PRACTICAL GUIDE TO ICP-MS
A Tutorial for Beginners

SECOND EDITION

PRACTICAL SPECTROSCOPY
A SERIES

1. Infrared and Raman Spectroscopy (in three parts), *edited by Edward G. Brame, Jr., and Jeanette G. Grasselli*
2. X-Ray Spectrometry, *edited by H. K. Herglotz and L. S. Birks*
3. Mass Spectrometry (in two parts), *edited by Charles Merritt, Jr., and Charles N. McEwen*
4. Infrared and Raman Spectroscopy of Polymers, *H. W. Siesler and K. Holland-Moritz*
5. NMR Spectroscopy Techniques, *edited by Cecil Dybowski and Robert L. Lichter*
6. Infrared Microspectroscopy: Theory and Applications, *edited by Robert G. Messerschmidt and Matthew A. Harthcock*
7. Flow Injection Atomic Spectroscopy, *edited by Jose Luis Burguera*
8. Mass Spectrometry of Biological Materials, *edited by Charles N. McEwen and Barbara S. Larsen*
9. Field Desorption Mass Spectrometry, *László Prókai*
10. Chromatography/Fourier Transform Infrared Spectroscopy and Its Applications, *Robert White*
11. Modern NMR Techniques and Their Application in Chemistry, *edited by Alexander I. Popov and Klaas Hallenga*
12. Luminescence Techniques in Chemical and Biochemical Analysis, *edited by Willy R. G. Baeyens, Denis De Keukeleire, and Katherine Korkidis*
13. Handbook of Near-Infrared Analysis, edited by *Donald A. Burns and Emil W. Ciurczak*
14. Handbook of X-Ray Spectrometry: Methods and Techniques, *edited by René E. Van Grieken and Andrzej A. Markowicz*
15. Internal Reflection Spectroscopy: Theory and Applications, *edited by Francis M. Mirabella, Jr.*
16. Microscopic and Spectroscopic Imaging of the Chemical State, *edited by Michael D. Morris*
17. Mathematical Analysis of Spectral Orthogonality, *John H. Kalivas and Patrick M. Lang*
18. Laser Spectroscopy: Techniques and Applications, *E. Roland Menzel*
19. Practical Guide to Infrared Microspectroscopy, *edited by Howard J. Humecki*
20. Quantitative X-ray Spectrometry: Second Edition, *Ron Jenkins, R. W. Gould, and Dale Gedcke*
21. NMR Spectroscopy Techniques: Second Edition, Revised and Expanded, *edited by Martha D. Bruch*
22. Spectrophotometric Reactions, *Irena Nemcova, Ludmila Cermakova, and Jiri Gasparic*

23. Inorganic Mass Spectrometry: Fundamentals and Applications, *edited by Christopher M. Barshick, Douglas C. Duckworth, and David H. Smith*

24. Infrared and Raman Spectroscopy of Biological Materials, *edited by Hans-Ulrich Gremlich and Bing Yan*

25. Near-Infrared Applications In Biotechnology, *edited by Ramesh Raghavachari*

26. Ultrafast Infrared and Raman Spectroscopy, *edited by M. D. Fayer*

27. Handbook of Near-Infrared Analysis: Second Edition, Revised and Expanded, *edited by Donald A. Burns and Emil W. Ciurczak*

28. Handbook of Raman Spectroscopy: From the Research Laboratory to the Process Line, *edited by Ian R. Lewis and Howell G. M. Edwards*

29. Handbook of X-Ray Spectrometry: Second Edition, Revised and Expanded, *edited by René E. Van Grieken and Andrzej A. Markowicz*

30. Ultraviolet Spectroscopy and UV Lasers, *edited by Prabhakar Misra and Mark A. Dubinskii*

31. Pharmaceutical and Medical Applications of Near-Infrared Spectroscopy, *Emil W. Ciurczak and James K. Drennen III*

32. Applied Electrospray Mass Spectrometry, *edited by Birendra N. Pramanik, A. K. Ganguly, and Michael L. Gross*

33. Practical Guide to ICP-MS, *Robert Thomas*

34. NMR Spectroscopy of Biological Solids, *edited by A. Ramamoorthy*

35. Handbook of Near-Infrared Analysis, Third Edition, *edited by Donald A. Burns and Emil W. Ciurczak*

36. Coherent Vibrational Dynamics, *edited by Sandro De Silvestri, Giulio Cerullo, and Guglielmo Lanzani*

37. Practical Guide to ICP-MS: A Tutorial for Beginners, Second Edition, *Robert Thomas*

PRACTICAL SPECTROSCOPY SERIES VOLUME 37

PRACTICAL GUIDE TO ICP-MS
A Tutorial for Beginners

SECOND EDITION

ROBERT THOMAS

CRC Press
Taylor & Francis Group
Boca Raton   London   New York

CRC Press is an imprint of the
Taylor & Francis Group, an **informa** business

CRC Press
Taylor & Francis Group
6000 Broken Sound Parkway NW, Suite 300
Boca Raton, FL 33487-2742

© 2008 by Taylor & Francis Group, LLC
CRC Press is an imprint of Taylor & Francis Group, an Informa business

Library of Congress Cataloging-in-Publication Data

Thomas, Robert.
    Practical guide to ICP-MS : a tutorial for beginners / Robert Thomas. -- 2nd ed.
        p. cm. -- (Practical spectroscopy)
    Includes bibliographical references and index.
    ISBN 978-1-4200-6786-6 (alk. paper)
    1. Inductively coupled plasma mass spectrometry. I. Title. II. Series.

QD96.I47T46 2008
543'.65--dc22   2008013041

Visit the Taylor & Francis Web site at
http://www.taylorandfrancis.com

and the CRC Press Web site at
http://www.crcpress.com

Dedication

A great deal has happened to me since I published the first edition of a *Practical Guide to ICP-MS* in 2004. I compiled and wrote a cookbook, which has been a dream of mine for a long time because of my love of cooking. I became a U.S. citizen after living here for almost 20 years, which means I can now finally vote for any one of the 3 presidential candidates still left in the primary race. Oh, and I forgot to mention, I also had a heart attack. I must admit, the first two events were exciting and extremely rewarding, but the third just wasn't as much fun. In fact, there wouldn't have been a second edition of the book if I hadn't made it through sextuple (yes, that's six) cardiac bypass surgery in August 2005. The U.S. health care system has many undesirable characteristics, but there is no doubt in my mind that it is the best in the world. Having been born and raised in the United Kingdom, I can speak with some authority that if I had experienced my heart attack there, I might not have been so fortunate. So I dedicate this book to the medical team from Washington Adventist Hospital in Tacoma Park, Maryland, including Dr. Gregory Kumkumian, who carried out the initial catheterization procedure, and Dr. Anjum Qazi, who did the bypass surgery. There's no doubt that their skill saved my life and gave me the opportunity to write the second edition of my book. So, since this is a dedication to them, please allow me a few minutes of your time to fill you in on the details. If it encourages just one person to change their lifestyle for the better, it will be worth it.

Monday, August 15, 2005, began like any other summer's day. It was extremely hot and humid, which you come to expect for the middle of August in Laytonsville, Maryland. My wife's family was in town for our annual reunion. It was something we did every year … get together for a week of activities, plenty of good food, drinking more than we should, and just having a great time together.

Sporting activities have always been a big part of our family get-togethers. This year was no exception. We had 10 of her immediate family staying with us, so we decided to play a game of soccer. It started at a frantic pace. It was clear we couldn't keep this going for the entire game. Feet were flying everywhere and bodies were constantly tumbling to the ground. There was no plan or strategy. It was basically kick and rush and shoulder-charge whoever got in your way. My team quickly went into a 3–0 lead, but then after about 20 minutes the game slowed to a crawl.

Everyone was feeling the effects of the heat so a time-out was called. Anyway, after a few sips of water I started to feel dizzy. This was unusual for me because I was used to running in the heat. I then began to feel slightly nauseous and experienced some discomfort in my chest. I thought that maybe it was related to the extreme heat and that it would just pass. After about 5 minutes, everyone headed back to the field. At this point, I began to experience severe indigestion. I knew something was wrong, so I told them I had to go to the bathroom. When I got to the house, I immediately took an antacid tablet and lay down on the sofa.

I was having difficulty in breathing. The panic was setting in. I knew I had to go back out and tell someone. I got as far the picnic table when the nausea hit me again. I had to sit down. Fortunately, my wife had walked back to the house to see where I was. I told her I felt really strange. She called to the others for help, as I was perspiring rather heavily. Their first thought was that it was heat exhaustion, so they suggested I cool off in the pool. I slowly walked there and sat down on the step with my body immersed under the water. That had no effect so someone started to hose me down with extremely cold well water (ouch). It wasn't helping either as I then started to shiver violently. When I told them I was getting chest pains, there was a realization that I could be having a heart attack.

A decision was then made to drive me to the local hospital (Montgomery General in Olney) and not wait for an ambulance. It all became a blur when we got to the ER. I just remember getting hooked up to an EKG machine and being intravenously pumped with nitroglycerine, which eventually relieved the chest pain somewhat. After about an hour I was medevaced to Washington Adventist Hospital … not what I imagined my first ride in a helicopter would be like! That evening I underwent an emergency cardiac catheterization procedure, where they inserted a metal stent through my groin into a blocked artery on the right side of my heart. It was during this procedure that they found that arteries on the other side of my heart were 75–90% blocked. The next morning I had open-heart surgery, where they carried out six bypasses. Unfortunately, there were problems, and I went back into surgery twice more with bleeding complications (I required 6 units of blood). I eventually came out of the anesthetic early the next morning, 17 hours after they began the surgery. I will not go into the fine details, but the next 3 months were extremely traumatic, both physically and mentally, as I slowly recovered from the ordeal.

Anyway, I recently passed the second-year anniversary of my brush with the "after-life." I look back on the experience and realize I was very lucky that my family was with me when it happened. If I had been out jogging on my own, who knows what the outcome would have been? There was no indication beforehand that I had such serious blockages. I was 56 years old, but had maintained a high fitness level. Running had been a part of my life for 40 years. (I had run 2 marathons, 5 half-marathons, and numerous 5 and 10 Ks.) I had never smoked and always ate a reasonably healthy diet. I thought I was doing everything right.

Thankfully, I made a complete recovery. I started an exercise program a few weeks after I got home from the hospital. The doctors said that my active lifestyle played a vital role in getting back to health so quickly. I am probably in better shape now than before the heart attack. I am eating more wisely and exercising (running or lifting weights) every other day of the week. If there is anything to learn from this life-changing experience, it is that you never know when something like this is going to happen. It is absolutely essential to get checked out on a regular basis, make intelligent dietary choices, don't ever think about smoking, and probably the most important decision—make exercise a part of your lifestyle. Trust me, these decisions could possibly save your life one day … and maybe give you the opportunity to write a chemistry textbook!

Robert Thomas
November 2007

Contents

Foreword xvii
Preface xix
Author xxv

Chapter 1   An Overview of ICP Mass Spectrometry 1
Principles of Operation 1

Chapter 2   Principles of Ion Formation 7
Ion Formation 7
Natural Isotopes 8

Chapter 3   Sample Introduction 13
Aerosol Generation 13
Droplet Selection 15
Nebulizers 16
      Concentric Design 16
      Cross-Flow Design 18
      Microflow Design 18
Spray Chambers 20
      Double-Pass Spray Chamber 20
      Cyclonic Spray Chamber 21
References 22

Chapter 4   Plasma Source 23
The Plasma Torch 24
Formation of an ICP Discharge 26
The Function of the RF Generator 26
Ionization of the Sample 28
References 29

Chapter 5   Interface Region 31
Capacitive Coupling 32
Ion Kinetic Energy 34
Benefits of a Well-Designed Interface 36
References 37

Chapter 6   Ion-Focusing System 39

Role of the Ion Optics 39
Dynamics of Ion Flow 41
Commercial Ion Optic Designs... 43
References... 46

Chapter 7   Mass Analyzers: Quadrupole Technology 47

Quadrupole Mass Filter Technology.. 47
Basic Principles of Operation 48
Quadrupole Performance Criteria 49
  Resolution 50
  Abundance Sensitivity 52
  Benefit of Good Abundance Sensitivity 53
References... 54

Chapter 8   Mass Analyzers: Double-Focusing Magnetic Sector Technology 57

Magnetic Sector Mass Spectroscopy: A Historical Perspective............................. 57
Use of Magnetic Sector Technology for ICP-MS 58
Principles of Operation of Magnetic Sector Systems 59
  Resolving Power 60
Other Benefits of Magnetic Sector Instruments 63
References... 63

Chapter 9   Mass Analyzers: Time-of-Flight Technology 65

Basic Principles of TOF Technology 65
Commercial Designs... 66
Differences Between Orthogonal and On-Axis TOF 68
Benefits of TOF Technology for ICP-MS 70
  Rapid Transient Peak Analysis 70
  Improved Precision 70
  Rapid Data Acquisition.. 71
References... 72

Chapter 10   Mass Analyzers: Collision/Reaction Cell and
     Interface Technology... 73

Basic Principles of Collision/Reaction Cells 74
Different Collision/Reaction Cell Approaches 75
  Collisional Mechanisms Using Nonreactive Gases and Kinetic Energy
    Discrimination 76
  Reaction Mechanisms with Highly Reactive Gases and Discrimination
    by Selective Bandpass Mass Filtering 80

The Collision/Reaction Interface...84
Using Reaction Mechanisms in a Collision Cell 85
Detection Limit Comparison 89
Summary...89
References 91

Chapter 11 Ion Detectors 93

Channel Electron Multiplier 93
Faraday Cup..94
Discrete Dynode Electron Multiplier...95
Extending the Dynamic Range...96
    Filtering the Ion Beam...96
    Using Two Detectors..96
    Using Two Scans with One Detector 96
    Using One Scan with One Detector 97
Extending the Dynamic Range Using Pulse-Only Mode...98
References 100

Chapter 12 Peak Measurement Protocol..101

Measurement Variables...101
Measurement Protocol 102
Optimization of Measurement Protocol 106
Multielement Data Quality Objectives 107
References 113

Chapter 13 Methods of Quantitation 115

Quantitative Analysis..115
    External Standardization 116
    Standard Additions 117
    Addition Calibration 118
Semiquantitative Analysis 118
Isotope Dilution 120
Isotope Ratios..123
Internal Standardization 123
References 124

Chapter 14 Review of Interferences..125

Spectral Interferences 125
    Oxides, Hydroxides, Hydrides, and Doubly Charged Species 127
    Isobaric Interferences 128

Ways to Compensate for Spectral Interferences 128
        Mathematical Correction Equations 128
        Cool/Cold Plasma Technology 130
        Collision/Reaction Cells 131
        High-Resolution Mass Analyzers 132
Matrix Interferences 132
        Compensation Using Internal Standardization 133
        Space-Charge-Induced Matrix Interferences 134
References 135

Chapter 15 Contamination Issues Associated with Sample Preparation 137

Collecting the Sample 137
Preparing the Sample 138
Grinding the Sample 138
Sample Dissolution Methods 139
Choice of Reagents and Standards 141
Vessels, Containers, and Sample Preparation Equipment 142
The Environment 145
The Analyst 146
Instrument and Methodology 147
References 149

Chapter 16 Routine Maintenance 151

Sample Introduction System 152
        Peristaltic Pump Tubing 152
        Nebulizers 152
        Spray Chamber 154
        Plasma Torch 155
Interface Region 156
Ion Optics 157
Roughing Pumps 158
Air Filters 159
Other Components to Be Periodically Checked 159
        The Detector 159
        Turbomolecular Pumps 160
        Mass Analyzer 160

Chapter 17 Alternative Sample Introduction Techniques 163

Laser Ablation 164
        Commercial Systems for ICP-MS 165
        Excimer Lasers 165
        Benefits of Laser Ablation for ICP-MS 166
        Optimum Laser Design Based on Application Requirements 167

Flow Injection Analysis 171
Electrothermal Vaporization.. 174
Chilled Spray Chambers and Desolvation Devices 178
    Water-Cooled and Peltier-Cooled Spray Chambers 178
    Ultrasonic Nebulizers 179
    Specialized Microflow Nebulizers with Desolvation Techniques............. 181
Direct Injection Nebulizers... 183
Rapid Sampling Procedures... 184
References 185

Chapter 18 Coupling ICP-MS with Chromatographic Techniques for Trace
                  Element Speciation Studies... 187

HPLC Coupled with ICP-MS 190
Chromatographic Separation Requirements.. 191
    Ion Exchange Chromatography (IEC).. 191
    Reversed-Phase Ion Pair Chromatography (RP-IPC)............................. 192
    Column Material.. 193
    Isocratic or Gradient Elution 193
Sample Introduction Requirements 195
Optimization of ICP-MS Parameters... 196
    Compatibility with Organic Solvents 197
    Collision/Reaction Cell or Interface Capability 197
    Optimization of Peak Measurement Protocol 199
    Full Software Control and Integration...200
Summary.. 201
References 202

Chapter 19 ICP-MS Applications 203

Environmental..204
Biomedical 208
    Sample Preparation...209
    Interference Corrections 210
    Calibration 210
    Stability... 211
Geochemical 211
    Determination of Rare Earth Elements 212
    Analysis of Digested Rock Samples Using Flow Injection 214
    Geochemical Prospecting 215
    Isotope Ratio Studies 216
    Laser Ablation.. 218
Semiconductor 219
Nuclear..224
    Applications Related to the Production of Nuclear Materials....................226
    Applications in the Characterization of High-Level Nuclear Waste 227

        Applications Involving the Monitoring of the Nuclear Industry's
            Impact on the Environment 227
        Applications Involving Human Health Studies 228
Other Applications 229
        Metallurgical Applications 229
        Petrochemical and Organic-Based Samples 231
        Food and Agriculture 234
Summary 236
References 236

Chapter 20 Comparing ICP-MS with Other Atomic Spectroscopic
                Techniques 241

Flame Atomic Absorption 242
Electrothermal Atomization 243
Radial-View ICP Optical Emission 243
Axial-View ICP Optical Emission 243
Inductively Coupled Plasma Mass Spectrometry 243
        Define the Objective 244
        Establish Performance Criteria 244
        Define the Application Task 244
                Application 244
                Installation 245
                User 245
                Financial 245
        Comparison of Techniques 245
                Detection Limits 245
                Analytical Working Range 246
                Sample Throughput 249
                Interferences 251
                Usability 252
                Cost of Ownership 252
                Cost per Sample 257
Conclusion 259
References 259

Chapter 21 How to Select an ICP Mass Spectrometer: Some Important
                Analytical Considerations 261

Evaluation Objectives 261
        Analytical Performance 262
                Detection Capability 263
                Precision 267
                Isotope Ratio Precision 269
                Accuracy 271
                Dynamic Range 272

Interference Reduction..274
Reduction of Matrix-Induced Interferences....................................281
Sample Throughput...284
Transient Signal Capability..285
Usability Aspects...286
Ease of Use...286
Routine Maintenance...287
Compatibility with Alternative Sampling Accessories....................288
Installation of Instrument 288
Technical Support 289
Training..289
Reliability Issues...289
Service Support...290
Financial Considerations..291
The Evaluation Process: A Summary...292
References...293

Chapter 22 Glossary of ICP-MS Terms 295

Chapter 23 Useful Contact Information..333

Index...341

Foreword

Much has changed in the field of inductively coupled plasma mass spectrometry (ICP-MS) since the publication of the first edition of this book as illustrated by the recent program of the 2008 Winter Conference on Plasma Spectrochemistry (www.icpinformation.org). Basic ICP and MS instrumentation has changed little, however. The major growth areas of ICP-MS have been its applications. So what's new to justify another edition of a "Practical Guide to ICP-MS: A tutorial for beginners"? Rob Thomas has taken his very successful 1st edition and updated the majority of the chapters to reflect the development of the technology over the past four years. In addition, he has included new material, based on published application work being carried by the ICP-MS analytical community and written it in a way a novice audience can understand.

For example, he has added a chapter on elemental speciation, which has probably been the most rapidly developing application area of ICP-MS during the past four years, especially for biological and environmental materials. The complementary nature of elemental (with ICP and glow discharge ion sources) and molecular (e.g., electrospray ion sources) mass spectrometry has strengthened the analyst's capability to identify and quantitate many different species of interest. Elemental speciation of solids lags other applications, and these analyses typically require surface and other techniques that don't use ICP-MS. However, spatial distribution studies and quantitative mapping (imaging) of elemental distributions in hard and soft tissues, gels, or membranes, for example, has developed rapidly during the past few years with progress in matrix assisted laser desorption/ionization (MALDI), secondary ionization MS and laser ablation ICP-MS (LA-ICP-MS). With that in mind, Rob has put together a very informative chapter on laser ablation systems, together with a very good overview of all the new sampling devices and desolvation systems currently available, in the updated chapter on alternative sample introduction approaches.

His expanded chapter on collision/reaction cells and interfaces for resolving mass spectral interferences shows how the appeal of this technology has grown significantly in the past four years. His assessment of the different commercial approaches, particularly regarding their application strengths and weaknesses is a much needed addition to the often confusing information about collision/reaction cells in the public domain for the novice user. With real-world analyses in mind, Rob's expanded chapter on applications gives a very good overview of what is currently being carried out by routine laboratories, as well as giving the reader a glimpse into what novel applications will emerge in the not-to-distant future. In particular, geochemical applications of the high-end ICP-MS systems (i.e., high resolution and/or multi-collector MS) have opened significant new directions in geochronology, for example, but accurate isotope ratio determinations still require careful chemical pretreatment and stable ion signals and/or simultaneous detection. Research and development in microplasma sources and their applications, modern materials analyses, and provenance of everything from food to ancient art objects are stimulating areas of ICP-MS waiting for new users to explore. Instrumental developments of solid state (focal-plane array) systems for simultaneous multichannel detection (by Hieftje et al.), and novel experiments with ICP ion trap MS especially a linear ion trap/orbital trapping arrangement for very high-resolution measurements (by Koppenaal et al.) potentially hold the future to significant improvements in commercial ICP-MS systems. However, these topics are beyond the scope of this introductory text. On the other hand, a critical need exists for novel academic, commercial, and industrial research to extend the ICP source, as well as sample introduction and MS systems. We encourage readers to assist their spectroscopy societies, universities, and research funding programs by attending meetings, contributing to funding appeals, and expressing the need for research support.

Rob has also put together an excellent "Glossary of Terms" for the ICP-MS beginner. His straight forward A-Z descriptions and explanations of terms used in ICP-MS are quite refreshing amongst the multitude of articles and papers we see published today. He takes complex topics and attempts to simplify and demystify them for someone who is new to the technique. Continuing on the theme of educating the user, numerous educational courses on ICP-MS topics are now offered commercially by instrument manufacturers and major national and international spectroscopy conferences such as the Pittsburgh Conference on Analytical Chemistry and Applied Spectroscopy and the Winter Conference on Plasma Spectrochemistry in Asia, Europe, and America. For example, more than 30 introductory and advanced courses were presented in January 2008 at the Winter Conference in Temecula, California. Readers are encouraged to continue their ICP-MS exposure and education after reading this tutorial by enrolling in one of these training programs.

On a personal note, Rob's recovery from a heart attack is a blessing to his family and friends, and we are especially fortunate that he has completed the second edition of his book. We wish him continued good health, so that future revisions and editions will appear at regular intervals as ICP-MS continues its progress.

Ramon M. Barnes
Amherst, Massachusetts
January 26, 2008

Preface

I cannot believe that it has been 4 years since I published the first edition of this textbook. What was originally intended as a series of tutorials on the basic principles of ICP-MS for *Spectroscopy* in 2001 quickly grew into a textbook focusing on the practical side of the technique. With over fifteen hundred copies of the English version sold and a Chinese (Mandarin) version now in print, I'm very honored that the book has gained the reputation as being the reference book of choice for novices and beginners to the technique. Sales of the book have exceeded my wildest expectations. Of course, it helps when it is "recommended reading" for a short course I teach twice a year on "How to Select an ICP-MS." It also helps when you get the visibility of your book being displayed at 15 different vendors' booths at the Pittsburgh Conference every year. But there is no question in my mind that the major reason for its success is that it presents ICP-MS in a way that is very easy to understand for beginners, and also shows the practical benefits of the technique for carrying out routine trace element analysis.

However, 4 years is a long time for a book to remain current, even if sales of the book in its fourth year were almost as high as they were in its first year in 2004. For that reason, it made sense to not only write an updated version to represent the current state of the technology, but also to incorporate all the great feedback I received from users and vendors over the past few years. Still, it should not lose sight of the fact that its target audience was going to be users who had just started with ICP-MS or analytical chemists who were in the process of thinking about investing in the technique. So with that in mind, I present to you the second edition of *Practical Guide to ICP-MS: A Tutorial for Beginners*. Below is a summary of the major changes from the first edition.

I have included two new chapters:

- **Coupling ICP-MS with Chromatographic Techniques for Trace Elemental Speciation Studies:** The demand for trace element speciation studies, particularly using liquid chromatography coupled with ICP-MS, has sky-rocketed in the past few years, especially in the environmental, toxicological, and clinical fields. This new chapter attempts to capture the current state of this exciting hyphenated technique, with particular emphasis on the most common applications being carried out and the optimum hardware and software configurations for both the liquid chromatographic separation and ICP-MS detection, to ensure the most successful speciation analysis.
- **Glossary of ICP-MS Terms:** I regret not putting this chapter in the first edition of my book, based on feedback I received. So I knew when I got around to writing the next edition, a basic dictionary or glossary of ICP-MS terms was an absolute necessity. I have therefore put together a list of explanations and definitions of the most common words, expressions, and terms used in this book. It is aimed specifically at beginners to use as a quick

reference guide, without having to go looking for a more detailed explanation of the subject matter somewhere else in the book.

There are also major rewrites and significant additions to the following chapters:

- **Collision/Reaction Cell and Interface Technology:** Collision and reaction cells and interfaces have revolutionized ICP-MS analysis since their commercialization in 1999. They are allowing determinations that were previously difficult if not impossible to carry out using quadrupole-based instrumentation. However, as more and more applications are getting developed, it is clear that all the different commercial designs on the market approach the analytical determinations differently. This extended chapter on collision/reaction cells will attempt to clarify the strengths and weaknesses of each design and how they impact the strategy for developing a method and the analysis of unknown samples.

- **Alternative Sample Introduction Techniques:** Nonstandard sampling accessories like laser ablation systems, flow injection analyzers, electrothermal vaporizers, cooled spray chambers, desolvation equipment, direct injection nebulizers, and automated sample delivery systems and dilutors are considered critical to enhancing the practical capabilities of the technique. Their use has increased significantly over the past few years as ICP-MS is being asked to solve more and more diverse application problems. This chapter reflects the increased interest in sampling accessories, especially in the area of specialized sample introduction and desolvation devices to reduce the impact of common interferences.

- **ICP-MS Applications:** As ICP-MS is getting less expensive, it is being installed in more and more routine labs and, as a result, is being asked to solve more diverse application problems every year. In addition, the power of collision/reaction cell/interface technology is taking the technique into application areas that previously required high-resolution instrumentation to carry out the analysis. This chapter, while still emphasizing the most common routine-type applications, also gives insight into the emerging new application areas being addressed.

- **Comparison of ICP-MS with Other Atomic Spectroscopy Techniques:** This chapter has been updated to compare the performance capabilities of modern ICP-MS (with collision/reaction cell/interface capabilities) with the latest ICP-OES, GFAA, and FAA instrumentation. In addition, the section on "cost of analysis" has been updated to reflect the price of gases, electricity, and instrument consumables in 2007–2008, compared to 4 years ago.

- **How to Select an ICP-MS System: The Most Important Analytical Considerations:** My experience of teaching two short courses a year on this subject has given me a unique insight into the kinds of questions that most novices have when evaluating commercial instrumentation. This chapter focuses on the most important analytical considerations when going through the evaluation process, with particular emphasis on what I have learned from teaching my course over the past four years.

In addition, I have made minor modifications to many of the other chapters in the book to reflect the advancement of the technology over the past 4 years.

ICP-MS MARKETPLACE

Before I go on to talk about the technique in greater detail, it is definitely worth reiterating what I said in the preface of my first book and give you an update on the current size of the ICP-MS marketplace. In 2007, 24 years after ICP-MS was first commercialized, there are approximately 8000 systems installed worldwide. If this is compared with ICP-OES, first commercialized in 1974, the difference is quite significant. In 1998, 24 years after ICP-OES was introduced, about 20,000 units had been sold, and if this is compared with the same time period that ICP-MS has been available, the difference is even more staggering. From 1983 to the present day, approximately 40,000 ICP-OES systems have been installed—about five times more than the number of ICP-MS systems. If the comparison is made with all atomic spectroscopy (AS) instrumentation (ICP-MS, ICP-OES, electrothermal atomization [ETA], and FAA), the annual sales for ICP-MS is less than 7.5% of the total AS market—900 units compared to approximately 12,000 AS systems. It is worth emphasizing that the global ICP-MS market is growing at about 7% annually, compared to approximately 3% for ICP-OES and a virtually flat growth for AA instrumentation (1). This makes the comparison a little more positive for ICP-MS as compared to the numbers I presented in my first book, but it is still unclear to me as to why ICP-MS is not growing at a faster rate. It is even more surprising when one considers that the technique offers so much more than the other AS techniques, including superb detection limits, rapid multielement analysis, and isotopic measurement capabilities.

ICP-MS: RESEARCH OR ROUTINE?

Clearly, one of the many reasons that ICP-MS has not become more popular is its relatively high price-tag—an ICP mass spectrometer still costs 2x more than ICP-OES and 3x more than ETA. But in a competitive world, the "street price" of an ICP-MS system is much closer to a top-of-the-line ICP-OES with sampling accessories or an ETA system that has all the "bells and whistles" on it. So if ICP-MS is not significantly more expensive than ICP-OES and ETA, why has it not been more widely accepted by the analytical community? It is still my firm opinion that the major reason ICP-MS has not gained the popularity of the other trace element techniques lies in the fact that it is still considered a complicated research-type technique, requiring a very skilled person to operate it. Manufacturers of ICP-MS equipment are constantly striving to make the systems easier to operate, the software easier to use, and the hardware easier to maintain, but even after 25 years, it is still not perceived as a mature, routine technique like FAA or ICP-OES. The picture is even fuzzier now that most instruments are sold with collision/reaction cells/interfaces. This means that even though this exciting new technology is making ICP-MS more powerful and flexible, the method development process for unknown samples is generally still a little more complex. In addition, vendors of this type of equipment are very skilled at inflating the capabilities of their technology while at the same time pointing out

the limitations of other approaches, making it even more confusing for the inexperienced user.

The bottom line is that ICP-MS has still not gained the reputation as a technique
that you can allow a complete novice to use with no supervision, for the fear of generating erroneous data. This makes for all the more reason why there is still a need for
a good textbook explaining the basic principles and application benefits of ICP-MS
in a way that is interesting, unbiased, and easy to understand for a novice who has
limited knowledge of the technique. There is no question that there are some excellent books out there (2,3,4,5,6,7), but they are mainly written or edited by scientists
who are not approaching the subject from a beginner's perspective. So they tend to
be technically "heavy" and more biased towards fundamental principles and less on
how ICP-MS is being applied to solve real-world application problems.

ICP-MS FOR DUMMIES?

I would hesitate to call my book *ICP-MS for Dummies,* but it is definitely intended
for analytical chemists who might be termed "ionically-challenged" (if you do fall
into this category, the glossary of ICP-MS terms was written especially for you). So
for those of you who think an ion is used to put a crease in your pants, this book is
yours. Inside you will find chapters not only on the fundamentals and basics of the
technique, but also on practical issues like contamination control, routine maintenance, and when best to use the many kinds of sampling accessories. I also felt it was
important to compare ICP-MS with other trace element techniques, like FAA, ETA,
and ICP-OES, focusing on criteria like elemental range, detection capability, sample throughput, analytical working range, interferences, sample preparation, maintenance issues, operator skill level, and running costs. This kind of head-to-head
comparison will enable the reader to relate both the advantages and disadvantages of
ICP-MS to other atomic spectroscopy instrumentation they are more familiar with. I
included this because there is still a role for the other techniques, and some vendors
who do not offer the full range of AS instrumentation might embellish the benefits
of ICP-MS over other techniques. In addition, in order to fully understand its practical capabilities, it is important to give an overview of the most common applications
currently being carried out by ICP-MS and its sampling accessories. This will give
you a flavor of the different industries and markets that are benefiting from the technique's enormous potential, especially the newer application segments, such as trace
element speciation analysis using HPLC coupled with ICP-MS. And for those of
you who might be interested in purchasing the technique, I have included a chapter
on the most important selection criteria. This is a critical ingredient in presenting
ICP-MS to a novice, because there is very little information in the public domain to
help someone carry out an evaluation of commercial instrumentation. Very often,
people go into this evaluation process completely unprepared and as a result may
end up with an instrument that is not ideally suited for their needs...something I am
very well aware of, based on teaching my Short Course at the Pittsburgh Conference
for the past 5 years.

Hopefully, after having completed the book, there is still a serious interest in
investing in ICP-MS instrumentation—unless, of course, you have purchased one

before reading the final chapter! Even though this might sound a little ambitious, the main objective is to make ICP-MS a little more compelling to purchase and ultimately open up its potential to the vast majority of the trace element community who have not yet realized the full benefits of its capabilities. So with this in mind, please feel free to come in and share my thoughts on a *Practical Guide to ICP-MS: A Tutorial for Beginners...*edition number two.

ACKNOWLEDGMENT

Having worked in the field of ICP-MS for over 20 years, my incentive for writing this book was based on a realization that there were no textbooks being written specifically for beginners with a very limited knowledge of the technique. I quickly came to the conclusion that the only way this was going to happen was to write it myself. So in 2002, I set myself the objective of putting together a reference book that could be used by both analytical chemists and senior management who were experienced in the field of trace metal analysis, but only had a limited knowledge of ICP-MS. The first edition of the book, published in 2004, represented the fruits of my effort. As I mentioned earlier, it has definitely reached its target audience, having sold over fifteen hundred copies worldwide.

About 12 months ago, we got a request from a Chinese publisher to get it translated into Mandarin. Knowing China's record on copyright infringement, this was a major achievement. As a result, the book became "legitimately" available to the Chinese marketplace in the summer of 2007 and has sold over three hundred copies already. I want to personally thank my U.S. publisher, CRC Press/Taylor and Francis, and the Chinese publisher, Atomic Industry Press of Beijing, for making this happen.

So now in 2008, as the second edition hits the "book stands," I would like to take this opportunity once again to thank some of the people and organizations that have helped me put the book together. First, I would like to thank the editorial staff of *Spectroscopy* magazine, who gave me the opportunity to write a monthly tutorial on ICP-MS back in the spring of 2001...this was most definitely the spark I needed to start the original project. They also allowed me to use many of the figures from the series, together with material from other ICP-MS articles I wrote for the magazine.

Second, I would like to thank all the manufacturers of ICP-MS instrumentation, ancillary equipment, sampling accessories, consumables, calibration standards, chemical reagents, and high-purity gases, who supplied me with the information, data, drawings, figures and schematics, etc., and particularly their willingness over the past 4 years to display the book at their Pittsburgh Conference exhibition booths. In fact, at last year's show in Chicago, we had 15 vendors showing the book. This alone has made a huge difference to the visibility of the book, and its success would not have been possible without their help.

Third, I would like to thank Dr. Ramon Barnes, Director of the Research Institute for Analytical Chemistry in Amherst, MA and the driving force behind the Winter Conference on Plasma Spectrochemistry, for the kind and complimentary words he wrote in the Foreword of both the first and second editions of my book.

Finally, I would like to thank my wife, Donna Marie, and two daughters, Glenna and Deryn for their encouragement and enthusiasm as I agonized over the decision to do a second edition, knowing the time, effort, and commitment needed to write the first edition. Anyone who lives in a house full of females knows how persuasive a wife and two teenage daughters can be!

REFERENCES

1. *Analytical Instruments Global Assessment Report, Ninth Edition*, Strategic Directions International, Inc., Los Angeles, July 2006.
2. *Inorganic Mass Spectrometry*: F. Adams, R. Gijbels, R. Van Grieken, Eds., Wiley and Sons, New York, 1988.
3. K. E. Jarvis, A. L. Gray, R. S. Houk, *Handbook of Inductively Coupled Plasma Mass Spectrometry*, Blackie, Glasgow, 1992.
4. A. Montasser, *Inductively Coupled Plasma Mass Spectrometry*, Wiley-VCH, New York, 1998.
5. H. E. Taylor, *Inductively Coupled Plasma Mass Spectrometry: Practices and Techniques*, Elsevier Science, Amsterdam, 2001.
6. *Inductively Coupled Plasma Mass Spectrometry Handbook*: S. Nelmes, Ed., CRC Press, Boca Raton, 2005.
7. *Inorganic Mass Spectrometry: Principles and Applications*: Sabine Becker J., John Wiley and Sons, UK, 2008.

Author

Robert J. Thomas is principal of Scientific Solutions, a consulting company based in Gaithersburg, Maryland, that serves the application, training, and technical writing needs of the trace element analysis user community. He has worked in the field of atomic spectroscopy (AS) for more than 30 years, with almost 20 years' experience in ICP-MS applications, product development, and marketing support at Perkin-Elmer Life and Analytical Sciences. He has written over 60 technical publications covering a wide variety of AS subjects, including the fundamental principles of solving real-world application problems with analytical instrumentation. He received his advanced degree in analytical chemistry from The University of Wales, Newport, Ghent in the United Kingdom, and is also a graduate member of the Royal Society of Chemistry and a Fellow of the Chemical Society of the United Kingdom.

1 An Overview of ICP Mass Spectrometry

ICP-MS not only offers extremely low detection limits in the sub parts per trillion (ppt) range, but also enables quantitation at the high parts per million (ppm) level. This unique capability makes the technique very attractive compared to other trace metal techniques such as ETA, which is limited to determinations at the trace level, or FAA and ICP-OES, which are traditionally used for the detection of higher concentrations. In Chapter 1 we will present an overview of ICP-MS and explain how its characteristic low detection capability is achieved.

Inductively coupled plasma mass spectrometry (ICP-MS) is undoubtedly the fastest growing trace element technique available today. Since its commercialization in 1983, approximately 8000 systems have been installed worldwide for many varied and diverse applications. The most common ones, which represent approximately 80% of the ICP-MS analysis being carried out today, include environmental, geological, semiconductor, biomedical, and nuclear application fields. There is no question that the major reason for its unparalleled growth is its ability to carry out rapid multielement determinations at the ultratrace level. Even though it can broadly determine the same suite of elements as other atomic spectroscopic techniques, such as flame atomic absorption (FAA), electrothermal atomization (ETA), and inductively coupled plasma optical emission (ICP-OES), ICP-MS has clear advantages in its multielement characteristics, speed of analysis, detection limits, and isotopic capability. Figure 1.1 shows approximate detection limits of all the elements that can be detected by ICP-MS, together with their isotopic abundance. For actual elemental detection limits and isotopic abundances, please refer to Table 20.1 and Figure 2.5 respectively.

PRINCIPLES OF OPERATION

There are a number of different ICP-MS designs available today that share many similar components, such as nebulizer, spray chamber, plasma torch, and detector, but can differ quite significantly in the design of the interface, ion-focusing system, mass separation device, and vacuum chamber. Instrument hardware is described in greater detail in the subsequent chapters, but let us begin here by giving an overview of the principles of operation of ICP-MS. Figure 1.2 shows the basic components that make up an ICP-MS system. The sample, which usually must be in a liquid form, is pumped at 1 mL/min, usually with a peristaltic pump into a nebulizer, where it is converted into a fine aerosol with argon gas at about 1 L/min. The fine droplets of the aerosol, which represent only 1–2% of the sample, are separated from larger droplets

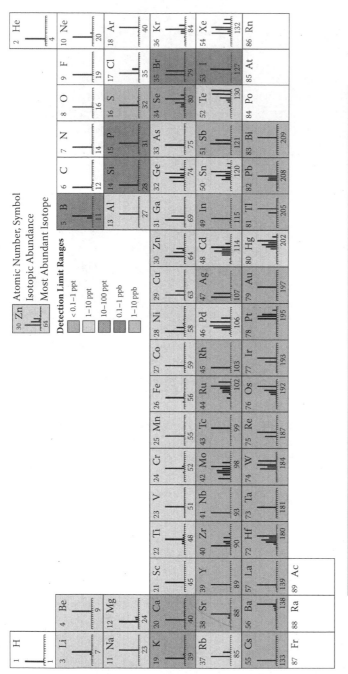

FIGURE 1.1 Approximate detection capability of ICP-MS, together with elemental isotropic abundance (copyright © 2003–2007, all rights reserved, PerkinElmer Inc.).

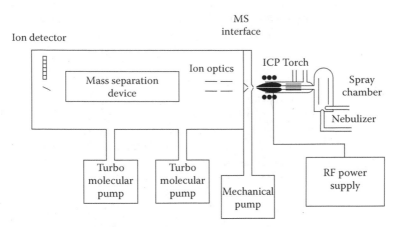

FIGURE 1.2 Basic instrumental components of an ICP mass spectrometer.

by means of a spray chamber. The fine aerosol then emerges from the exit tube of the spray chamber and is transported into the plasma torch via a sample injector.

It is important to differentiate between the roles of the plasma torch in ICP-MS compared to ICP-OES. The plasma is formed in exactly the same way, by the interaction of an intense magnetic field (produced by radio frequency (RF) passing through a copper coil) on a tangential flow of gas (normally argon), at about 15 L/min flowing through a concentric quartz tube (torch). This has the effect of ionizing the gas, which when seeded with a source of electrons from a high-voltage spark, forms a very-high-temperature plasma discharge (~10,000 K) at the open end of the tube. However, this is where the similarity ends. In ICP-OES, the plasma, which is normally vertical, is used to generate photons of light by the excitation of electrons of a ground-state atom to a higher energy level. When the electrons "fall" back to ground state, wavelength-specific photons are emitted that are characteristic of the element of interest. In ICP-MS the plasma torch, which is positioned horizontally, is used to generate positively charged ions and not photons. In fact, every attempt is made to stop the photons from reaching the detector because they have the potential to increase signal noise. It is the production and detection of large quantities of these ions that gives ICP-MS its characteristic low-ppt detection capability—about three to four orders of magnitude better than ICP-OES.

Once the ions are produced in the plasma, they are directed into the mass spectrometer via the interface region, which is maintained at a vacuum of 1–2 torr with a mechanical roughing pump. This interface region consists of two metallic cones (usually nickel), called the sampler and a skimmer cone, each with a small orifice (0.6–1.2 mm) to allow the ions to pass through to the ion optics, where they are guided into the mass separation device.

The interface region is one of the most critical areas of an ICP mass spectrometer, because the ions must be transported efficiently and with electrical integrity from the plasma, which is at atmospheric pressure (760 torr), to the mass spectrometer analyzer region, which is at approximately 10^{-6} torr. Unfortunately, there is capacitive coupling between the RF coil and the plasma, producing a potential difference of a few hundred volts. If this is not eliminated, an electrical discharge (called a secondary discharge or pinch effect) between the plasma and the sampler cone would occur. This discharge would increase the formation of interfering species and also dramatically affect the kinetic energy of the ions entering the mass spectrometer, making optimization of the ion optics very erratic and unpredictable. For this reason, it is absolutely critical that the secondary charge be eliminated by grounding the RF coil. There have been a number of different approaches used over the years to achieve this, including a grounding strap between the coil and the interface, balancing the oscillator inside the RF generator circuitry, a grounded shield or plate between the coil and the plasma torch, or the use of a double interlaced coil where RF fields go in opposing directions. They all work differently but basically achieve a similar result, which is to reduce or eliminate the secondary discharge.

Once the ions have been successfully extracted from the interface region, they are directed into the main vacuum chamber by a series of electrostatic lens, called ion optics. The operating vacuum in this region is maintained at about 10^{-3} torr with a turbomolecular pump. There are many different designs of the ion optic region, but they serve the same function, which is to electrostatically focus the ion beam toward the mass separation device, while stopping photons, particulates, and neutral species from reaching the detector.

The ion beam containing all the analyte and matrix ions exits the ion optics and now passes into the heart of the mass spectrometer—the mass separation device, which is kept at an operating vacuum of approximately 10^{-6} torr with a second turbomolecular pump. There are many different mass separation devices, all with their strengths and weaknesses. Three of the most common types are discussed in this book—quadrupole, magnetic sector, and time-of-flight technology—but they basically serve the same purpose, which is to allow analyte ions of a particular mass-to-charge ratio through to the detector and to filter out all the nonanalyte, interfering, and matrix ions. Depending on the design of the mass spectrometer, this is either a scanning process where the ions arrive at the detector in a sequential manner, or a simultaneous process where the ions are either sampled or detected at the same time. Most quadrupole instruments nowadays are also sold with collision/reaction cells or interfaces. This technology offers a novel way of minimizing polyatomic spectral interferences by bleeding a gas into the cell or interface and using ion–molecule collision and reaction mechanisms to reduce the impact of the ionic interference.

The final process is to convert the ions into an electrical signal with an ion detector. The most common design used today is called a discrete dynode detector, which contains a series of metal dynodes along the length of the detector. In this design, when the ions emerge from the mass filter, they impinge on the first dynode and are converted into electrons. As the electrons are attracted to the next dynode, electron multiplication takes place, which results in a very high stream of electrons emerging from the final dynode. This electronic signal is then processed by the data-handling system in the conventional way and converted into analyte concentration using ICP-MS calibration standards. Most detection systems can handle up to eight orders of dynamic range, which means they can be used to analyze samples from ppt levels up to a few hundred ppm.

It is important to emphasize that because of the enormous interest in the technique, most ICP-MS instrument companies have very active R&D programs in place, in order to get an edge in a very competitive marketplace. This is obviously very good for the consumer, because not only does it drive down instrument prices, but the performance, applicability, usability, and flexibility of the technique is being improved at a dramatic rate. Although this is extremely beneficial for the ICP-MS user community, it can pose a problem for a textbook writer who is attempting to present a snapshot of instrument hardware and software components at a particular moment in time. Hopefully, I have struck the right balance in not only presenting the fundamental principles of ICP-MS to a beginner, but also making them aware of what the technique is capable of achieving and where new developments might be taking it.

2 Principles of Ion Formation

Chapter 2 gives a brief overview of the fundamental principles used in ICP-MS—the use of a high-temperature argon plasma to generate positive ions. The highly energized argon ions that make up the plasma discharge are used to first produce analyte ground-state atoms from the dried sample aerosol, and then to interact with the atoms to remove an electron and generate positively charged ions, which are then steered into the mass spectrometer for detection and measurement.

In ICP-MS, the sample, which is usually in liquid form, is delivered into the sample introduction system, comprising a spray chamber and nebulizer. It emerges as an aerosol, where it eventually finds its way, via a sample injector, into the base of the plasma. As it travels through the different heating zones of the plasma torch, it is dried, vaporized, atomized, and ionized. During this time, the sample is transformed from a liquid aerosol to solid particles, and then into a gas. When it finally arrives at the analytical zone of the plasma, at approximately 6000–7000 K, it exists as ground-state atoms and ions, representing the elemental composition of the sample. The excitation of the outer electron of a ground-state atom to produce wavelength-specific photons of light is the fundamental basis of atomic emission. However, there is also enough energy in the plasma to remove an electron from its orbital to generate a free ion. The energy available in an argon plasma is ~15.8 eV, which is high enough to ionize most of the elements in the periodic table (the majority have first ionization potentials on the order of 4–12 eV). It is the generation, transportation, and detection of significant numbers of positively charged ions that gives ICP-MS its characteristic ultratrace detection capabilities. It is also important to mention that although ICP-MS is predominantly used for the detection of positive ions, negative ions are also produced in the plasma. However, because the extraction and transportation of negative ions is different from that of positive ions, most commercial instruments are not designed to measure them. The process of the generation of positively charged ions in the plasma is conceptually shown in greater detail in Figure 2.1.

ION FORMATION

The actual process of conversion of a neutral ground-state atom to a positively charged ion is shown in Figures 2.2 and 2.3. Figure 2.2 shows a very simplistic view of the chromium atom Cr^0, consisting of a nucleus with 24 protons (p^+) and 28 neutrons (n), surrounded by 24 orbiting electrons (e^-). (It must be emphasized that this is not meant to be an accurate representation of the electron's shells and subshells, but just a conceptual explanation for the purpose of clarity.) From this we can conclude

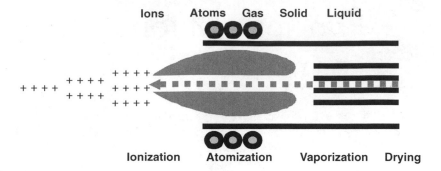

FIGURE 2.1   Generation of positively charged ions in the plasma.

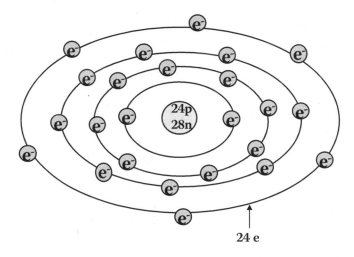

FIGURE 2.2   Simplified schematic of a chromium ground-state atom (Cr^0).

that the atomic number of chromium is 24 (number of protons) and its atomic mass is 52 (number of protons + neutrons).

If energy is then applied to the chromium ground-state atom in the form of heat from a plasma discharge, one of the orbiting electrons will be stripped off the outer shell. This will result in only 23 electrons left orbiting the nucleus. Because the atom has lost a negative charge (e^-) but still has 24 protons (p^+) in the nucleus, it is converted into an ion with a net positive charge. It still has an atomic mass of 52 and an atomic number of 24, but is now a positively charged ion and not a neutral ground-state atom. This process is shown in Figure 2.3.

NATURAL ISOTOPES

This is a very basic look at the process, because most elements occur in more than one form (isotope). In fact, chromium has four naturally occurring isotopes, which means that the chromium atom exists in four different forms, all with the same

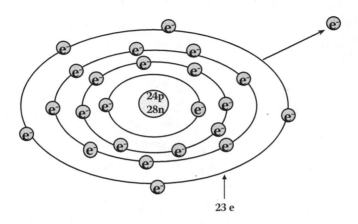

23 e

FIGURE 2.3 Conversion of a chromium ground-state atom (Cr^0) to an ion (Cr^+).

atomic number of 24 (number of protons), but with different atomic masses (numbers of neutrons).

To make this a little easier to understand, let us take a closer look at an element such as copper, which only has two different isotopes—one with an atomic mass of 63 (^{63}Cu) and another with an atomic mass of 65 (^{65}Cu). They both have the same number of protons and electrons, but differ in the number of neutrons in the nucleus. The natural abundances of ^{63}Cu and ^{65}Cu are 69.1% and 30.9%, respectively, which gives copper a nominal atomic mass of 63.55—the value you see for copper in atomic weight reference tables. Details of the atomic structure of the two copper isotopes are shown in Table 2.1.

TABLE 2.1

Breakdown of the Atomic Structure of Copper Isotopes

	^{63}Cu	^{65}Cu
Number of protons (p^+)	29	29
Number electrons (e^-)	29	29
Number of neutrons (n)	34	36
Atomic mass ($p^+ + n$)	63	65
Atomic number (p^+)	29	29
Natural abundance	69.17%	30.83%
Nominal atomic weight	63.55[a]	

[a] The nominal atomic weight of copper is calculated using the formula 0.6917n (^{63}Cu) + 0.3083n (^{65}Cu) + p^+ and referenced to the atomic weight of carbon.

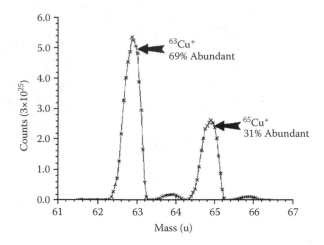

FIGURE 2.4   Mass spectra of the two copper isotopes—$^{63}Cu^+$ and $^{65}Cu^+$.

When a sample containing naturally occurring copper is introduced into the plasma, two different ions of copper, $^{63}Cu^+$, and $^{65}Cu^+$ are produced that generate two different mass spectra—one at mass 63 and another at mass 65. This can be seen in Figure 2.4, which is an actual ICP-MS spectral scan of a sample containing copper, showing a peak for the $^{63}Cu^+$ ion on the left, which is 69.17% abundant, and a peak for $^{65}Cu^+$ at 30.83% abundance on the right. You can also see small peaks for two Zn isotopes at mass 64 ($^{64}Zn^+$) and mass 66 ($^{66}Zn^+$). (Zn has a total of five isotopes at masses 64, 66, 67, 68, and 70.) In fact, most elements have at least two or three isotopes, and many elements including zinc and lead, have four or more isotopes. Figure 2.5 is a chart showing the relative abundance of the naturally occurring isotopes of all the elements.

Relative Abundance of the Natural Isotopes

Isotope	El	%	El	%	El	%
1	H	99.985				
2	H	0.015				
3	He	0.000137				
4	He	99.999863				
6	Li	7.5				
7	Li	92.5				
9	Be	100				
10	B	19.9				
11	B	80.1				
12	C	98.90				
13	C	1.10				
14	N	99.643				
15	N	0.366				
16	O	99.762				
17	O	0.038				
18	O	0.200				
19	F	100				
20	Ne	90.48				
21	Ne	0.27				
22	Ne	9.25				
23	Na	100				
24	Mg	78.99				
25	Mg	10.00				
26	Mg	11.01				
27	Al	100				
28	Si	92.23				
29	Si	4.67				
30	Si	3.10				
31	P	100				
32	S	95.02				
33	S	0.75				
34	S	4.21				
35	Cl	75.77				
36	S	0.02	Ar	0.337		
37	Cl	24.23				
38	Ar	0.063				
39	K	93.2581				
40	K	0.0117	Ar	99.600	Ca	96.941
41	K	6.7302				
42	Ca	0.647				
43	Ca	0.135				
44	Ca	2.086				
45	Sc	100				
46	Ca	0.004				
47	Ca	0.187				
48	Ca	0.250				
50	V	0.250	Cr	4.345		
51	V	99.750				
52	Cr	83.789				
53	Cr	9.501				
54	Cr	2.365	Fe	5.8		
55	Mn	100				
56	Fe	91.72				
57	Fe	2.2				
58	Fe	0.28	Ni	68.077		
59	Co	100				
60	Ni	26.223				

Isotope	El	%	El	%	El	%
61	Ni	1.140				
62	Ni	3.634				
63	Cu	69.17				
64	Ni	0.926	Zn	48.6		
65	Cu	30.83				
66	Zn	27.9				
67	Zn	4.1				
68	Zn	18.8				
69	Ga	60.108				
70	Ge	21.23	Zn	0.6		
71	Ga	39.892				
72	Ge	27.66				
73	Ge	7.73				
74	Ge	35.94	Se	0.89		
75	As	100				
76	Ge	7.44	Se	9.36		
77	Se	7.63				
78	Se	23.78	Kr	0.35		
79	Br	50.69				
80	Se	49.61	Kr	2.25		
81	Br	49.31				
82	Se	8.73	Kr	11.6		
83	Kr	11.5				
84	Kr	57.0				
85	Rb	72.165				
86	Kr	17.3	Sr	9.86		
87	Rb	27.835	Sr	7.00		
88	Sr	82.58				
89	Y	100				
90	Zr	51.45				
91	Zr	11.22				
92	Zr	17.15	Mo	14.84		
93	Nb	100				
94	Zr	17.38	Mo	9.25		
95	Mo	15.92				
96	Zr	2.80	Mo	16.68	Ru	5.52
97	Mo	9.55				
98	Mo	24.13	Ru	1.88		
99	Ru	12.7				
100	Mo	9.63	Ru	12.6		
101	Ru	17.0				
102	Pd	1.02	Ru	31.6		
103	Rh	100				
104	Pd	11.14	Ru	18.7		
105	Pd	22.33				
106	Pd	27.33	Cd	1.25		
107	Ag	51.839				
108	Pd	26.46	Cd	0.89		
109	Ag	48.161				
110	Pd	11.72	Cd	12.49		
111	Cd	12.80				
112	Sn	0.97	Cd	24.13		
113	In	4.3	Cd	12.22		
114	Sn	0.65	Cd	28.73		
115	Sn	0.34	In	95.7		
116	Sn	14.53	Cd	7.49		
117	Sn	7.68				
118	Sn	24.23				
119	Sn	8.59				
120	Sn	32.59	Te	0.096		

Isotope	El	%	El	%	El	%
121	Sb	57.36				
122	Sn	4.63	Te	2.603		
123	Sb	42.64	Te	0.908		
124	Sn	5.79	Te	4.816	Xe	0.10
125	Te	7.139				
126	Te	18.95	Xe	0.09		
127	I	100				
128	Te	31.69	Xe	1.91		
129	Xe	26.4				
130	Ba	0.106	Te	33.80	Xe	4.1
131	Xe	21.2				
132	Ba	0.101	Xe	26.9		
133	Cs	100				
134	Ba	2.417	Xe	10.4		
135	Ba	6.592				
136	Ba	7.854	Ce	0.19	Xe	8.9
137	Ba	11.23				
138	Ba	71.70	La	0.0902	Ce	0.25
139	La	99.9098				
140	Ce	88.48				
141	Pr	100				
142	Nd	27.13	Ce	11.08		
143	Nd	12.18				
144	Nd	23.80	Sm	3.1		
145	Nd	8.30				
146	Nd	17.19				
147	Sm	15.0				
148	Nd	5.76	Sm	11.3		
149	Sm	13.8				
150	Nd	5.64	Sm	7.4		
151	Eu	47.8				
152	Gd	0.20	Sm	26.7		
153	Eu	52.2				
154	Gd	2.18	Sm	22.7		
155	Gd	14.80				
156	Gd	20.47	Dy	0.06		
157	Gd	15.65				
158	Gd	24.84	Dy	0.10		
159	Tb	100				
160	Gd	21.86	Dy	2.34		
161	Dy	18.9				
162	Er	0.14	Dy	25.5		
163	Dy	24.9				
164	Er	1.61	Dy	28.2		
165	Ho	100				
166	Er	33.6				
167	Er	22.95				
168	Er	26.8	Yb	0.13		
169	Tm	100				
170	Er	14.9	Yb	3.05		
171	Yb	14.3				
172	Yb	21.9				
173	Yb	16.12				
174	Yb	31.8	Hf	0.162		
175	Lu	97.41				
176	Lu	2.59	Yb	12.7	Hf	5.206
177	Hf	18.606				
178	Hf	27.297				
179	Hf	13.629				
180	Ta	0.012	W	0.13	Hf	35.100

Isotope	El	%	El	%	El	%
181	Ta	99.988				
182	W	26.3				
183	W	14.3				
184	Os	0.02	W	30.67		
185	Re	37.40				
186	Os	1.58	W	28.6	Re	62.60
187	Os	1.6				
188	Os	13.3				
189	Os	16.1				
190	Os	26.4	Pt	0.01		
191	Ir	37.3				
192	Os	41.0	Ir	62.7	Pt	0.79
194	Pt	32.9				
195	Pt	33.8				
196	Hg	0.15	Pt	25.3		
197	Au	100				
198	Hg	9.97	Pt	7.2		
199	Hg	16.87				
200	Hg	23.10				
201	Hg	13.18				
202	Hg	29.86				
203	Tl	29.524				
204	Hg	6.87	Pb	1.4		
205	Tl	70.476				
206	Pb	24.1				
207	Pb	22.1				
208	Pb	52.4				
209	Bi	100				
231	Pa	100				
232	Th	100				
234	U	0.0055				
235	U	0.7200				
238	U	99.2745				

"Isotopic Compositions of the Elements 1989" Pure Appl. Chem., Vol. 63, No. 7, pp. 991–1002, 1991. © 1991 IUPAC

FIGURE 2.5 Relative abundance of the naturally occurring isotopes of the elements. (From UIPAC Isotopic Composition of the Elements, *Pure and Applied Chemistry* **75**[6], 683–799, 2003.)

3 Sample Introduction

Chapter 3 examines one of the most critical areas of the instrument—the sample introduction system. It discusses the fundamental principles of converting a liquid into a fine-droplet aerosol suitable for ionization in the plasma, and presents an overview of the different types of commercially available nebulizers and spray chambers. Although this chapter briefly touches upon some of the newer sampling components introduced in the past few years, the new breed of desolvating nebulizers and chilled spray chambers are specifically addressed in Chapter 17.

The majority of current ICP-MS applications involve the analysis of liquid samples. Even though the technique has been adapted over the years to handle solids and slurries, it was developed in the early 1980s primarily to analyze solutions. There are many different ways of introducing a liquid into an ICP mass spectrometer, but they all basically achieve the same result, which is to generate a fine aerosol of the sample so that it can be efficiently ionized in the plasma discharge. The sample introduction area has been called the Achilles' heel of ICP-MS, because it is considered the weakest component of the instrument. Only 2–5% of the sample finds its way into the plasma, depending on the matrix and method of introducing the sample.[1] Although there has recently been significant innovation in this area, particularly in instrument-specific components custom built by third-party vendors, the fundamental design of a traditional ICP-MS sample introduction system has not dramatically changed since the technique was first introduced in 1983.

Before I discuss the mechanics of aerosol generation in greater detail, let us look at the basic components of a sample introduction system. Figure 3.1 shows the location of the sample introduction area relative to the rest of the ICP mass spectrometer, whereas Figure 3.2 represents a more detailed view showing the individual components.

The traditional way of introducing a liquid sample into an analytical plasma can be considered as two separate events: aerosol generation using a nebulizer and droplet selection using a spray chamber.[2]

AEROSOL GENERATION

As mentioned previously, the main function of the sample introduction system is to generate a fine aerosol of the sample. It achieves this with a nebulizer and a spray chamber. The sample is normally pumped at about 1 mL/min via a peristaltic pump into the nebulizer. A peristaltic pump is a small pump with lots of minirollers that all rotate at the same speed. The constant motion and pressure of the rollers on the pump tubing feeds the sample through to the nebulizer. The benefit of a peristaltic pump

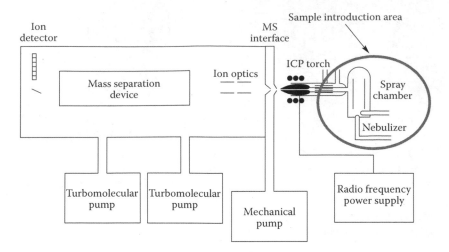

FIGURE 3.1 Location of the ICP-MS sample introduction area.

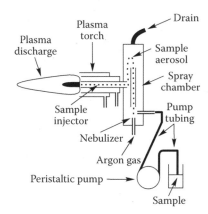

FIGURE 3.2 More detailed view of the ICP-MS sample introduction area.

is that it ensures a constant flow of liquid, irrespective of differences in viscosity between samples, standards, and blanks. Once the sample enters the nebulizer, the liquid is then broken up into a fine aerosol by the pneumatic action of a flow of gas (~1 L/min) "smashing" the liquid into tiny droplets, very similar to the spray mechanism in a can of deodorant. It should be noted that although pumping the sample is the most common approach to introducing the sample, some pneumatic designs such as concentric nebulizers do not require a pump, because they rely on the natural "venturi effect" of the positive pressure of the nebulizer gas to suck the sample through the tubing. Solution nebulization is conceptually represented in Figure 3.3, which shows aerosol generation using a cross-flow-designed nebulizer.

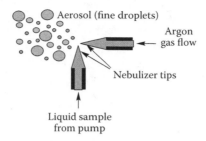

FIGURE 3.3   Conceptual representation of aerosol generation using a cross-flow nebulizer.

DROPLET SELECTION

Because the plasma discharge is not very efficient at dissociating large droplets, the function of the spray chamber is primarily to allow only the small droplets to enter the plasma. Its secondary purpose is to smooth out pulses that occur during the nebulization process, owing mainly to the peristaltic pump. Spray chambers are discussed in greater detail later in this chapter, but the most common type is the double-pass design, where the aerosol from the nebulizer is directed into a central tube running the entire length of the chamber. The droplets then travel the length of this tube, where the large droplets (greater than ~10 μm dia.) will fall out by gravity and exit through the drain tube at the end of the spray chamber. The fine droplets (<10 μm dia.) then pass between the outer wall and the central tube, where they eventually emerge from the spray chamber and are transported into the sample injector of the plasma torch.[3] Although there are many different designs available, the spray chamber's main function is to allow only the smallest droplets into the plasma for dissociation, atomization, and, finally, ionization of the sample's elemental components. A simplified schematic of this process using a double-pass designed spray chamber is shown in Figure 3.4.

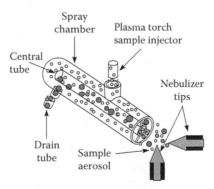

FIGURE 3.4   Simplified representation of the separation of large and fine droplets in a double-pass spray chamber.

Let us now look at the most common nebulizers and spray chamber designs used in ICP-MS. We cannot cover every conceivable design, because over the past few years there has been a huge demand for application-specific solutions, which has generated a number of third-party manufacturers that sell sample introduction components directly to ICP-MS users.

NEBULIZERS

By far the most common design used for ICP-MS is the pneumatic nebulizer, which uses mechanical forces of a gas flow (normally argon at a pressure of 20–30 psi) to generate the sample aerosol. Some of the most popular designs of pneumatic nebulizers include the concentric, microconcentric, microflow, and cross-flow. They are usually made from glass, but other nebulizer materials, such as various kinds of polymers, are becoming more popular, particularly for highly corrosive samples and specialized applications. It should be emphasized at this point that nebulizers designed for use with ICP-OES are far from ideal for use with ICP-MS. This is the result of a limitation in the quantity of total dissolved solids (TDS) that can be put into the ICP-MS interface area. Because the orifice sizes of the sampler and skimmer cones used in ICP-MS are so small (~0.6–1.2 mm), the matrix components must generally be kept below 0.2%, although higher concentrations of some matrices can be tolerated (refer to Chapter 5).[4] This means that general-purpose ICP-OES nebulizers that are designed to aspirate 1–2% dissolved solids, or high-solids nebulizers such as the Babbington, V-groove, or cone-spray, which are designed to handle up to 20% dissolved solids, are not ideally suited to analyzing solutions by ICP-MS. Some researchers have attempted to analyze slurries by ICP-MS using this approach. This is not recommended for high-throughput, routine work because of the potential of blocking the interface cones, but as long as the particle size of the slurry is kept below 10 μm in diameter, some success has been achieved using these types of nebulizers.[5] The most common of the pneumatic nebulizers used in commercial ICP mass spectrometers are the concentric and cross-flow design types. The concentric design is the most widely used nebulizer for clean samples, whereas the cross-flow is generally more tolerant to samples containing higher solids and particulate matter. However, recent advances in the concentric design have allowed for the aspiration of these types of samples.

CONCENTRIC DESIGN

In traditional concentric nebulization, a solution is introduced through a fine-bore capillary tube, where it comes into contact with a rapidly moving flow of argon gas at a pressure of approximately 30–50 psi. The high-speed gas and the lower-pressure sample combine to create a venturi effect, which results in the sample being sucked through to the end of the capillary, where it is broken up into a fine-droplet aerosol. Most concentric nebulizers being used today are manufactured from borosilicate glass or quartz. However, polymer-based materials are now being used for applications that require corrosion resistance. Typical sample flow rates for a standard concentric nebulizer are on the order of 1–3 mL/min, although lower flows can be

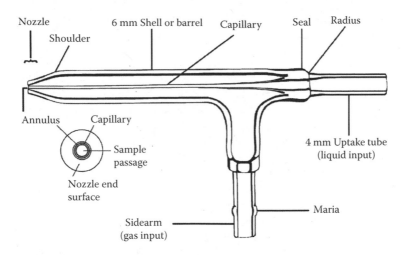

FIGURE 3.5 Schematic of a glass concentric nebulizer (courtesy of Meinhard Glass Products).

FIGURE 3.6 Aerosol generated by a concentric nebulizer (courtesy of Meinhard Glass Products).

used to accommodate more volatile sample matrices, such as organic solvents. A schematic of a glass concentric nebulizer with the different parts labeled is shown in Figure 3.5, and the aerosol generated by the nebulization process is shown in Figure 3.6.

The standard concentric pneumatic nebulizer will give excellent sensitivity and stability, particularly with clean solutions. However, the narrow capillary can be plagued by blockage problems, especially if heavier-matrix samples are being aspirated. For that reason, manufacturers of concentric nebulizers offer modifications to the basic design utilizing different size capillary tubing and recessed tips to allow aspiration of samples with higher dissolved solids and particulate matter. There are even specially designed concentric nebulizers with a smaller-bore input capillary to significantly reduce the dead volume for better coupling of a high-performance liquid chromatography (HPLC) system to the ICP-MS when carrying out trace element speciation studies.

CROSS-FLOW DESIGN

For the routine analysis of samples that contain a heavier matrix, or maybe small amounts of undissolved matter, the cross-flow design is probably the more rugged design. With this nebulizer, the argon gas is directed at right angles to the tip of a capillary tube, in contrast to the concentric design, where the gas flow is parallel to the capillary. The solution is either drawn up through the capillary tube via the pressure created by the high-speed gas flow, or, as is most common with cross-flow nebulizers, fed through the tube with a peristaltic pump. In either case, contact between the high-speed gas and the liquid stream causes the liquid to break up into an aerosol. Cross-flow nebulizers are generally not as efficient as concentric nebulizers at creating the very small droplets needed for ionization in the plasma. However, the larger-diameter liquid capillary and longer distance between liquid and gas injectors reduces the potential for clogging problems. Many analysts feel that the small penalty to be paid in analytical sensitivity and precision with cross-flow nebulizers compared to the concentric design is compensated by the fact that they are better suited for high-throughput, routine applications. In addition, they are typically manufactured from plastic materials, which makes them far more rugged than a glass concentric nebulizer. A cross section of a cross-flow nebulizer is shown in Figure 3.7.

MICROFLOW DESIGN

A new breed of nebulizers has been developed for ICP-MS called microflow or high-efficiency nebulizers, which are designed to operate at much lower sample flows. Whereas conventional nebulizers have a sample uptake rate of about 1 mL/min, microflow and high-efficiency nebulizers typically run at less than 0.1 mL/min. They are based on the concentric principle, but usually operate at higher gas pressure to accommodate the lower sample flow rates. The extremely low uptake rate makes them ideal for applications where sample volume is limited or where the sample or analyte is prone to sample introduction memory effects. The additional benefit of

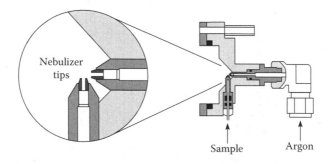

FIGURE 3.7   Schematic of a cross-flow nebulizer (copyright © 2003–2007, all rights reserved, PerkinElmer Inc.).

FIGURE 3.8   The OpalMist™ microflow concentric nebulizer made from PFA (courtesy of Glass Expansion Inc.).

this design is that it produces an aerosol with smaller droplets, and as a result it is generally more efficient than a conventional concentric nebulizer.

These nebulizers and their components are typically constructed from polymer materials, such as polytetrafluoroethylene (PTFE), perfluoroalkoxy (PFA), or polyvinylfluoride (PVF), although some designs are available in borosilicate glass or quartz. The excellent corrosion resistance of the polymer nebulizers means they have naturally low blank levels. This characteristic, together with their ability to handle small sample volumes found in applications such as vapor phase decomposition (VPD), makes them an ideal choice for semiconductor laboratories that are carrying out ultratrace element analysis.[6,7] A microflow concentric nebulizer made from PFA is shown in Figure 3.8, and a typical spray pattern of the nebulization process is shown in Figure 3.9.

The disadvantage of microconcentric nebulizers is that they use an extremely fine capillary, which makes them not very tolerant to high concentrations of dissolved solids or suspended particles. Their high efficiency also means that most of the sample makes it into the plasma, and as a result can cause more severe matrix suppression problems. In addition, the higher level of matrix components entering the interface has the potential to cause cone blockage problems over extended periods of operation. For these reasons, they have been found to be most applicable for the analysis of samples containing low levels of dissolved solids.

One of the application areas that high-efficiency nebulizers are very well suited to is in the handling of extremely small volumes being eluted from an HPLC or flow injection analyzer (FIA) system into an ICP-MS for doing speciation/microsampling work. The analysis of discrete sample volumes encountered in these types of applications allows for detection limits equivalent to a standard concentric nebulizer, while consuming 10–20 times less sample.

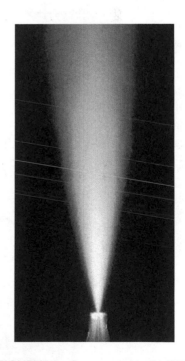

FIGURE 3.9   Spray pattern of a PFA microflow concentric nebulizer (courtesy of Elemental Scientific Inc.).

SPRAY CHAMBERS

Let us now turn our attention to spray chambers. There are basically two designs that are used in today's commercial ICP-MS instrumentation: double-pass and cyclonic spray chambers. The double-pass is by far the most common, with the cyclonic type rapidly gaining in popularity. As mentioned earlier, the function of the spray chamber is to reject the larger aerosol droplets and also to smooth out nebulization pulses produced by the peristaltic pump, if it is used. In addition, some ICP-MS spray chambers are externally cooled for thermal stability of the sample and to reduce the amount of solvent going into the plasma. This can have a number of beneficial effects, depending on the application, but the main advantages are to reduce oxide species, minimize signal drift, and reduce the solvent loading when aspirating volatile organic solvents.

DOUBLE-PASS SPRAY CHAMBER

By far the most common design of the double-pass spray chamber is the Scott design, which selects the small droplets by directing the aerosol into a central tube. The larger droplets emerge from the tube, and by gravity exit the spray chamber via a drain tube. The liquid in the drain tube is kept at positive pressure (usually by way of a loop), which forces the small droplets back between the outer wall and the central tube and emerges from the spray chamber into the sample injector of the plasma torch. Double-pass spray chambers come in a variety of shapes, sizes, and materials, and are generally considered the most rugged design for routine use. Figure 3.10 shows a Scott double-pass spray chamber made of a polysulfide-type material, coupled to a cross-flow nebulizer.

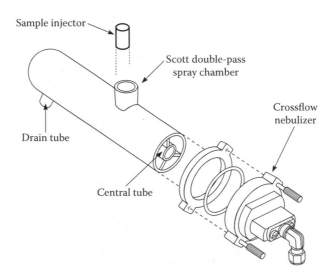

FIGURE 3.10   A Scott double-pass spray chamber with cross-flow nebulizer (copyright © 2003–2007, all rights reserved, PerkinElmer Inc.).

CYCLONIC SPRAY CHAMBER

The cyclonic spray chamber operates by centrifugal force. Droplets are discriminated according to their size by means of a vortex produced by the tangential flow of the sample aerosol and argon gas inside the chamber. Smaller droplets are carried with the gas stream into the ICP-MS, whereas the larger droplets impinge on the walls and fall out through the drain. It is generally accepted that a cyclonic spray chamber has a higher sampling efficiency, which for clean samples translates into higher sensitivity and lower detection limits. However, the droplet size distribution appears to be different from a double-pass design, and for certain types of samples can give slightly inferior precision. Beres and coworkers published a very useful study of the capabilities of a cyclonic spray chamber.[8] Figure 3.11 shows a cyclonic spray chamber connected to a concentric nebulizer.

The cyclonic spray chamber is definitely growing in popularity, particularly as its potential is getting realized in more and more application areas. Just as there is a wide selection of nebulizers available for different applications, there is also a wide choice of customized cyclonic spray chambers, manufactured from glass, quartz, and different polymer materials. Depending on the application being carried out, modifications to the cyclonic design are available for low sample flows, high dissolved solids, fast sample washout, corrosion resistance, and organic solvents. Figure 3.12 shows one of the many variations of cyclonic spray chamber, called the jacketed Cinnabar™, which is a water-cooled borosilicate glass spray chamber optimized for aspirating small sample volumes with a microflow concentric nebulizer.

It is worth emphasizing that cooling the spray chamber is generally beneficial in ICP-MS, because it reduces the solvent loading on the plasma. This has three major benefits. First, because very little plasma energy is wasted vaporizing the solvent, more is available to excite and ionize the analytes. Second, if there is less water being delivered to the plasma, there is less chance of forming oxide and hydroxide species,

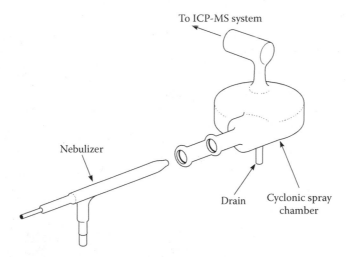

To ICP-MS system

Nebulizer

Drain   Cyclonic spray chamber

FIGURE 3.11 A cyclonic spray chamber (shown with a concentric nebulizer). (From S. A. Beres, P. H. Bruckner, and E. R. Denoyer, *Atomic Spectroscopy,* **15**[2], 96–99, 1994.)

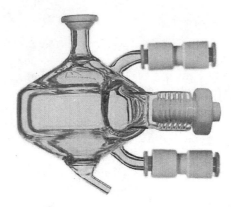

FIGURE 3.12 The low-flow Cinnabar™, a water-cooled cyclonic spray chamber for use with a microflow concentric nebulizer (courtesy of Glass Expansion Inc.).

which can potentially interfere with other analytes. Finally, if the spray chamber is kept at a constant temperature, it leads to better long-term signal stability, especially if there are environmental temperature changes over the time period of the analysis. For these reasons, some manufacturers supply cooled spray chambers as a standard, whereas others offer the capability as an option. There is also a wide variety of cooled and chilled spray chambers available from third-party vendors.

There are many other nonstandard sample introduction devices such as laser ablation, ultrasonic nebulizers, desolvation devices, direct injection nebulizers, flow injection systems, and electrothermal vaporization, which are not described in this chapter. However, because they are becoming more and more important, particularly as ICP-MS users are demanding higher performance and more flexibility, they are covered in greater detail in Chapter 17.

REFERENCES

1. R. A. Browner and A. W. Boorn, *Analytical Chemistry*, **56**, 786–798A, 1984.
2. B. L. Sharp, *Analytical Atomic Spectrometry*, **3**, 613, 1980.
3. L. C. Bates and J. W. Olesik, *Journal of Analytical Atomic Spectrometry*, **5**(3), 239, 1990.
4. R. S. Houk, *Analytical Chemistry*, **56**, 97A, 1986.
5. J. G. Williams, A. L. Gray, P. Norman, and L. Ebdon, *Journal of Analytical Atomic Spectrometry*, **2**, 469–472, 1987.
6. E. Debrah, S. A. Beres, T. J. Gluodennis, R. J. Thomas, and E. R. Denoyer, *Atomic Spectroscopy*, **16**(7), 197–202, 1995.
7. R. A. Aleksejczyk and D. Gibilisco, *Micro,* September 1997.
8. S. A. Beres, P. H. Bruckner, and E. R. Denoyer, *Atomic Spectroscopy,* **15**(2), 96–99, 1994.

4 Plasma Source

Chapter 4 takes a look at the region where the ions are generated—the plasma discharge. It gives a brief historical perspective of some of the common analytical plasmas used over the years and discusses the components used to create the inductively coupled plasma (ICP). It then goes on to explain the fundamental principles of formation of a plasma discharge and how it is used to convert the sample aerosol into a stream of positively charged ions of low kinetic energy required by the ion-focusing system and the mass spectrometer.

ICPs are by far the most common type of plasma sources used in today's commercial ICP optical emission (ICP-OES) and ICP mass spectrometric (ICP-MS) instrumentation. However, it was not always that way. In the early days, when researchers were attempting to find the ideal plasma source to use for spectrometric studies, it was not clear which approach would prove to be the most successful. In addition to ICPs, some of the other novel plasma sources developed were direct current plasmas (DCPs) and microwave-induced plasmas (MIPs). Before I go on to describe the ICP, let us first take a closer look at these other two excitation sources.

A DCP is formed when a gas (usually argon) is introduced into a high current flowing between two or three electrodes. Ionization of the gas produces a Y-shaped plasma. Unfortunately, early DCP instrumentation was prone to interference effects and also had some usability and reliability problems. For these reasons, the technique never became widely accepted by the analytical community.[1] However, its one major benefit was that it could aspirate high dissolved or suspended solids because there was no restrictive sample injector for the solid material to block. This feature alone made it very attractive for some laboratories, and once the initial limitations of DCPs were better understood, the technique became more accepted. In fact, a DCP excitation source coupled to an optical emission instrument today, using an Echelle-based grating and a solid-state detector, has been commercially available for a number of years.[2]

Limitations in the DCP approach led to the development of electrodeless plasma, of which the MIP was the simplest form. In this system, microwave energy (typically 100–200 W) is supplied to the plasma gas from an excitation cavity around a glass/quartz tube. The plasma discharge in the form of a ring is generated inside the tube. Unfortunately, even though the discharge achieves a very high power density, the high excitation temperatures only exist along a central filament. The bulk of the MIP never goes above 2000–3000 K, which means it is prone to very severe matrix effects. In addition, they are easily extinguished when aspirating liquid samples. For these reasons, they have had limited success as an emission source because they are not considered robust enough for the analysis of real-world solution-based samples.

However, they have gained acceptance as an ion source for mass spectrometry (MS),[3] and also as emission-based detectors for gas chromatography.

Because of the limitations of the DCP and MIP approaches, ICPs became the dominant area of research for both optical emission and mass spectrometric studies. As early as 1964, Greenfield and coworkers reported that an atmospheric pressure ICP coupled with OES could be used for elemental analysis.[4] Although crude by today's standards, it showed the enormous possibilities of the ICP as an excitation source and most definitely opened the door in the early 1980s to the even more exciting potential of using the ICP to generate ions.[5]

THE PLASMA TORCH

Before we take a look at the fundamental principles behind the creation of an ICP used in ICP-MS, let us take a look at the basic components used to generate the source—a plasma torch, radio-frequency (RF) coil, and power supply. Figure 4.1 shows their proximity compared to the rest of the instrument, and Figure 4.2 is a more detailed view of the plasma torch and RF coil relative to the MS interface.

The plasma torch consists of three concentric tubes, which are normally made from quartz. In Figure 4.2, these are shown as the outer tube, middle tube, and sample injector. The torch can either be one piece, in which all three tubes are connected, or it can employ a demountable design in which the tubes and the sample injector are separate. The gas (usually argon) that is used to form the plasma (plasma gas) is passed between the outer and middle tubes at a flow rate of ~12–17 L/min. A second gas flow (auxiliary gas) passes between the middle tube and the sample injector at ~1 L/min and is used to change the position of the base of the plasma relative to the tube and the injector. A third gas flow (nebulizer gas), also at ~1 L/min, brings the sample, in the form of a fine-droplet aerosol, from the sample introduction system

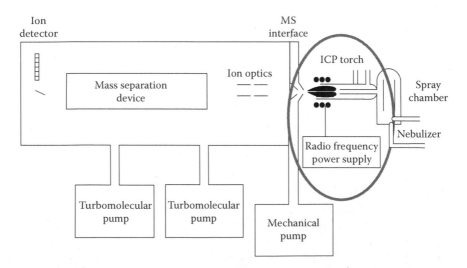

FIGURE 4.1 Inductively coupled plasma mass spectrometry (ICP-MS) system showing location of the plasma torch and RF power supply.

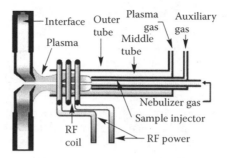

FIGURE 4.2 Detailed view of plasma torch and radio-frequency (RF) coil relative to the inductively coupled plasma mass spectrometry (ICP-MS) interface.

(for details refer to Chapter 3) and physically punches a channel through the center of the plasma. The sample injector is often made from other materials besides quartz, such as alumina, platinum, and sapphire—if highly corrosive materials need to be analyzed. It is worth mentioning that although argon is the most suitable gas to use for all three flows, there are analytical benefits in using other gas mixtures, especially in the nebulizer flow.[6] The plasma torch is mounted horizontally and positioned centrally in the RF coil, approximately 10–20 mm from the interface. This can be seen in Figure 4.3, which shows a photograph of a plasma torch mounted in an instrument.

FIGURE 4.3 Photograph of a plasma torch mounted in an instrument (copyright © 2003–2007, all rights reserved, PerkinElmer Inc.).

It must be emphasized that the coil used in an ICP-MS plasma is slightly different from the one used in ICP-OES, the reason being that in a plasma discharge, there is a potential difference of a few hundred volts produced by capacitive coupling between the RF coil and the plasma. In an ICP mass spectrometer, this would result in a secondary discharge between the plasma and the interface cone, which can negatively affect the performance of the instrument. To compensate for this, the coil must be grounded to keep the interface region as close to zero potential as possible. The full implications of this are discussed in greater detail in Chapter 5.

FORMATION OF AN ICP DISCHARGE

Let us now discuss the mechanism of formation of the plasma discharge in greater detail. First, a tangential (spiral) flow of argon gas is directed between the outer and middle tube of a quartz torch. A load coil (usually copper) surrounds the top end of the torch and is connected to an RF generator. When RF power (typically 750–1500 W, depending on the sample) is applied to the load coil, an alternating current oscillates within the coil at a rate corresponding to the frequency of the generator. In most ICP generators, this frequency is either 27 or 40 MHz (commonly known as megahertz or million cycles/second). This RF oscillation of the current in the coil causes an intense electromagnetic field to be created in the area at the top of the torch. With argon gas flowing through the torch, a high-voltage spark is applied to the gas, causing some electrons to be stripped from their argon atoms. These electrons, which are caught up and accelerated in the magnetic field, then collide with other argon atoms, stripping off still more electrons. This collision-induced ionization of the argon continues in a chain reaction, breaking down the gas into argon atoms, argon ions, and electrons, forming what is known as an inductively coupled plasma (ICP) discharge. The ICP discharge is then sustained within the torch and load coil as RF energy is continually transferred to it through the inductive coupling process. The amount of energy required to generate argon ions in this process is on the order of 15.8 eV (first ionization potential), which is enough energy to ionize the majority of the elements in the periodic table. The sample aerosol is then introduced into the plasma through a third tube called the sample injector. The entire process is conceptually shown in Figure 4.4.[7]

THE FUNCTION OF THE RF GENERATOR

Although the principles of an RF power supply have not changed since the work of Greenfield, the components have become significantly smaller. Some of the early generators that used nitrogen or air required 5–10 kW of power to sustain the plasma discharge—and literally took up half the room. Most of today's generators use solid-state electronic components, which means that vacuum power amplifier tubes are no longer required. This makes modern instruments significantly smaller and, because vacuum tubes were notoriously unreliable and unstable, far more suitable for routine operation.

As mentioned previously, two frequencies have typically been used for ICP RF generators—27 and 40 MHz. These frequencies have been set aside specifically for

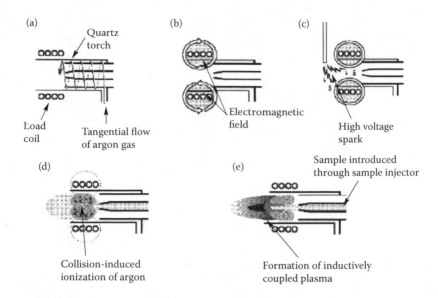

FIGURE 4.4   Schematic of an ICP torch and load coil showing how the inductively coupled plasma (ICP) is formed. (a) A tangential flow of argon gas is passed between the outer and middle tube of the quartz torch. (b) RF power is applied to the load coil, producing an intense electromagnetic field. (c) A high-voltage spark produces free electrons. (d) Free electrons are accelerated by the RF field, causing collisions and ionization of the argon gas. (e) The ICP is formed at the open end of the quartz torch. The sample is introduced into the plasma via the sample injector. (From C. B. Boss and K. J. Fredeen, *Concepts, Instrumentation and Techniques in Inductively Coupled Plasma Optical Emission Spectrometry*, 2nd edition, Perkin Elmer Corporation, 1997.)

RF applications of this kind, so that they will not interfere with other communication-based frequencies. There has been much debate over the years as to which frequency gives the best performance.[8,9] I think it is fair to say that although there have been a number of studies, no frequency appears to give a significant analytical advantage over the other. In fact, of all the commercially available ICP-MS systems, there seems to be roughly an equal number of 27 and 40 MHz generators.

The more important consideration is the coupling efficiency of the RF generator to the coil. The majority of modern solid-state RF generators are on the order of 70–75% efficient, which means that 70–75% of the delivered power actually makes it into the plasma. This was not always the case, and some of the older vacuum-tube-designed generators were notoriously inefficient, with some of them experiencing over a 50% power loss. Another important criterion to consider is the way the matching network compensates for changes in impedance (a material's resistance to the flow of an electric current) produced by the sample's matrix components or differences in solvent volatility, or both. In earlier-designed crystal-controlled generators, this was usually done with servo-driven capacitors. They worked very well with most sample types but, because they were mechanical devices, struggled to compensate for very rapid impedance changes produced by some samples. As a result, it was fairly easy to extinguish the plasma, particularly when aspirating volatile organic solvents.

These problems were partially overcome by the use of free-running RF generators, in which the matching network was based on electronic tuning of small changes in frequency brought about by the sample solvent or matrix components or both. The major benefit of this approach was that compensation for impedance changes was virtually instantaneous, because there were no moving parts. This allowed for the successful analysis of many sample types, which would most probably have extinguished the plasma of a crystal-controlled generator. However, because of improvements in electronic components over the years, the later crystal-controlled generators appear to be as responsive as free-running designs.

IONIZATION OF THE SAMPLE

To better understand what happens to the sample on its journey through the plasma source, it is important to understand the different heating zones within the discharge. Figure 4.5 shows a cross-sectional representation of the discharge along with the approximate temperatures for different regions of the plasma.

As mentioned previously, the sample aerosol enters the injector via the spray chamber. When it exits the sample injector, it is moving at such a velocity that it physically punches a hole through the center of the plasma discharge. It then goes through a number of physical changes, starting at the preheating zone and continuing through the radiation zone, before it eventually becomes a positively charged ion in the analytical zone. To explain this in a very simplified way, let us assume that the element exists as a trace metal salt in solution. The first step that takes place is desolvation of the droplet. With the water molecules stripped away, it then becomes a very small solid particle. As the sample moves further into the plasma, the solid particle changes first into gaseous form and then into a ground-state atom. The final process of conversion of an atom to an ion is achieved mainly by collisions of energetic argon electrons (and to a lesser extent by argon ions) with the ground-state atom.[10] The ion then emerges from the plasma and is directed into the interface of the mass spectrometer (for details on the mechanisms of ion generation, refer to Chapter 2). This process of conversion of droplets into ions is represented in Figure 4.6.

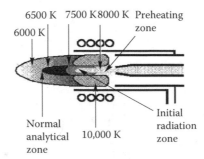

FIGURE 4.5 Different temperature zones in the plasma. (From C. B. Boss and K. J. Fredeen, *Concepts, Instrumentation and Techniques in Inductively Coupled Plasma Optical Emission Spectrometry*, 2nd edition, Perkin Elmer Corporation, 1997.)

Droplet   (Desolvation) Solid (Vaporization) Gas (Atomization) Atom (Ionization) Ion

$$M(H_2O)^1X^2 \longrightarrow (MX)_n \longrightarrow MX \longrightarrow M \longrightarrow M^1$$

From sample injector ———————————————————→ To mass spectrometer

FIGURE 4.6 Mechanism of conversion of a droplet to a positive ion in the inductively coupled plasma (ICP).

REFERENCES

1. A. L. Gray, *Analyst*, **100**, 289–299, 1975.
2. G. N. Coleman, D. E. Miller, and R. W. Stark, *American Lab.*, **30**(4), 33R, 1998.
3. D. J. Douglas and J. B. French, *Analytical. Chemistry*, **53**, 37–41, 1981.
4. S. Greenfield, I. L. Jones, and C. T. Berry, *Analyst*, **89**, 713–720, 1964.
5. R. S. Houk, V. A. Fassel, and H. J. Svec, *Dynamic Mass Spectrometry*, **6**, 234, 1981.
6. J. W. Lam and J. W. McLaren, *Journal of Analytical Atomic Spectrometry*, **5**, 419–424, 1990.
7. C. B. Boss and K. J. Fredeen, *Concepts, Instrumentation and Techniques in Inductively Coupled Plasma Optical Emission Spectrometry*, 2nd edition, Perkin Elmer Corporation, 1997.
8. K. E. Jarvis, P. Mason, T. Platzner, and J. G. Williams, *Journal of Analytical Atomic Spectrometry*, **13**, 689–696, 1998.
9. G. H. Vickers, D. A. Wilson, and G. M. Hieftje, *Journal of Analytical Atomic Spectrometry* **4**, 749–754, 1989.
10. T. Hasegawa and H. Haraguchi, *ICPs in Analytical Atomic Spectrometry*, A. Montasser, D. W. Golightly (eds), 2nd edition, VCH, New York, 1992.

5 Interface Region

Chapter 5 takes a look at the interface region, which is probably the most critical area of the entire ICP-MS system. It gave the early pioneers of the technique the most problems to overcome. Although we take all the benefits of ICP-MS for granted, the process of taking a liquid sample, generating an aerosol that is suitable for ionization in the plasma and then sampling a representative number of analyte ions, transporting them through the interface, focusing them via the ion optics into the mass spectrometer, and finally ending up with detection and conversion to an electronic signal is not a trivial task. Each part of the journey has its own unique problems to overcome, but probably the most challenging is the movement of the ions from the plasma into the mass spectrometer.

The role of the interface region, which is shown in Figure 5.1, is to transport the ions efficiently, consistently, and with electrical integrity from the plasma, which is at atmospheric pressure (760 torr), to the mass spectrometer analyzer region at approximately 10^{-6} torr.

This is first achieved by directing the ions into the interface region. The interface consists of two metallic cones with very small orifices, which are maintained at a vacuum of ~1–2 torr with a mechanical roughing pump. After the ions are generated in the plasma, they pass into the first cone, known as the sampler cone, which has an orifice of 0.8–1.2 mm i.d. From there, they travel a short distance to the skimmer cone, which is generally smaller and more pointed than the sampler cone. The skimmer also has a much smaller orifice (typically 0.4–0.8 mm i.d.) than the sampler cone. Both cones are usually made of nickel, but can be made of other materials such as platinum, which is far more tolerant to corrosive liquids. To reduce the effects of high-temperature plasma on the cones, the interface housing is water cooled and made from a material that dissipates heat easily, such as copper or aluminum. The ions then emerge from the skimmer cone, where they are directed through the ion optics and, finally, guided into the mass separation device. Figure 5.2 shows the interface region in greater detail, and Figure 5.3 shows a close-up of a platinum sampler cone on the left and a platinum skimmer cone on the right.

It should be noted that for most sample matrices, it is desirable to keep the TDS below 0.2%, because of the possibility of deposition of the matrix components around the sampler cone orifice. This is not such a serious problem with short-term use, but it can lead to long-term signal instability if the instrument is being run for extended periods of time. The TDS levels can be higher (0.5–1%) when analyzing a matrix that forms a volatile oxide such as sodium chloride because, once deposited on the cones, the volatile sodium oxide tends to revaporize without forming a significant layer that

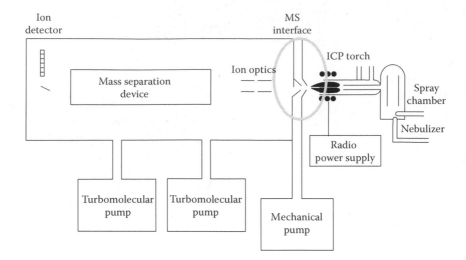

FIGURE 5.1 Schematic of an inductively coupled plasma mass spectrometer (ICP-MS), showing proximity of the interface region.

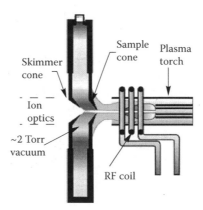

FIGURE 5.2 Detailed view of the interface region.

could potentially affect the flow through the cone orifice. In fact, some researchers have reported running a 1:1 dilution of seawater (1.5% NaCl) for extended periods of time with good stability and no significant cone blockage—by careful optimization of the plasma RF power, sampling depth, and extraction lens voltage.[1]

CAPACITIVE COUPLING

This process sounds fairly straightforward, but proved to be very problematic during the early development of ICP-MS because of an undesired electrostatic (capacitive) coupling between the voltage on the load coil and the plasma discharge, producing a potential difference of 100–200 V. Although this potential is a physical characteristic of all ICP discharges, it was more serious in an ICP mass spectrometer, because the

FIGURE 5.3 Close-up of a platinum sampler cone (left) and a platinum skimmer cone (right) (courtesy of Spectron Inc.).

capacitive coupling created an electrical discharge between the plasma and the sampler cone. This discharge, commonly called the pinch effect or secondary discharge, shows itself as arcing in the region where the plasma is in contact with the sampler cone.[2] This is shown in a simplified manner in Figure 5.4.

If not taken care of, this arcing can cause all kinds of problems, including an increase in doubly charged interfering species, a wide kinetic energy spread of sampled ions, formation of ions generated from the sampler cone, and decreased orifice lifetime. These were all problems reported by many of the early researchers into the technique.[3,4] In fact, because the arcing increased with sampler cone orifice size, the source of the secondary discharge was originally thought to be the result of an electro-gas-dynamic effect, which produced an increase in electron density at the orifice.[5] After many experiments, it was eventually realized that the secondary discharge was a result of electrostatic coupling of the load coil to the plasma. The problem was first eliminated by grounding the induction coil at the center, which had the effect of reducing the RF potential to a few volts. This can be seen in Figure 5.5,

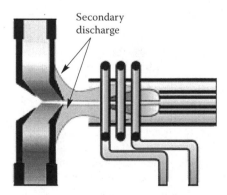

Secondary
discharge

FIGURE 5.4 Interface showing area affected by a secondary discharge.

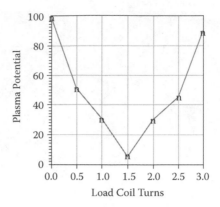

FIGURE 5.5 Reduction in plasma potential as the load coil is grounded at different positions (turns) along its length.[6]

which is taken from one of the early papers and shows the reduction in plasma potential as the coil is grounded at different positions (turns) along its length.[6]

This work has since been supported by other researchers who carried out Langmuir probe measurements, the results indicating that plasma potential was lowest with a center-tapped coil as opposed to the grounding being elsewhere on the coil.[7,8] In today's instrumentation, the "grounding" is implemented in a number of different ways, depending on the design of the interface. Some of the most popular designs include balancing the oscillator inside the circuitry of the RF generator,[9] positioning a grounded shield or plate between the coil and the plasma torch,[10] and using two interlaced coils where the RF fields go in opposite directions.[11] They all work differently, but many experts believe that the center-tapped coil and the interlaced coil achieve the lowest plasma potential compared to the other designs. However, they all appear to work equally well when it comes to using cool plasma conditions requiring higher RF power and lower nebulizer gas flow. Further details about cool and cold plasma technology can be found in Chapter 14.

ION KINETIC ENERGY

The impact of a secondary discharge cannot be overemphasized with respect to its effect on the kinetic energy of the ions being sampled. It is well documented that the energy spread of the ions entering the mass spectrometer must be as low as possible to ensure they can all be focused efficiently and with full electrical integrity by the ion optics and the mass separation device. When the ions emerge from the argon plasma, they will all have different kinetic energies, depending on their mass-to-charge ratio. Their velocities should all be similar, because they are controlled by rapid expansion of the bulk plasma, which will be neutral as long as it is maintained at zero potential. As the ion beam passes through the sampler cone into the skimmer cone, expansion will take place, but its composition and integrity will be maintained, assuming the plasma is neutral. This can be seen in Figure 5.6.

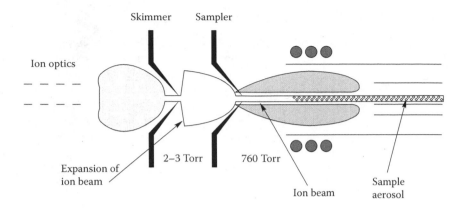

FIGURE 5.6 The composition of the ion beam is maintained as it passes through the interface, a neutral plasma being assumed.

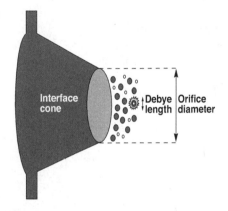

FIGURE 5.7 Electrodynamic forces do not affect the composition of the ion beam entering the sampler or the skimmer cone. ·

Electrodynamic forces do not play a role as the ions enter the sampler or the skimmer, because the distance over which the ions exert an influence on one another (known as the Debye length) is small (typically 10^{-3}–10^{-4} mm) compared to the diameter of the orifice (0.5–1.0 mm),[6] as shown in Figure 5.7.

It is therefore clear that maintaining a neutral plasma is of paramount importance to guarantee the electrical integrity of the ion beam as it passes through the interface region. If a secondary discharge is present, the electrical characteristics of the plasma change, which will affect the kinetic energy of the ions differently, depending on their mass-to-charge ratio. If the plasma is at zero potential, the ion energy spread is on the order of 5–10 eV. However, if a secondary discharge is present, a much wider spread of ion energies entering the mass spectrometer (typically 20–40 eV) results, which makes ion focusing far more complicated.[6]

BENEFITS OF A WELL-DESIGNED INTERFACE

The benefits of a well-designed interface are not readily obvious if simple aqueous samples are being analyzed using only one set of operating conditions. However, it becomes more apparent when many different sample types are being analyzed, requiring different operating parameters. The design of the interface is really put to the test when plasma conditions need to be changed, when the sample matrix changes, or when ICP-MS is being used to analyze solid materials. Analytical scenarios such as these have the potential to induce a secondary discharge, change the kinetic energy of the ions entering the mass spectrometer, and affect the tuning of the ion optics. It is therefore critical that the interface grounding mechanism be able to handle these types of real-world analytical situations, including the following:

- **Using cool plasma conditions:** All instruments today have the ability to use cool plasma conditions (500–700 W power and 1.0–1.3 L/min nebulizer gas flow) to lower the plasma temperature and reduce argon-based polyatomic interferences such as $^{40}Ar^{16}O^+$, $^{40}Ar^+$, and $^{38}ArH^+$ in the determination of difficult elements such as $^{56}Fe^+$, $^{40}Ca^+$, and $^{39}K^+$. Such dramatic deviations from normal operating conditions (1000 W, 0.8 L/min) will affect the electrical characteristics of the plasma.
- **Running organic solvents:** Analyzing oil or organic-based samples requires a chilled spray chamber (typically −20°C) or a membrane desolvation system to reduce the solvent loading on the plasma. In addition, higher RF power (~1300–1500 W) and lower nebulizer gas flow (~0.4–0.8 L/min) are required to dissociate the organic components in the sample. A reduction in the amount of solvent entering the plasma combined with higher power and lower nebulizer gas flow translates into a hotter plasma, and a change in its ionization mechanism.
- **Optimizing conditions for low oxides:** The formation of oxide species can be problematic in some sample types. For example, in geochemical applications it is quite common to sacrifice sensitivity by lowering the nebulizer gas flow and increasing the RF power to reduce the formation of rare earth oxides, which can spectrally interfere with the determination of other analytes. Unfortunately, these conditions will change the electrical characteristics of the plasma, which can induce a secondary discharge.
- **Using sampling accessories:** Sampling accessories such as membrane desolvators, laser ablation systems, and electrothermal vaporization devices are being used more routinely to enhance the flexibility of ICP-MS. The major difference between these sampling devices and a conventional liquid sample introduction system is that they generate a "dry" sample aerosol, which requires completely different operating conditions compared to a conventional "wet" plasma. An aerosol that contains no solvent can have a dramatic effect on the ionization conditions in the plasma.

Even though most modern ICP-MS interfaces have been designed to minimize the effects of the secondary discharge, it should not be taken for granted that they can all handle changes in operating conditions and matrix components with the same ease. The most noticeable problems that have been reported include spectral peaks of the cone material appearing in the blank, erosion/discoloration of the sampling cones, widely different optimum plasma conditions (neb flow/RF power) for different masses, and frequent retuning of the ion optics.[12,13] Chapter 21 goes into this subject in greater detail, but there is no question that the plasma discharge, interface region, and ion optics have to be designed in concert to ensure the instrument can handle a wide range of operating conditions and sample types.

REFERENCES

1. M. Plantz and S. Elliott, *Application Note* # ICP-MS 17, Varian Instruments, 1998.
2. A. L. Gray and A. R. Date, *Analyst,* **108**, 1033, 1983.
3. R. S. Houk, V. A. Fassel, and H. J. Svec, *Dynamic Mass Spectrometry,* **6**, 234, 1981.
4. A. R. Date and A. L. Gray, *Analyst,* **106**, 1255, 1981.
5. A. L. Gray and A. R. Date, *Dynamic Mass Spectrometry,* **6**, 252, 1981.
6. D. J. Douglas and J. B. French, *Spectrochimica Acta,* **41B**(3), 197, 1986.
7. A. L. Gray, R. S. Houk, and J. G. Williams, *Journal of Analytical Atomic Spectrometry,* **2**, 13–20, 1987.
8. R. S. Houk, J. K. Schoer, and J. S. Crain, *Journal of Analytical Atomic Spectrometry,* **2**, 283–286, 1987.
9. S. D. Tanner, *Journal of Analytical Atomic Spectrometry,* **10**, 905, 1995.
10. K. Sakata and K. Kawabata, *Spectrochimica Acta,* **49B**, 1027, 1994.
11. S. Georgitus and M. Plantz, *Winter Conference on Plasma Spectrochemistry,* **FP4**, Fort Lauderdale, 1996.
12. D. J. Douglas, *Canadian Journal of Spectroscopy,* **34**, 2, 1989.
13. J. E. Fulford and D. J. Douglas, *Applied Spectroscopy,* **40**, 7, 1986.

6 Ion-Focusing System

Chapter 6 takes a detailed look at the ion-focusing system—a crucial area of the ICP mass spectrometer—where the ion beam is focused before it enters the mass analyzer. Sometimes known as the ion optics, it comprises one or more ion lens components, which electrostatically steer the analyte ions in an axial (straight) or orthogonal (right-angled) direction from the interface region into the mass separation device. The strength of a well-designed ion-focusing system is its ability to produce a flat signal response over the entire mass range, low background levels, good detection limits, and stable signals in real-world sample matrices.

Although the detection capability of ICP-MS is generally recognized as being superior to any of the other atomic spectroscopic techniques, it is probably most susceptible to the sample's matrix components. The inherent problem lies in the fact that ICP-MS is relatively inefficient—out of a million ions generated in the plasma, only one actually reaches the detector. One of the main contributing factors to the low efficiency is the higher concentration of matrix elements compared to the analyte, which has the effect of defocusing the ions and altering the transmission characteristics of the ion beam. This is sometimes referred to as a space charge effect, and can be particularly severe when the matrix ions are of a heavier mass than the analyte ions.[1] The role of the ion-focusing system is therefore to transport the maximum number of analyte ions from the interface region to the mass separation device, while rejecting as many of the matrix components and non-analyte-based species as possible. Let us now discuss this process in greater detail.

ROLE OF THE ION OPTICS

The ion optics, shown in Figure 6.1, are positioned between the skimmer cone and mass separation device and consist of one or more electrostatically controlled lens components, maintained at a vacuum of approximately 10^{-3} torr with a turbomolecular pump. They are not traditional optics that we associate with ICP emission or atomic absorption, but are made up of a series of metallic plates, barrels, or cylinders, which have a voltage placed on them. The function of the ion optic system is to take ions from the hostile environment of the plasma at atmospheric pressure via the interface cones and steer them into the mass analyzer, which is under high vacuum. The nonionic species such as particulates, neutral species, and photons are prevented from reaching the detector by using some kind of physical barrier, positioning the mass analyzer off axis relative to the ion beam, or electrostatically bending the ions by 90° into the mass analyzer.

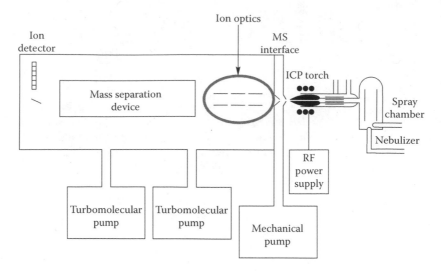

FIGURE 6.1   Position of ion optics relative to the plasma torch and interface region.

As mentioned in Chapters 4 and 5, the plasma discharge and interface region have to be designed in concert with the ion optics. It is absolutely critical that the composition and electrical integrity of the ion beam be maintained as it enters the ion optics. For this reason, it is essential that the plasma be at zero potential to ensure that the magnitude and spread of ion energies are as low as possible.[2]

A secondary, but also very important, role of the ion optic system is to stop particulates, neutral species, and photons from getting through to the mass analyzer and the detector. These species cause signal instability and contribute to background levels, which ultimately affect the performance of the system. For example, if photons or neutral species reach the detector, they will elevate the noise of the background and therefore degrade detection capability. In addition, if particulates from the matrix penetrate further into the mass spectrometer region, they have the potential to deposit on lens components, and in extreme cases, get into the mass analyzer. In the short term, this will cause signal instability, and in the long term, increase the frequency of cleaning and routine maintenance.

There are basically three different approaches to reducing the chances that these undesirable species will enter the mass spectrometer. The first method is to place a grounded metal stop (disk) behind the skimmer cone. This stop allows the ion beam to move around it and physically block the particulates, photons, and neutral species from traveling "downstream."[3] The second approach is to set the mass analyzer off axis to the ion lens system (in some systems this is called a chicane design). The positively charged ions are then steered with the lens components into the mass analyzer, while the photons, neutral, and nonionic species are ejected out of the ion beam.[4] The third and most recent development is to reflect the ion beam 90° with a "hollow" ion mirror.[5] This allows the photons, neutrals, and solid particles to pass through, whereas the ions are reflected at right angles into an off-axis mass analyzer that incorporates curved fringe rod technology.[6] The principle of this design is shown schematically in Figure 6.2.

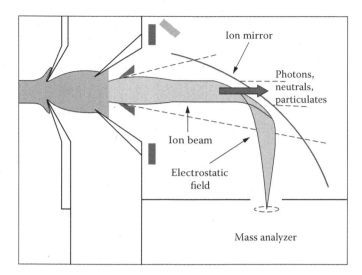

FIGURE 6.2   An ion-focusing system that uses a hollowed-out ion mirror to deflect the ion beam 90° to the mass analyzer, while allowing photons, neutrals, and solid particles to pass through (courtesy of Varian Inc.).

It is also worth mentioning that some lens systems incorporate an extraction lens after the skimmer cone to electrostatically "pull" the ions from the interface region. This has the benefit of improving the transmission and detection limits of the low-mass elements (which tend to be pushed out of the ion beam by the heavier elements), resulting in a more uniform response across the full mass range. In an attempt to reduce these space charge effects, some older designs have utilized lens components to accelerate the ions downstream. Unfortunately, this can have the effect of degrading the resolving power and abundance sensitivity (ability to differentiate an analyte peak from the wing of an interference) of the instrument, because of the much higher kinetic energy of the accelerated ions as they enter the mass analyzer.[7]

DYNAMICS OF ION FLOW

To fully understand the role of the ion optics in ICP-MS, it is important to get an appreciation of the dynamics of ion flow from the plasma through the interface region into the mass spectrometer. When the ions generated in the plasma emerge from the skimmer cone, there is a rapid expansion of the ion beam as the pressure is reduced from 760 torr (atmospheric pressure) to approximately 10^{-3} to 10^{-4} torr in the lens chamber with a turbomolecular pump. The composition of the ion beam immediately behind the cone is the same as in front of the cone because the expansion at this stage is controlled by normal gas dynamics and not by electrodynamics. One of the main reasons for this is that in the ion-sampling process, the Debye length (the distance over which ions exert influence on one another) is small compared to the orifice diameter of the sampler or skimmer cone. Consequently, there is little electrical interaction between the ion beam and the cone, and relatively little interaction between the individual ions in the beam. In this way, the compositional integrity

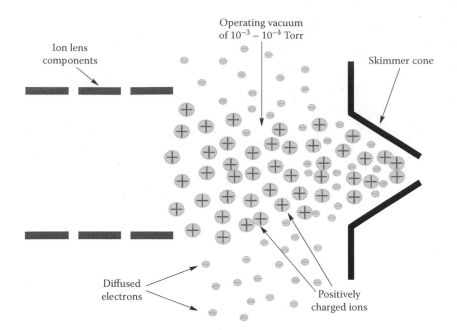

FIGURE 6.3   Extreme pressure drop in the ion optic chamber produces diffusion of electrons, resulting in a positively charged ion beam.

of the ion beam is maintained throughout the interface region.[8] With this rapid drop in pressure in the lens chamber, electrons diffuse out of the ion beam. Because of the small size of the electrons relative to the positively charged ions, the electrons diffuse further from the beam than the ions, resulting in an ion beam with a net positive charge. This is represented schematically in Figure 6.3.

The generation of a positively charged ion beam is the first stage in the charge separation process. Unfortunately, the net positive charge of the ion beam means that there is now a natural tendency for the ions to repel each other. If nothing is done to compensate for this, ions of higher mass-to-charge ratio will dominate the center of the ion beam and force the lighter ions to the outside. The degree of loss will depend on the kinetic energy of the ions—those with high kinetic energy (high-mass elements) will be transmitted in preference to ions with medium (midmass elements) or low kinetic energy (low-mass elements). This is shown in Figure 6.4. The second stage of charge separation, therefore, consists of electrostatically steering the ions of interest back into the center of the ion beam by placing voltages on one or more ion lens components. It should be emphasized that this is only possible if the interface is kept at zero potential, which ensures a neutral gasdynamic flow through the interface, maintaining the compositional integrity of the ion beam. It also guarantees that the average ion energy and energy spread of each ion entering the lens systems are at levels optimum for mass separation. If the interface region is not grounded correctly, stray capacitance will generate a discharge between the plasma and sampler cone and increase the kinetic energy of the ion beam, making it very difficult to optimize the ion lens voltages (refer to Chapter 5 for details).

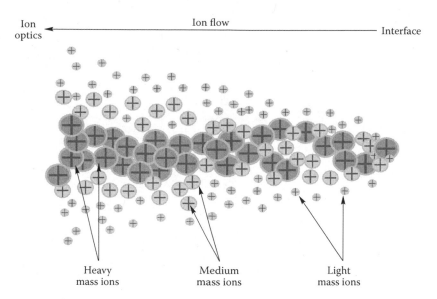

Ion optics ← Ion flow → Interface

Heavy mass ions   Medium mass ions   Light mass ions

FIGURE 6.4 The degree of ion repulsion will depend on the kinetic energy of the ions—those with high kinetic energy (heavy masses) will be transmitted in preference to those with medium (medium masses) or low kinetic energy (light masses).

COMMERCIAL ION OPTIC DESIGNS

Over the years, there have been many different ion optic designs. Although they have their own individual characteristics, they perform the same basic function of discriminating between undesirable matrix- or solvent-based ions, so that only the analyte ions are transmitted to the mass analyzer. The oldest and most mature design of ion optics in use today consists of several lens components, all of which have a specific role to play in the transmission of the analyte ions with a minimum of mass discrimination. With these multicomponent lens systems, the voltage can be optimized on every lens of the ion optics to achieve the desired ion specificity. This type of lens configuration has been used in commercial instrumentation for almost 25 years and has proved to be very durable. One of its main benefits is that it produces a uniform response across the mass range with very low background levels, particularly when combined with an off-axis mass analyzer.[9] A schematic of a commercially available multicomponent lens systems is shown in Figure 6.5. It should be emphasized that because of the interactive nature of parameters that affect the signal response, the more complex the lens system, the more the variables that have to be optimized. For this reason, if many different sample types are being analyzed, extensive lens optimization procedures have to be carried out for each matrix or group of elements. This is not such a major problem, because most of the lens voltages are computer controlled and methods can be stored for every new sample scenario. However, it could be a factor if the instrument is being used for the routine analysis of many diverse sample types, all requiring different lens settings.

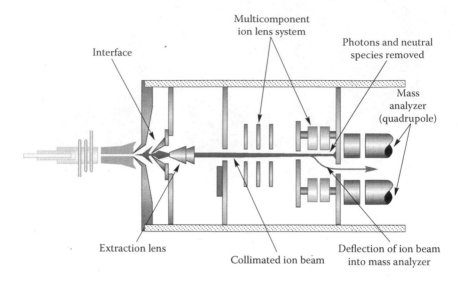

FIGURE 6.5 Schematic of a multicomponent lens system. (From Y. Kishi, *Agilent Technologies Application Journal*, August 1997.)

A more novel approach is to use just one cylinder lens, combined with a grounded stop—positioned just inside the skimmer cone. With this design, the voltage is dynamically ramped "on the fly," in concert with the mass scan of the analyzer (typically, a quadrupole). The benefit of this approach is that the optimum lens voltage is placed on every mass in a multielement run to allow the maximum number of analyte ions through, while keeping the matrix ions down to an absolute minimum.[10] This is represented in Figure 6.6, which shows a lens voltage scan of six elements, Li, Co, Y, In, Pb, and U at 7, 59, 89, 115, 208, and 238 amu, respectively. It can be seen that each element has its own optimum value, which is then used to calibrate the system so the lens can be ramp-scanned across the full mass range. This type of approach is typically used in conjunction with a grounded stop to act as a physical

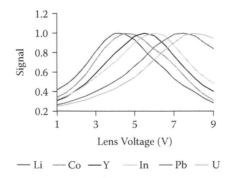

FIGURE 6.6 A calibration of optimum lens voltages is used to ramp-scan the ion lens in concert with the mass scan of the analyzer. (From E.R. Denoyer, D. Jacques, E. Debrah, S. D. Tanner, *Atomic Spectroscopy*, **16**[1], 1, 1995.)

barrier to reduce the chances that particulates, neutral species, and photons will reach the mass analyzer and detector. Although this design does not generate such a uniform mass response across the full range as an off-axis multilens system with an extraction lens, it appears to offer better long-term stability with real-world samples. It works well for many sample types, but is most effective when low-mass elements are being determined in the presence of high-mass matrix elements.

Another approach is to use a high-efficiency, multipole-based ion guide. This simplified version of a collision cell is driven by mass-dependent RF voltages to optimize ion transmission equally across the entire mass range. In this mode, the cell is not used with a traditional collision gas, but instead utilizes the multipole to act as an ion-focusing guide to direct the ions into the mass analyzer. The design of this type of ion lens system is usually incorporated with an off-axis quadrupole and a chicane-type deflector. The major advantage of this design is that it gives extremely low background levels.

A more recent development in ion-focusing optics utilizes a parabolic electrostatic field created with an ion mirror to reflect and refocus the ion beam at 90° to the ion source.[5] The ion mirror incorporates a hollow structure, which allows photons, neutrals, and solid particles to pass through it, while allowing ions to be reflected at right angles into the mass analyzer. The major benefit of this design is the very efficient way the ions are refocused, offering the capability of extremely high sensitivity across the mass range, with very little sacrifice in oxide performance. In addition, there is very little contamination of the ion optics, because a vacuum pump sits behind the ion mirror to immediately remove these particles before they have a chance to penetrate further into the mass spectrometer. Removing these undesirable species and photons before they reach the detector, in addition to incorporating curved fringe rods prior to an off-axis mass analyzer, means that background levels are very low. Figure 6.7 shows a schematic of a quadrupole-based ICP-MS that utilizes a 90° ion optic design.[6,11]

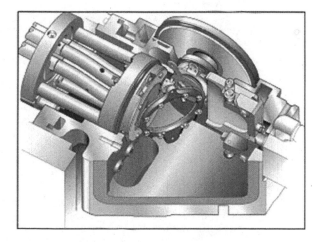

FIGURE 6.7 A 90° ion optic design used with curved fringe rods and an off-axis quadrupole mass analyzer (courtesy of Varian Inc.).

It is also worth emphasizing that a number of ICP-MS systems offer what is called a high-sensitivity option. All these work slightly differently but share similar components. By using a combination of slightly different cone geometry, higher vacuum at the interface, one or more extraction lenses, and slightly modified ion optic design, they offer up to 10 times the sensitivity of a traditional interface. However, in some systems, this increased sensitivity sometimes comes with slightly worse stability and an increase in background levels, particularly for samples with a heavy matrix. To get around this, these kinds of samples typically need to be diluted before analysis—which has somewhat limited their applicability to real-world samples with high dissolved solids.[12] However, they have found a use in non-liquid-based applications in which high sensitivity is crucial—for example, in the analysis of small spots on the surface of a geological specimen using laser ablation ICP-MS. For this application, the instrument must offer high sensitivity, because a single laser pulse is often used to ablate very small amounts of the sample, which is then swept into the ICP-MS for analysis.

The importance of the ion-focusing system cannot be overemphasized, because it has a direct bearing on the number of ions that find their way to the mass analyzer. In addition to affecting background levels and instrument response across the entire mass range, it has a huge impact on both long- and short-term signal stability, especially in real-world samples. However, there are many different ways of achieving this. It is almost irrelevant whether the design of the ion optics is based on a dynamically scanned single ion lens or a multicomponent lens system, whether a grounded stop, an off-axis mass analyzer, or a right-angled bend is used to stop photons, particulates, and neutral species from hitting the detector, or even whether an extraction lens is used. The most important consideration when evaluating an ion lens system is not the actual design but its ability to perform well with real sample matrices.

REFERENCES

1. J. A. Olivares and R. S. Houk, *Analytical Chemistry*, **58**, 20, 1986.
2. D. J. Douglas and J. B. French, *Spectrochimica Acta*, **41B**(3), 197, 1986.
3. S. D. Tanner, L. M. Cousins, and D. J. Douglas, *Applied Spectroscopy*, **48**, 1367, 1994.
4. D. Potter, *American Lab*, July, 1994.
5. I. Kalinitchenko, Ion Optical System for a Mass Spectrometer, Patent Number 750860, 1999.
6. S. Elliott, M. Plantz, and L. Kalinitchenko, Oral Paper 1360–8, *Pittsburgh Conference*, Orlando, FL, 2003.
7. P. Turner, Paper at *2nd International Conference on Plasma Source Mass Spec.* Durham, UK, 1990.
8. S. D. Tanner, D. J. Douglas, and J. B. French, *Applied Spectroscopy*, **48**, 1373, 1994.
9. Y. Kishi, *Agilent Technologies Application Journal*, August, 1997.
10. E. R. Denoyer, D. Jacques, E. Debrah, and S. D. Tanner, *Atomic Spectroscopy*, **16**(1), 1, 1995.
11. I. Kalinitchenko, Mass Spectrometer Including a Quadrupole Mass Analyzer Arrangement, Patent applied for—WO 01/91159 A1.
12. B. C. Gibson, Paper at *Surrey International Conference on ICP-MS*, London, UK, 1994.

7 Mass Analyzers
Quadrupole Technology

The next four chapters deal with the heart of the system—the mass separation device. Sometimes called the mass analyzer, it is the region of the ICP mass spectrometer that separates the ions according to their mass-to-charge ratio. This selection process is achieved in a number of different ways, depending on the mass separation device, but they all have one common goal, which is to separate the ions of interest from all other nonanalyte, matrix, solvent, and argon-based ions. Quadrupole mass filters are described in this chapter, followed by magnetic sector systems, time-of-flight mass spectrometers, and finally, collision/reaction cell and interface technology.

Although ICP-MS was commercialized in 1983, the first 10 years of its development utilized a traditional quadrupole mass analyzer to separate the ions of interest. These worked exceptionally well for most applications, but proved to have limitations when determining difficult elements or dealing with more complex sample matrices. This led to the development of alternative mass separation devices that allowed ICP-MS to be used for applications that were previously beyond the capabilities of quadrupole-based technology. Before we discuss these different mass spectrometers in greater detail, let us take a look at the proximity of the mass analyzer in relation to the ion optics and detector. Figure 7.1 shows this in greater detail.

As can be seen, the mass analyzer is positioned between the ion optics and detector, and it is maintained at a vacuum of approximately 10^{-6} torr with an additional turbomolecular pump to the one that is used for the lens chamber. Assuming the ions are emerging from the ion optics at the optimum kinetic energy, they are ready to be separated according to their mass-to-charge ratio (m/z) by the mass analyzer. There are basically four different kinds of commercially available mass analyzers: quadrupole mass filters, double-focusing magnetic sectors, time-of-flight mass spectrometers, and collision/reaction cell technology. They all have their own strengths and weaknesses, which is discussed in greater detail over the next four chapters. Let us first begin with the most common type of mass separation device used in ICP-MS—the quadrupole mass filter.

QUADRUPOLE MASS FILTER TECHNOLOGY

Developed in the early 1980s, quadrupole-based systems represent approximately 85% of all ICP mass spectrometers used today. This design was the first to be commercialized, and as a result, today's quadrupole ICP-MS technology is considered a very mature, routine trace element technique. A quadrupole consists of four cylindrical or hyperbolic metallic rods of the same length and diameter. They are typically made of stainless steel or molybdenum and sometimes coated with a ceramic coating

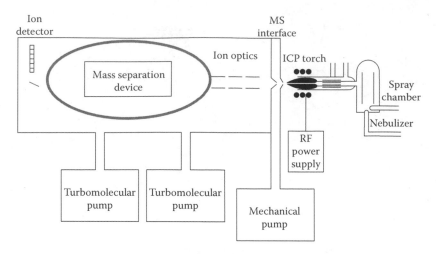

FIGURE 7.1   The mass separation device is positioned between the ion optics and the detector.

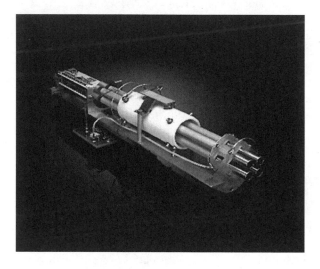

FIGURE 7.2   Photograph of a quadrupole system mounted in its housing (copyright © 2003–2007, all rights reserved, PerkinElmer Inc.).

for corrosion resistance. Quadrupoles used in ICP-MS are typically 15–25 cm in length, about 1 cm in diameter, and operate at a frequency of 2–3 MHz. Figure 7.2 shows a photograph of a quadrupole system mounted in its housing.

BASIC PRINCIPLES OF OPERATION

A quadrupole operates by placing both a direct current (DC) field and a time-dependent alternating current (AC) of radio frequency on opposite pairs of the four rods. By selecting the optimum AC/DC ratio on each pair of rods, ions of a selected mass

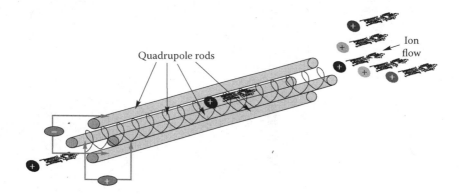

FIGURE 7.3 Schematic representation showing principles of mass separation using a quadrupole mass filter.

are allowed to pass through the rods to the detector, whereas the others are unstable and ejected from the quadrupole. Figure 7.3 shows this in greater detail.

In this simplified example, the analyte ion (black) and four other ions (shades of gray) have arrived at the entrance to the four rods of the quadrupole. When a particular AC/DC potential is applied to the rods, the positive or negative bias on the rods will electrostatically steer the analyte ion of interest down the middle of the four rods to the end, where it will emerge and be converted to an electrical pulse by the detector. The other ions of different m/z values will be unstable, pass through the spaces between the rods, and be ejected from the quadrupole. This scanning process is then repeated for another analyte with a completely different mass-to-charge ratio until all the analytes in a multielement analysis have been measured. The process for the detection of one particular mass in a multielement run is represented in Figure 7.4. It shows a $^{63}Cu^+$ ion emerging from the quadrupole and being converted to an electrical pulse by the detector. As the AC/DC voltage of the quadrupole—corresponding to $^{63}Cu^+$—is repeatedly scanned, the ions are stored and counted by a multichannel analyzer as electrical pulses. This multichannel data acquisition system typically has 20 channels per mass. As the electrical pulses are counted in each channel, a profile of the mass is built up over the 20 channels, corresponding to the spectral peak of $^{63}Cu^+$. In a multielement run, repeated scans are made over the entire suite of analyte masses as opposed to just one mass represented in this example.

Quadrupole scan rates are typically on the order of 2500 amu per second and can cover the entire mass range of 0–300 amu in about one tenth of a second. However, real-world analysis speeds are much slower than this, and in practice, 25 elements can be determined in duplicate with good precision in 1–2 min, depending on the analytical requirements.

QUADRUPOLE PERFORMANCE CRITERIA

There are two very important performance specifications of a mass analyzer that govern its ability to separate an analyte peak from a spectral interference. The first is the resolving power (R), which in traditional mass spectrometry is represented by

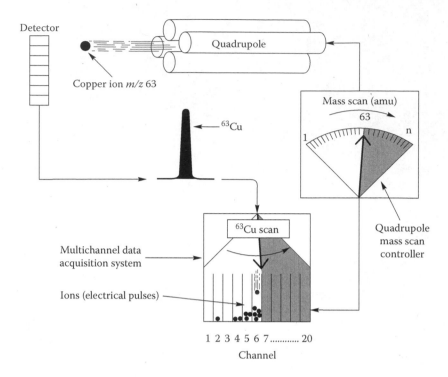

FIGURE 7.4   Profiles of different masses are built up using a multichannel data acquisition system (copyright © 2003–2007, all rights reserved, PerkinElmer Inc.).

the equation $R = m/\Delta m$, where m is the nominal mass at which the peak occurs and Δm is the mass difference between two resolved peaks.[1] However, for quadrupole technology, the term *resolution* is more commonly used and is normally defined as the width of a peak at 10% of its height. The second specification is abundance sensitivity, which is the signal contribution of the tail of an adjacent peak at one mass lower and one mass higher than the analyte peak.[2] Even though they are somewhat related and both define the quality of a quadrupole, the abundance sensitivity is probably the most critical. If a quadrupole has good resolution but poor abundance sensitivity, it will often prohibit the measurement of an ultratrace analyte peak next to a major interfering mass.

RESOLUTION

Let us now discuss this area in greater detail. The ability to separate different masses with a quadrupole is determined by a combination of factors, including shape, diameter, and length of the rods; frequency of quadrupole power supply; operating vacuum; applied RF/DC voltages; and the motion and kinetic energy of the ions entering and exiting the quadrupole. All these factors will have a direct impact on the stability of the ions as they travel down the middle of the rods, and therefore, the quadrupole's ability to separate ions with differing m/z values. This is represented

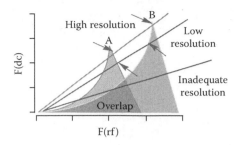

FIGURE 7.5 Simplified Mathieu stability diagram of a quadrupole mass filter, showing separation of two different masses A and B. (From *Quadrupole Mass Spectrometry and Its Applications*: P. H. Dawson, Ed., Elsevier, Amsterdam, 1976; reissued by AIP Press, Woodbury, NY, 1995.)

in Figure 7.5, which shows a simplified version of the Mathieu mass stability plot of two separate masses (A and B) entering the quadrupole at the same time.[3]

Any of the RF/DC conditions shown under the peak on the left will only allow mass A to pass through the quadrupole, whereas any combination of RF/DC voltages under the peak on the right will only allow mass B to pass through the quadrupole. If the slope of the RF/DC scan rate is steep, represented by the top line (high resolution), the spectral peaks will be narrow and masses A and B will be well separated. However, if the slope of the scan is shallow, represented by the middle line (low resolution), the spectral peaks will be wide and masses A and B will not be well separated. On the other hand, if the slope of the scan is too shallow, represented by the lower line (inadequate resolution), the peaks will overlap each other and both masses A and B will pass through the quadrupole without being separated. Theoretically, the resolution of a quadrupole mass filter can be varied between 0.3 and 3.0 amu, but is normally kept at 0.7–1.0 amu for most applications. However, improved resolution is always accompanied by a sacrifice in sensitivity, as seen in Figure 7.6, which shows a comparison of the same mass at resolutions of 3.0, 1.0, and 0.3 amu.

It can be seen that the peak height at 3.0 amu is much larger than that at 0.3 amu, but as expected, it is also much wider. This would prohibit using a resolution of

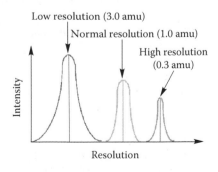

FIGURE 7.6 Sensitivity comparison of a quadrupole operated at 3.0, 1.0, and 0.3 amu resolutions.

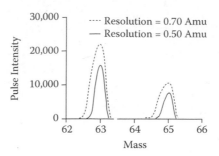

FIGURE 7.7 Sensitivity comparison of two copper isotopes—$^{63}Cu^+$ and $^{65}Cu^+$— at resolution settings of 0.70 and 0.50 amu.

3.0 amu with spectrally complex samples. Conversely, the peak width at 0.3 amu is very narrow, but the sensitivity is low. For this reason, a compromise between peak width and sensitivity normally has to be reached, depending on the application. This can clearly be seen in Figure 7.7, which shows a spectral overlay of two copper isotopes—$^{63}Cu^+$ and $^{65}Cu^+$—at resolution settings of 0.70 and 0.50 amu. In practice, the quadrupole is normally operated at a resolution of 0.7–1.0 amu for the majority of applications.

It is worth mentioning that most quadrupoles are operated in the first stability region, where resolving power is typically on the order of 500–600. If the quadrupole is operated in the second or third stability regions, resolving powers of 4000[4] and 9000,[5] respectively, have been achieved. However, improving resolution using this approach has resulted in a significant loss of signal. Although there are ways of improving sensitivity, other problems have been encountered. As a result, to date there are no commercial quadrupole instruments available based on using higher stability regions.

Some instruments can vary the peak width "on the fly," which means that the resolution can be changed between 3.0 and 0.3 amu for every analyte in a multi-element run. Although this appears to offer some benefits, in reality they are few and far between, and for the vast majority of applications it is adequate to use the same resolution setting for every analyte. Even though quadrupoles can be operated at a higher resolution (in the first stability region), up to now the slight improvement has not been shown to be of practical benefit for most routine applications.

ABUNDANCE SENSITIVITY

It can be seen in Figure 7.7 that the tail of the spectral peaks drops off more rapidly at the high-mass end of the peak compared to the low-mass end. The overall peak shape, particularly its low-mass and high-mass tail, is determined by the abundance sensitivity of the quadrupole, which is impacted by a combination of factors, including the design of the rods, frequency of the power supply, and operating vacuum.[6] Even though they are all important, probably the biggest impact on abundance sensitivity is the motion and kinetic energy of the ions as they enter and exit the quadrupole. If one looks at the Mathieu stability plot in Figure 7.5, it can be seen that the stability boundaries of each mass are less defined (not so sharp) on the low-mass

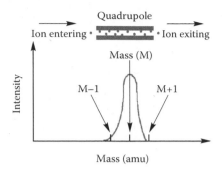

FIGURE 7.8 Ions entering the quadrupole are slowed down by the filtering process and produce peaks with a pronounced tail or shoulder at the low-mass end (M − 1) compared to the high-mass end (M + 1).

side than on the high-mass side.[3] As a result, the characteristic of ion motion at the low-mass boundary is different from that at the high-mass boundary and is therefore reflected in poorer abundance sensitivity at the low-mass side compared to the high-mass side. The velocity, and therefore the kinetic energy, of the ions entering the quadrupole will affect the ion motion and, as a result, will have a direct impact on the abundance sensitivity. For that reason, factors that affect the kinetic energy of the ions, such as high plasma potential and the use of lenses to accelerate the ion beam, will have a negative effect on the instrument's abundance sensitivity.[7]

These are the fundamental reasons why the peak shape is not symmetrical with a quadrupole and explains why there is always a pronounced shoulder at the low-mass side of the peak compared to the high-mass side—as represented in Figure 7.8, which shows the theoretical peak shape of a nominal mass M. It can be seen that the shape of the peak at one mass lower (M − 1) is slightly different from the other side of the peak at one mass higher (M + 1) than the mass M. For this reason, the abundance sensitivity specification for all quadrupoles is always worse on the low-mass side than the high-mass side and is typically 1×10^{-6} at M − 1 and 1×10^{-7} at M + 1. In other words, an interfering peak of 1 million counts per second (mcps) at M − 1 would produce a background of 1 cps at M, whereas it would take an interference of 10 million cps at M + 1 to produce a background of 1 cps at M.

BENEFIT OF GOOD ABUNDANCE SENSITIVITY

An example of the importance of abundance sensitivity is shown in Figure 7.9. Figure 7.9a is a spectral scan of 50 ppm of the doubly charged europium ion, $^{151}Eu^{++}$, at 75.5 amu (a doubly charged ion is one with two positive charges, as opposed to a normal singly charged positive ion, and exhibits an m/z peak at half its mass). It can be seen that the intensity of the peak is so great that its tail overlaps the adjacent mass at 75 amu, which is the only available mass for the determination of arsenic. This is highlighted in Figure 7.9b, which shows an expanded view of the tail of the $^{151}Eu^{++}$ together with a scan of 1 ppb of As at mass 75. It can be seen very clearly that the $^{75}As^+$ signal lies on the sloping tail of the $^{151}Eu^{++}$ peak. Measurement on a sloping background similar to this would result in a significant degradation in the

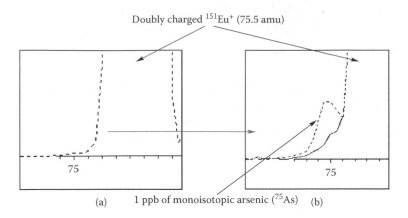

FIGURE 7.9 A low abundance sensitivity specification is critical to minimize spectral interferences, as shown by (a), which represents a spectral scan of 50 ppb of $^{151}Eu^{++}$ at 75.5 amu, and (b), which shows how the tail of the $^{151}Eu^{++}$ elevates the spectral background of 1 ppb of As at mass 75 (copyright © 2003–2007, all rights reserved, PerkinElmer Inc.).

arsenic detection limit, particularly as the element is monoisotopic and no alternative mass is available. In this particular example, a slightly higher resolution setting was also used (0.5 amu instead of 0.7 amu) to enhance the separation of the arsenic peak from the europium peak, but nevertheless still emphasizes the importance of good abundance sensitivity in ICP-MS.

There are many different designs of quadrupole used in ICP-MS, all made from different materials with varied dimensions, shape, and physical characteristics. In addition, they are all maintained at a slightly different vacuum chamber pressure and operate at different frequencies. Theoretically, these hyperbolic rods should generate a better hyperbolic (elliptical) field than cylindrical rods, resulting in higher transmission of ions at higher resolution. It also tells us that a higher operating frequency means a higher rate of oscillation—and therefore separation—of the ions as they travel down the quadrupole. Finally, it is very well accepted that a higher vacuum produces fewer collisions between gas molecules and ions, resulting in a narrower spread in kinetic energy of the ions, and therefore, a reduction in the tail at the low-mass side of a peak. Given all these theoretical differences, in reality the practical capabilities of most modern quadrupoles used in ICP-MS are very similar. However, there are some subtle differences in each instrument's measurement protocol and the software's approach to peak quantitation. This is such an important area that it will be discussed in greater detail in Chapter 12.

REFERENCES

1. F. Adams, R. Gijbels, and R. Van Grieken, *Inorganic Mass Spectrometry,* John Wiley and Sons, New York, 1988.
2. *Inductively Coupled Plasma Mass Spectrometry*: E. Montasser, Ed., Wiley-VCH, Berlin, 1998.
3. *Quadrupole Mass Spectrometry and Its Applications*: P. H. Dawson, Ed., Elsevier, Amsterdam, 1976; reissued by AIP Press, Woodbury, NY, 1995.

4. Z. Du, T. N. Olney, and D. J. Douglas, *Journal of American Society of Mass Spectrometry*, **8**, 1230–1236, 1997.
5. P. H. Dawson and Y. Binqi, *International Journal of Mass Spectrometry*, Ion Proc., **56**, 25, 1984.
6. D. Potter, Agilent Technologies Application Note: 228–349, January, 1996.
7. E. R. Denoyer, D. Jacques, E. Debrah, and S. D. Tanner, *Atomic Spectroscopy*, **16**(1), 1, 1995.

8 Mass Analyzers
Double-Focusing Magnetic Sector Technology

Although quadrupole mass analyzers represent approximately 85% of all ICP-MS systems installed worldwide, limitations in their resolving power has led to the development of high-resolution spectrometers based on the double-focusing magnetic sector design. In this chapter we take a detailed look at this very powerful mass separation device, which has found its niche in solving challenging application problems that require excellent detection capability, exceptional resolving power, and very high precision.

As discussed in Chapter 7, a quadrupole-based ICP-MS system typically offers a resolution of 0.7–1.0 amu. This is quite adequate for most routine applications, but has proved to be inadequate for many elements that are prone to argon-, solvent-, and/or sample-based spectral interferences. These limitations in quadrupoles drove researchers in the direction of traditional high-resolution magnetic sector technology to improve quantitation by resolving the analyte mass away from the spectral interference.[1] These ICP-MS instruments that were first commercialized in the late 1980s offered resolving power of up to 10,000, compared to that of a quadrupole, which was on the order of ~300. This dramatic improvement in resolving power allowed difficult elements such as Fe, K, As, V, and Cr to be determined with relative ease, even in complex sample matrices.

MAGNETIC SECTOR MASS SPECTROSCOPY: A HISTORICAL PERSPECTIVE

Mass spectrometers, using separation based on velocity focusing[2,3] and magnetic deflection,[4,5] were first developed over 80 years ago, primarily to investigate isotopic abundances and calculate atomic weights. Even though these designs were combined into one instrument in the 1930s to improve both sensitivity and resolving power,[6,7] they were still considered rather bulky and expensive to build. For that reason, in the late 1930s and 1940s, magnetic field technology, and in particular the small radius sector design of Nier,[8] became the preferred method of mass separation. Because Nier was a physicist, most of the early work carried out with this design was used for isotope studies in the disciplines of earth and planetary sciences. However, it was the oil industry that accelerated the commercialization of MS, because of its demand for the fast and reliable analysis of complex hydrocarbons in oil refineries.

Once scanning magnetic sector technology became the most accepted approach for high-resolution mass separation in the 1940s, the challenges that lay ahead for mass spectroscopists were in the design of the ionization source—especially as the technique was being used more and more for the analysis of solids. The gas discharge ion source that was developed for gases and high-vapor-pressure liquids proved to be inadequate for most solid materials. For this reason, one of the first successful methods of ionizing solids was carried out using the hot anode method,[9] where the previously dissolved material was deposited onto a strip of platinum foil and evaporated by passing an electric current through it. Unfortunately, although there were variations of this approach that all worked reasonably well, the main drawback of a thermal evaporation technique was selective ionization. In other words, because of the different volatilities of the elements, it could not be guaranteed that the ion beam properly represented the compositional integrity of the sample.

It was finally the work carried out by Dempster in 1946,[10] using a vacuum spark discharge and a high-frequency, high-voltage spark, that led researchers to believe that it could be applied to sample electrodes and used as a general-purpose source for the analysis of solids. The breakthrough came in 1954 with the development of the first modern spark source mass spectrometer (SSMS) based on the Mattauch-Herzog mass spectrometer design.[11] Using this design, Hannay and Ahearn showed that it was possible to determine sub-ppm impurity levels directly in a solid material.[12]

Over the years, as a result of a demand for more stable ionization sources, lower detection capability, and higher precision, researchers were led in the direction of other techniques such as secondary ion mass spectrometry (SIMS),[13] ion microprobe mass spectrometry (IMMS),[14] and laser induced mass spectrometry (LIMS).[15] Although they are considered somewhat complementary to SSMS, they all had their own strengths and weaknesses, depending on the analytical objectives for the solid material being analyzed. However, it should be emphasized that these techniques were predominantly used for microanalysis because only a very small area of the sample is vaporized. This meant that it could only provide meaningful analytical data of the bulk material if the sample was sufficiently homogeneous. For that reason, other ionization sources that sampled a much larger area, such as the glow discharge, became a lot more practical for the bulk analysis of solids by MS.[16]

USE OF MAGNETIC SECTOR TECHNOLOGY FOR ICP-MS

Even though magnetic sector technology was the most common mass separation device for the analysis of inorganic compounds using traditional ion sources, it lost out to quadrupole technology when ICP-MS was first developed in the early 1980s. However, it was not until the mid- to late-1980s that the analytical community realized that quadrupole ICP-MS suffered from serious limitations in its ability to resolve troublesome polyatomic spectral interferences, and began to look at double-focusing magnetic sector technology to eliminate these kinds of problems. Initially, it was found to be unsuitable as a separation device for an ICP because of the high voltage required to accelerate the ions into the mass analyzer. This high potential at the interface region dramatically changed the energy of the ions entering the mass spectrometer, and therefore made it very difficult to steer the ions through the ion optics and still maintain a narrow spread of ion kinetic energies. For this reason, basic changes had to be made to the ion acceleration mechanism for magnetic sector technology to

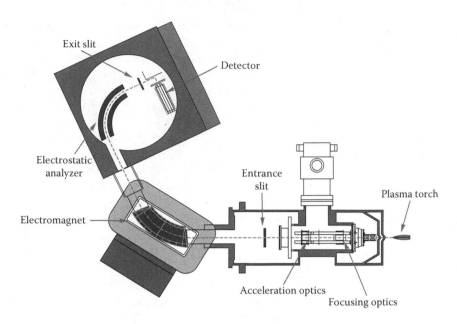

FIGURE 8.1 Schematic of a reverse Nier–Johnson double-focusing magnetic sector mass spectrometer (from U. Geismann and U. Greb, *Fresnius' Journal of Analytical Chemistry,* **350**, 186–193, 1994).

be successfully used as a separation device for ICP-MS. This was a significant challenge when magnetic sector systems were first developed in the late 1980s.

However, by the early 1990s, one instrument manufacturer solved this problem by moving the high-voltage components away from the plasma and positioning the interface closer to the mass spectrometer. Modern instrumentation has typically been based on two different approaches, the "standard" and "reverse" Nier–Johnson geometry. Both these designs, which use the same basic principles, consist of two analyzers: a traditional electromagnet and an electrostatic analyzer (ESA). In the standard (sometimes called forward) design, the ESA is positioned before the magnet, and in the reverse design it is positioned after the magnet. A schematic of the reverse Nier–Johnson spectrometer is shown in Figure 8.1.

PRINCIPLES OF OPERATION OF MAGNETIC SECTOR SYSTEMS

The original concept of magnetic sector technology was to scan over a large mass range by varying the magnetic field over time with a fixed acceleration voltage. During a small window in time, which was dependent on the resolution chosen, ions of a particular mass-to-charge ratio are swept past the exit slit to produce the characteristic flat-top peaks. As the resolution of a magnetic sector instrument is independent of mass, ion signals, particularly at low mass, are far apart. Unfortunately, this results in a relatively long time being spent scanning and settling the magnet. This was not such a major problem for qualitative analysis or mass spectral fingerprinting of unknown compounds, but proved to be impractical for rapid trace element analysis, where you had to scan to individual masses, slow down, settle the magnet, stop, take measurements, and then scan to the next mass.

However, by using the double-focusing approach, the ions are sampled from the plasma in a conventional manner and then accelerated in the ion optic region to a few kilovolts (kV) before they enter the mass analyzer. The magnetic field, which is dispersive with respect to ion energy and mass, then focuses all the ions with diverging angles of motion from the entrance slit. The ESA, which is only dispersive with respect to ion energy, then focuses all the ions onto the exit slit, where the detector is positioned. If the energy dispersion of the magnet and ESA are equal in magnitude but opposite in direction, they will focus both ion angles (first focusing) and ion energies (second or double focusing) when combined together. Changing the electric field in the opposite direction during the cycle time of the magnet (in terms of the mass passing the exit slit) has the effect of "freezing" the mass for detection. Then, as soon as a certain magnetic field strength is passed, the electric field is set to its original value and the next mass is "frozen." The voltage is varied on a per-mass basis, allowing the operator to scan only the mass peaks of interest rather than the full mass range.[17,18]

It should be pointed out that although this approach represents an enormous time savings over traditional magnet scanning technology, it is still slower than quadrupole-based instruments. The inherent problem lies in the fact that a quadrupole can be electronically scanned faster than a magnet. Typical speeds for a full mass scan (0–250 amu) of a magnet are on the order of 200 ms compared to 100 ms for a quadrupole. In addition, it takes much longer for a magnet to slow down, settle, and stop to take measurements—typically, 20 ms compared to 1–2 ms for a quadrupole. So, even though in practice the electric scan dramatically reduces the overall analysis time, modern double-focusing magnetic sector ICP-MS systems are still slower than state-of-the-art quadrupole instruments, which makes them less than ideal for rapid, high-throughput multielement applications.

RESOLVING POWER

As mentioned previously, most commercial magnetic sector ICP-MS systems offer up to 10,000 resolving power (10% valley definition), which is high enough to resolve the majority of spectral interferences. It is worth emphasizing that resolving power is represented by the equation $R = m/\Delta m$, where m is the nominal mass at which the peak occurs and Δm is the mass difference between two resolved peaks.[19] In a quadrupole, the resolution is selected by changing the ratio of the RF and DC voltages on the quadrupole rods.

However, because a double-focusing magnetic sector instrument involves focusing ion angles and ion energies, mass resolution is achieved by using two mechanical slits—one at the entrance to the mass spectrometer and another at the exit, prior to the detector. Varying resolution is achieved by scanning the magnetic field under different entrance and exit slit width conditions. Similar to optical systems, low resolution is achieved by using wide slits, whereas high resolution is achieved with narrow slits. Varying the width of both the entrance and exit slits effectively changes the operating resolution.

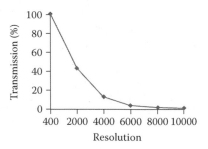

FIGURE 8.2 Ion transmission with a magnetic sector instrument decreases as the resolution increases.

However, it should be emphasized that, similar to optical spectrometry, as the resolution is increased, the transmission decreases. So even though extremely high resolution is available, detection limits will be compromised under these conditions. This can be seen in Figure 8.2, which shows a plot of resolution against ion transmission. It can be seen that a resolving power of 400 produces 100% transmission, but at a resolving power of 10,000, only ~2% is achievable. This dramatic loss in sensitivity could be an issue if low detection limits are required in spectrally complex samples that require the highest possible resolution. However, spectral demands of this nature are not very common. Table 8.1 shows the resolution required to resolve fairly common polyatomic interferences from a selected group of elemental isotopes, together with the achievable ion transmission.

TABLE 8.1

Resolution Required to Resolve Some Common Polyatomic Interferences from a Selected Group of Isotopes

Isotope	Matrix	Interference	Resolution	Transmission (%)
$^{39}K^+$	H_2O	$^{38}ArH^+$	5570	6
$^{40}Ca^+$	H_2O	$^{40}Ar^+$	199,800	0
$^{44}Ca^+$	HNO_3	$^{14}N^{14}N^{16}O^+$	970	80
$^{56}Fe^+$	H_2O	$^{40}Ar^{16}O^+$	2504	18
$^{31}P^+$	H_2O	$^{15}N^{16}O^+$	1460	53
$^{34}S^+$	H_2O	$^{16}O^{18}O^+$	1300	65
$^{75}As^+$	HCl	$^{40}Ar^{35}Cl^+$	7725	2
$^{51}V^+$	HCl	$^{35}Cl^{16}O^+$	2572	18
$^{64}Zn^+$	H_2SO_4	$^{32}S^{16}O^{16}O^+$	1950	42
$^{24}Mg^+$	Organics	$^{12}C^{12}C^+$	1600	50
$^{52}Cr^+$	Organics	$^{40}Ar^{12}C^+$	2370	20
$^{55}Mn^+$	HNO_3	$^{40}Ar^{15}N^+$	2300	20

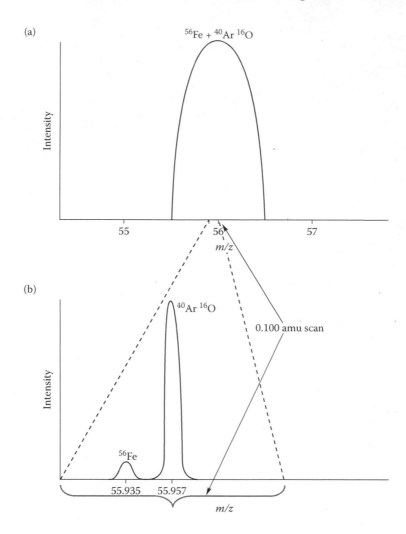

FIGURE 8.3 Comparison of resolution between (a) a quadrupole and (b) a magnetic sector instrument for the polyatomic interference of $^{40}Ar^{16}O^+$ on $^{56}Fe^+$. (From U. Greb and L. Rottman, *Labor Praxis*, August 1994.)

Figure 8.3 is a comparison between a quadrupole and a magnetic sector instrument of one of the most common polyatomic interference—$^{40}Ar^{16}O^+$ on $^{56}Fe^+$, which requires a resolution of 2504 to separate the peaks. Figure 8.3a shows a spectral scan of $^{56}Fe^+$ using a quadrupole instrument. What it does not show is the massive polyatomic interference $^{40}Ar^{16}O^+$ (produced by oxygen ions from the water combining with argon ions from the plasma) completely overlapping the $^{56}Fe^+$. It shows very clearly that these two masses are irresolvable with a quadrupole. If that same spectral scan is carried out on a magnetic sector–type instrument, the result is the scan shown in Figure 8.3b.[20] It should be pointed out that in order to see the spectral scan on the same scale, it is necessary to examine a much smaller range. For this reason, a 0.100 amu window was taken, as indicated by the dotted lines.

OTHER BENEFITS OF MAGNETIC SECTOR INSTRUMENTS

Besides high resolving power, another attractive feature of magnetic sector instruments is their very high sensitivity combined with extremely low background levels. High ion transmission in low-resolution mode translates into sensitivity specifications of up to 1 billion counts per second (bcps) per parts per million, whereas background levels resulting from extremely low dark current noise are typically 0.1–0.2 counts per second (cps). This compares to sensitivity levels of 10–50 mcps and background levels of ~10 cps for a quadrupole instrument. For this reason, detection limits, especially for high-mass elements such as uranium where high resolution is generally not required, are typically 5–10 times better than a quadrupole-based instrument.

Besides good detection capability, another of the recognized benefits of the magnetic sector approach is its ability to quantitate with excellent precision. Measurement of the characteristically flat-topped spectral peaks translates directly into high-precision data. As a result, in the low-resolution mode relative standard deviation (RSD) values of 0.01–0.05% are fairly common, which makes it an ideal approach for carrying out high-precision isotope ratio work.[21] Although precision is usually degraded as resolution is increased, modern instrumentation with high-speed electronics and low mass bias are still capable of precision values of <0.1% RSD in medium- or high-resolution mode.[22]

The demand for ultra-high-precision data, particularly in the field of geochemistry, has led to the development of instruments dedicated to isotope ratio analysis. These are based on the double-focusing magnetic sector design, but instead of using just one detector, these instruments use multiple detectors. Often referred to as multi-collector systems, they offer the capability of detecting and measuring multiple ion signals at exactly the same time. As a result of this simultaneous measurement approach, they are recognized as producing the ultimate in isotope ratio precision.[23]

There is no question that double-focusing magnetic sector ICP-MS systems are no longer a novel analytical technique. They have proved themselves to be a valuable addition to the trace element toolkit, particularly for challenging applications that require good detection capability, exceptional resolving power, and very high precision. Even though they are no competition for quadrupole instruments when it comes to rapid, high-sample-throughput applications, the scan speeds of modern systems have been improved considerably over the past few years. For that reason, they can now be considered a viable alternative to quadrupoles for carrying out multielement determinations on transient peaks using laser ablation[24] or chromatographic separation devices.[25]

REFERENCES

1. N. Bradshaw, E. F. H. Hall, and N. E. Sanderson, *Journal of Analytical Atomic Spectrometry,* **4**, 801–803, 1989.
2. F. W. Aston, *Philosophical Magazine,* **38**, 707, 1919.
3. J. L. Costa, *Annals of Physics,* **4**, 425, 1925.
4. A. J. Dempster, *Physical Review,* **11**, 316, 1918.
5. W. F. G. Swann, *Journal of the Franklin Institute,* **210**, 751, 1930.
6. A. J. Dempster, *Proceedings of the American Philosophical Society,* **75**, 755, 1935.

7. K. T. Bainbridge and E. B. Jordan, *Physical Review*, **50**, 282, 1936.
8. A. O. Nier, *Review of Scientific Instruments,* **11**, 252, 1940.
9. G. P. Thomson, *Philosophical Magazine,* **42**, 857, 1921.
10. A. J. Dempster, MDDC 370, U. S. Department of Commerce, Washington, DC, 1946.
11. J. Mattauch and R. Herzog, *Zeitschrift für Physik,* **89**, 786, 1934.
12. N. B. Hannay and A. J. Ahearn, *Analytical Chemistry,* **26**, 1056, 1954.
13. R. E. Honig, *Journal of Applied Physics,* **29**, 549, 1958.
14. R. Castaing and G. Slodzian, *Journal of Microscopy,* **1**, 395, 1962.
15. R. E. Honig and J. R. Wolston, *Applied Physics Letters*, **2**, 138, 1963.
16. J. W. Coburn, *Rev. Sci. Instrum.,* **41**, 1219, 1970.
17. R. Hutton, A. Walsh, D. Milton, and J. Cantle, *ChemSA*, **17**, 213–215, 1991.
18. U. Geismann and U. Greb, *Fresnius' Journal of Analytical Chemistry,* **350**, 186–193, 1994.
19. F. Adams, R. Gijbels, and R. Van Grieken, *Inorganic Mass Spectrometry,* John Wiley and Sons, New York, 1988.
20. U. Greb and L. Rottman, *Labor Praxis*, August 1994.
21. F. Vanhaecke, L. Moens, R. Dams, and R. Taylor, *Analytical Chemistry,* **68**, 567, 1996.
22. M. Hamester, D. Wiederin, J. Willis, W. Keri, and C. B. Douthitt, *Fresnius' Journal of Analytical Chemistry,* **364**, 495–497, 1999.
23. J. Walder and P. A. Freeman, *Journal of Analytical Atomic Spectrometry,* **7**, 571, 1992.
24. S. Shuttleworth and D. Kremser, *Journal of Analytical Atomic Spectrometry,* **13**, 697–699, 1998.
25. D. Klueppel, N. Jakubowski, J. Messerschmidt, D. Stuewer, and D. Klockow, *Journal of Analytical Atomic Spectrometry,* **13**, 255, 1998.

9 Mass Analyzers
Time-of-Flight Technology

Let us turn our attention to the most recent mass separation device to be commercialized: time-of-flight (TOF) technology. Although the first TOF mass spectrometer was first described in the literature in the late 1940s,[1] it has taken over 50 years to adapt it for use with a commercial ICP mass spectrometer. The recent growth in TOF ICP-MS instrument sales has come about because of its unique ability to sample all ions generated in the plasma at exactly the same time, which is ideally suited for multielement determinations of rapid transient signals, high-precision ratio analysis, and rapid data acquisition.

BASIC PRINCIPLES OF TOF TECHNOLOGY

The simultaneous nature of sampling ions in TOF offers distinct advantages over traditional scanning (sequential) quadrupole technology for ICP-MS applications in which large amounts of data need to be captured in a short span of time. To understand the benefits of this mass separation device, let us first take a look at its fundamental principles. All TOF mass spectrometers are based on the same principle: the kinetic energy (KE) of an ion is directly proportional to its mass (*m*) and velocity (*V*). This can be represented by the following equation:

$$KE = \tfrac{1}{2}\, mV^2$$

Therefore, if a population of ions—all with different masses—is given the same KE by an accelerating voltage (U), the velocities of the ions will all be different, depending on their masses. This principle is then used to separate ions of different mass-to-charge ratios (m/e) in the time (*t*) domain over a fixed flight path distance (*D*), represented by the following equation:

$$m/e = \frac{2Ut^2}{D^2}$$

This is schematically shown in Figure 9.1, which shows three ions of different mass-to-charge ratios being accelerated into a "flight tube" and arriving at the detector at different times. It can be seen that, depending on their velocities, the lightest ion arrives first, followed by the medium mass ion, and finally the heaviest one. Using flight tubes of 1 m length, even the heaviest ions typically take less than 50 µs to reach the detector. This translates into approximately 20,000 mass spectra per

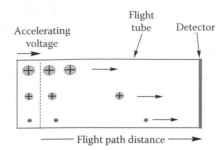

FIGURE 9.1 Principles of ion detection using time-of-flight technology, showing separation of three different masses in the time domain.

second—three orders of magnitude faster than the sequential scanning mode of a quadrupole system.

COMMERCIAL DESIGNS

Even though this process sounds fairly straightforward, it is not a trivial task to sample the ions in a simultaneous manner from a continuous source of ions being generated in the plasma discharge. When TOF ICP-MS technology was first commercialized, there were basically two different sampling approaches used—the orthogonal design,[2] in which the flight tube is positioned at right angles to the sampled ion beam, and the axial design,[3] in which the flight tube is along the same axis as the ion beam. However, the only approach that is commercially available today is the orthogonal design. The axial design was discontinued about 2 years ago. For the purpose of this tutorial, I will describe both approaches, but the reader should be aware that the only commercially available design is the orthogonal approach.

In both designs, all ions that contribute to the mass spectrum are sampled through the interface cones, but instead of being focused into the mass filter in the conventional way, packets (groups) of ions are electrostatically injected into the flight tube at exactly the same time. With the orthogonal approach, an accelerating potential is applied at right angles to the continuous ion beam from the plasma source. The ion beam is then "chopped" by using a pulsed voltage supply coupled to the orthogonal accelerator to provide repetitive voltage "slices" at a frequency of a few kilohertz. The "sliced" packets of ions, which are typically tall and thin in cross section (in the vertical plane), are then allowed to "drift" into the flight tube, where the ions are temporally resolved according to their differing velocities. This is shown schematically in Figure 9.2.

The axial approach is similar in design to the orthogonal approach, except that an accelerating potential is applied axially (in the same axis) to the incoming ion beam as it enters the extraction region. Because the ions are in the same plane as the detector, the beam has to be modulated using an electrode grid to repel the "gated" packet of ions into the flight tube. This kind of modulation generates an ion packet that is long and thin in cross section (in the horizontal plane). The different masses are then resolved in the time domain in a similar manner to the orthogonal design. An on-axis TOF system is schematically shown in Figure 9.3.

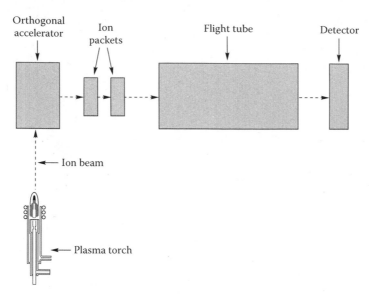

FIGURE 9.2   A schematic of an orthogonal acceleration TOF analyzer. (From Technical Note: 001-0877-00, GBC Scientific, February 1998.)

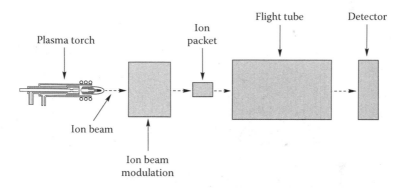

FIGURE 9.3   A schematic of an axial acceleration TOF analyzer. (From Technical Note: 001-0877-00, GBC Scientific, February 1998.)

Figures 9.2 and 9.3 offer a rather simplified explanation of the TOF principles of operation. In practice, there are many complex ion-focusing components in a TOF mass analyzer that ensure that the maximum number of analyte ions reach the detector, and also that undesired photons, neutral species, and interferences are ejected from the ion beam. Some of these components are seen in Figure 9.4, which shows a more detailed view of a commercial orthogonal TOF ICP-MS system. It can be seen in this design that an orthogonal accelerator is used to inject packets of ions at right angles from the ion beam emerging from the MS interface. They are then directed toward an ion blanker, where unwanted ions are rejected from the flight path by deflection plates. The packets of ions are then directed into an ion reflectron where

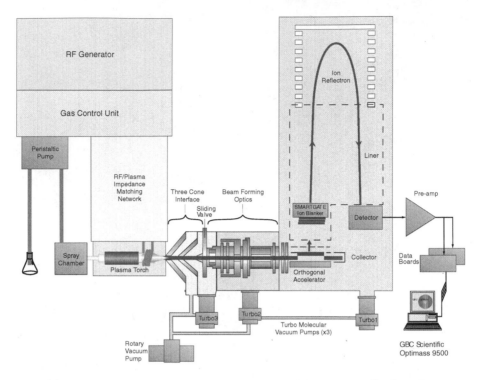

FIGURE 9.4 A more detailed view of an orthogonal (right-angled) TOF ICP-MS system, showing some of the ion-steering components (courtesy of GBC Scientific Equipment Pty Ltd.).

they do a U-turn and are deflected back 180°, where they are detected by a channel electron multiplier or discrete dynode detector. The reflectron, which is a type of ion mirror, functions as an energy compensation device so that different ions of the same mass arrive at the detector at the same time.

DIFFERENCES BETWEEN ORTHOGONAL AND ON-AXIS TOF

Although there are real benefits of using TOF over quadrupole technology for some ICP-MS applications, there are also subtle differences in the capabilities of each type of TOF design. Without getting into the advantages and disadvantages of different commercial instrumentation, it is well worth presenting the major differences between the orthogonal and on-axis approaches and comparing them with today's quadrupole-based instruments. Some of these differences include the following:

Sensitivity: The axial approach tends to produce higher ion transmission because the steering components are in the same plane as the ion generation system (plasma) and the detector. This means that the direction and magnitude of greatest energy dispersion is along the axis of the flight tube. In addition, when ions are extracted orthogonally, the energy dispersion can produce angular divergence of the ion beam, resulting in poor transmission efficiency. However, on the basis of current evidence, the sensitivity of both TOF designs is generally an order of magnitude lower than the latest commercial quadrupole instruments.

Background levels: The on-axis design tends to generate higher background levels because neutral species and photons stand a better chance of reaching the detector. This results in background levels on the order of 20–50 cps— approximately one order of magnitude higher than the orthogonal design. However, because the ion beam in the axial design has a smaller cross section, a smaller detector can be used, which generally has better noise characteristics. In comparison, most commercial quadrupole instruments offer background levels of 1–10 cps, depending on the design.

Duty cycle: This is usually defined as the fraction (percentage) of extracted ions that actually make it into the mass analyzer. Unfortunately, with a TOF ICP mass spectrometer, which has to use "pulsed" ion packets from a continuous source of ions generated in the plasma, this process is relatively inefficient. It should also be emphasized that even though the ions are sampled at the same time, detection is not simultaneous because of different masses arriving at the detector at different times. The difference between the sampling mechanisms of the orthogonal and axial TOF designs translates into subtle differences in their duty cycles.

With the orthogonal design, duty cycle is defined by the width of the extracted ion packets, which are typically tall and thin in cross section, as shown in Figure 9.2. In comparison, the duty cycle of an axial design is defined by the length of the extracted ion packet, which is typically wide and thin in cross section, as shown in Figure 9.3. The duty cycle can be improved by changing the cross-sectional area of the ion packet, but, depending on the design, it is generally at the expense of resolution. However, this is not a major issue, because TOF instruments are generally not used for high-resolution ICP-MS applications. In practice, the duty cycles for both orthogonal and axial designs are on the order of 15–20%.

Resolution: The resolution of the orthogonal approach is slightly better because of its two-stage extraction/acceleration mechanism. Because a pulse of voltage pushes the ions from the extraction area into the acceleration region, the major energy dispersion lies along the axis of ion generation. For this reason, the energy spread is relatively small in the direction of extraction compared to spread with the axial approach, resulting in better resolution. However, the resolving power of commercial TOF ICP-MS systems is typically on the order of 500–2000, depending on the mass region, which makes them inadequate to resolve many of the problematic polyatomic species encountered in ICP-MS. In comparison, commercial high-resolution systems based on the double-focusing magnetic sector design offer resolving power up to 10,000, whereas commercial quadrupoles achieve 300 to 400.

Mass bias: This is also known as mass discrimination and is the degree to which ion transport efficiency varies with mass. All instruments show some degree of mass bias, which is usually compensated for by measuring the difference between the theoretical and observed ratio of two different isotopes of the same element. In TOF, the velocity (energy) of the initial ion beam will affect the instrument's mass bias characteristics. In theory, it should be less with the axial design because the extracted ion packets do not have any velocity in a direction perpendicular to the axis of the flight tube, which could potentially impact their transport efficiency.

BENEFITS OF TOF TECHNOLOGY FOR ICP-MS

It should be emphasized that these performance differences between the two designs are subtle and should not detract from the overall benefits of the TOF approach for ICP-MS. As mentioned earlier, a scanning device such as a quadrupole can only detect one mass at a time, which means that there is always a compromise between number of elements, detection limits, precision, and the overall measurement time. However, with the TOF approach, the ions are sampled at the same moment in time, which means that multielement data can be collected with no significant deterioration in quality. The ability of a TOF system to capture a full mass spectrum, approximately three orders of magnitude faster than a quadrupole, translates into three major benefits—multielement determinations in a fast transient peak, improved precision, especially for isotope ratioing techniques, and rapid data acquisition for carrying out qualitative or semiquantitative scans. Let us look at these in greater detail.

RAPID TRANSIENT PEAK ANALYSIS

Probably, the most exciting potential for TOF ICP-MS is in the multielement analysis of a rapid transient signal generated by sampling accessories such as laser ablation (LA),[5] electrothermal vaporization (ETV),[6] and flow injection systems.[7] Even though a scanning quadrupole can be used for this type of analysis, it struggles to produce high-quality multielement data when the transient peak lasts only a few seconds. The simultaneous nature of TOF instrumentation makes it ideally suited for this type of analysis, because the entire mass range can be collected in less than 50 µs. In particular, when used with an ETV system, the high acquisition speed of TOF can help to reduce matrix-based spectral overlaps by resolving them from the analyte masses in the temperature domain.[6] There is no question that TOF technology is ideally suited (probably more than any other design of ICP mass spectrometer) for the analysis of transient peaks.

IMPROVED PRECISION

To better understand how TOF technology can help improve precision in ICP-MS, it is important to know the major sources of instability. The most common source of noise in ICP-MS is the flicker noise associated with the sample introduction process (peristaltic pump pulsations, nebulization mechanisms, plasma fluctuations, etc.) and the shot noise derived from photons, electrons, and ions hitting the detector. Shot noise is based on counting statistics and is directly proportional to the square root of the signal. It therefore follows that as the signal intensity gets larger, the shot noise has less of an impact on the precision (% RSD) of the signal. This means that at high ion counts, the most dominant source of imprecision in ICP-MS is derived from the flicker noise generated in the sample introduction area.

One of the most effective ways to reduce instability produced by flicker noise is to use a technique called internal standardization, where the analyte signal is compared and ratioed to the signal of an internal standard element (usually of similar mass or ionization characteristics) that is spiked into the sample. Even though a quadrupole-based system can do an adequate job of compensating for these signal fluctuations, it is ultimately limited by its inability to measure the internal standard at precisely the same time as the analyte isotope. So, in order to compensate for sample introduction– and plasma-based noise and achieve high precision, the analyte and internal standard isotopes need to be sampled and measured simultaneously. For this reason, the design of a TOF mass analyzer is perfect for simultaneous internal standardization required for high-precision work. It therefore follows that TOF is also well suited for high-precision isotope ratio analysis, where its simultaneous nature of measurement is capable of achieving precision values close to the theoretical limits of counting statistics. Also, unlike a scanning-quadrupole-based system, it can measure ratios for as many isotopes or isotopic pairs as needed—all with excellent precision.[8]

RAPID DATA ACQUISITION

As with a scanning ICP-OES system, the speed of a quadrupole ICP mass spectrometer is limited by its scanning rate. To determine 10 elements in duplicate with good precision and detection limits, an integration time of 3 s per mass is normally required. When overhead scanning and settling times are added for each mass and replicate, this translates into approximately 2 min per sample. With a TOF system, the same analysis would take significantly less time, because all the data are captured simultaneously. In fact, detection limit levels in a TOF instrument are typically achieved within a 10–30 s integration time, which translates into a 5–10-fold improvement in data acquisition time over a quadrupole instrument. The added benefit of a TOF instrument is that the speed of analysis is not impacted by the number of analytes being determined: it would not matter if the method contained 10 or 70 elements—the measurement time would be virtually the same. However, there is one point that must be stressed. A large portion of the overall analysis time is taken up for flushing an old sample out of and pumping a new sample into the sample introduction system. This can be as much as 2 min per sample for real-world matrices. So, when this is taken into account, the difference between the sample throughput of a quadrupole and a TOF ICP mass spectrometer is not so evident.

Another benefit of the fast acquisition time is that qualitative or semiquantitative analysis is relatively seamless compared to scanning quadrupole technology, because every multielement scan contains data for every mass. This also makes spectral identification much easier by comparing the spectral fingerprint of unknown samples against a known reference standard. This is particularly useful for forensic work, where the evidence is often an extremely small sample.

There is no question that TOF ICP-MS, with its rapid, simultaneous mode of measurement, excels at multielement applications that generate fast transient signals, such as laser ablation. It offers excellent precision, particularly for isotope ratioing techniques, and also has the potential for very fast data acquisition. As mentioned previously, only the orthogonal (right-angled) design is currently available on a commercial basis. However, this should not detract from its overall capabilities. Commercially, TOF ICP-MS technology is almost 10 years old; so, although it is not as mature as quadrupole technology, it definitely should be considered as an option if the application demands it.

REFERENCES

1. A. E. Cameron and D. F. Eggers, *The Review of Scientific Instruments,* **19**(9), 605, 1948.
2. D. P. Myers, G. Li, P. Yang, and G. M. Hieftje, *Journal of American Society of Mass Spectrometry,* **5**, 1008–1016, 1994.
3. D. P. Myers, *12th Asilomar Conference on Mass Spectrometry,* Pacific Grove, CA, September 20–24, 1996.
4. Technical Note: 001-0877-00, GBC Scientific, February 1998.
5. P. Mahoney, G. Li, and G. M. Hieftje, *Journal of American Society of Mass Spectrometry,* **11**, 401–406, 1996.
6. Technical Note: 001-0876-00, GBC Scientific, February 1998.
7. R. E. Sturgeon, J. W. H. Lam, and A. Saint, *Journal of Analytical Atomic Spectrometry,* **15**, 607–616, 2000.
8. F. Vanhaecke, L. Moens, R. Dams, L. Allen, and S. Georgitis, *Analytical Chemistry,* **71**, 3297, 1999.

10 Mass Analyzers
Collision/Reaction Cell and Interface Technology

The detection capability for some elements using traditional quadrupole mass analyzer technology is severely compromised because of the formation of polyatomic spectral interferences generated by a combination of argon, solvent, and matrix-derived ions. Although there are ways to minimize these interferences, including correction equations, cool plasma technology, and matrix separation, they cannot be completely eliminated. However, a novel approach using collision/reaction cell and interface technology has been developed that significantly reduces the formation of many of these harmful species before they enter the mass analyzer. This chapter takes a detailed look at this relatively new technique and the exciting potential it has to offer.

There are a small number of elements that are recognized as having poor detection limits by ICP-MS. These are predominantly elements that suffer from major spectral interferences generated by ions derived from the plasma gas, matrix components, or the solvent/acid used in sample preparation. Examples of these interferences include the following:

- $^{40}Ar^{16}O^+$ in the determination of $^{56}Fe^+$
- $^{38}ArH^+$ in the determination of $^{39}K^+$
- $^{40}Ar^+$ in the determination of $^{40}Ca^+$
- $^{40}Ar^{40}Ar^+$ in the determination of $^{80}Se^+$
- $^{40}Ar^{35}Cl^+$ in the determination of $^{75}As^+$
- $^{40}Ar^{12}C^+$ in the determination of $^{52}Cr^+$
- $^{35}Cl^{16}O^+$ in the determination of $^{51}V^+$

The cold/cool plasma approach, which uses a lower temperature to reduce the formation of the argon-based interferences, has been a very effective way to get around some of these problems.[1] However, this approach can sometimes be difficult to optimize, is only suitable for a few of the interferences, is susceptible to more severe matrix effects, and it can be time consuming to change back and forth between normal and cool plasma conditions. These limitations and the desire to improve performance have led to the commercialization of collision/reaction cells (CRC) and collision/reaction interfaces (CRI). Designs for CRC and CRI were based on the early work of Rowan and Houk, who used Xe and CH_4 in the late 1980s to reduce the formation of ArO^+ and Ar_2^+ species in the determination of Fe and Se with a modified tandem mass spectrometer.[2] This research was investigated further by

Koppenaal and coworkers in 1994, who carried out studies using an ion trap for the determination of Fe, V, As, and Se in a 2% hydrochloric acid matrix.[3] However, it was not until 1996 that studies describing the coupling of a collision/reaction cell with a traditional quadrupole ICP mass spectrometer were published. Eiden and coworkers experimented using hydrogen as a collision gas,[4] whereas Turner and coworkers based their investigations on using helium gas.[5] These studies and the work of other groups at the time[6,7] proved to be the basis for modern collision and reaction cells that are commercially available today. Let us take a look at the fundamental principles of both collision/reaction cells and interfaces.

BASIC PRINCIPLES OF COLLISION/REACTION CELLS

With all collision/reaction cells, ions enter the interface in the normal manner, then are directed into a collision/reaction cell positioned prior to the analyzer quadrupole. A collision/reaction gas (e.g., helium, hydrogen, ammonia, or oxygen, depending on the design) is then bled via an inlet aperture into the cell containing a multipole (a quadrupole, hexapole, or octapole), usually operated in the RF-only mode. The RF-only field does not separate the masses as a traditional quadrupole mass analyzer does, but instead, it has the effect of focusing the ions, which then collide and react with molecules of the collision/reaction gas. By a number of different ion–molecule collision and reaction mechanisms, polyatomic interfering ions such as $^{40}Ar^+$, $^{40}Ar^{16}O^+$, and $^{38}ArH^+$ will either be converted to harmless noninterfering species, or the analyte will be converted to another ion that is not interfered with. This process is exemplified by the equation below, which shows the use of hydrogen as a reaction gas to reduce the $^{40}Ar^+$ interference in the determination of $^{40}Ca^+$.

$$H_2 + {}^{40}Ar^+ = Ar + H_2^+$$

$$H_2 + {}^{40}Ca^+ = {}^{40}Ca^+ + H_2 \text{ (no reaction)}$$

It can be seen that the hydrogen molecule interacts with the argon interference to form atomic argon and the harmless H_2^+ ion. However, there is no interaction between the hydrogen and the calcium. As a result, the $^{40}Ca^+$ ions, free of the argon interference, emerge from the collision/reaction cell through the exit aperture where they are directed towards the quadrupole analyzer for normal mass separation. Other gases are better suited to reduce the $^{40}Ar^+$ interference, but this process at least demonstrates the principles of the reaction mechanisms in a collision/reaction cell. The layout of a typical collision/reaction cell within the instrument is shown in Figure 10.1.

The equation above is an example of an ion–molecule reaction using the process of charge transfer. By the transfer of a positive charge from the argon ion to the hydrogen molecule, an innocuous neutral Ar atom is formed, which is invisible to the mass analyzer. There are many other reaction and collisional mechanisms that can take place in the cell, depending on the nature of the analyte ion, the interfering species, the reaction/collision gas, and the type of multipole used. Other possible mechanisms that can occur in the cell, in addition to charge transfer, include the following:

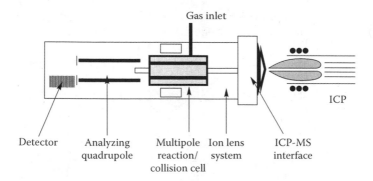

FIGURE 10.1 Layout of a typical collision/reaction cell instrument.

- *Proton transfer*—The interfering polyatomic species gives up a proton, which is then transferred to the reaction gas molecule to form a neutral atom.
- *Hydrogen atom transfer*—A hydrogen atom is transferred to the interfering ion, which is converted to an ion at one mass higher.
- *Molecular association reactions*—An interfering ion associates with a neutral species (atom or molecule) to form a molecular ion.
- *Collisional fragmentation*—The polyatomic ion is broken apart or fragmented by the process of multiple collisions with the gaseous atoms.
- *Collisional retardation*—The gas atoms/molecules undergo multiple collisions with the polyatomic interfering ion in order to retard or lower its kinetic energy. Because the interfering ion has a larger cross-sectional area than the analyte ion, it undergoes more collisions, and as a result, can be separated or discriminated from the analyte ion based on their kinetic energy differences.
- *Collisional focusing*—Analyte ions lose energy as they collide with the gaseous molecules, and depending on the molecular weight of the gas, will either enhance ion transmission as the ions migrate towards the central axis of the cell, or decrease sensitivity if ion scattering takes place.

The collision/reaction interface, which will be discussed later in this chapter, uses a slightly different principle to remove the interfering ions. It does not use a pressurized cell before the mass analyzer, but instead, injects a reaction/collision gas directly into the interface between the sampler and skimmer cones. The injection of the collision/reaction into this region of the ion beam produces collisions between the argon gas and the injected gas molecules, and as a result, argon-based polyatomic interferences are destroyed or removed before they are extracted into the ion optics.

DIFFERENT COLLISION/REACTION CELL APPROACHES

All these possible interactions between ions and molecules indicate that many complex secondary reactions and collisions can take place, which generate undesirable interfering species. If these species are not eliminated or rejected, they could potentially lead to additional spectral interferences. There are basically two different

approaches used to reduce the formation of polyatomic interferences and discriminate the products of these unwanted side reactions from the analyte ion. They are:

- **Collision mechanisms** using nonreactive gases and kinetic energy discrimination (KED)
- **Reaction mechanisms** using highly reactive gases and discrimination by
selective bandpass mass filtering

The major differences between the two approaches are how the gaseous molecules
interact with the interfering species and what type of multipole is used in the cell.
These dictate whether it is an ion–molecule collision or reaction mechanism taking
place. Let us take a closer look at each process because there are distinct differences
in the way the interference is rejected and separated from the analyte ion.

COLLISIONAL MECHANISMS USING NONREACTIVE GASES
AND KINETIC ENERGY DISCRIMINATION

The collisional mechanisms approach was adapted from collision-induced dissociation (CID) technology, which was first used in the early to mid-1990s in the study
of organic molecules using tandem mass spectrometry. The basic principle relies on
using a nonreactive gas in a hexapole collision cell to stimulate ion–molecule collisions. The more collision-induced daughter species that are generated, the better
the chance of identifying the structure of the parent molecule.[10,11] However, this very
desirable CID characteristic for identifying and quantifying biomolecules was a disadvantage in inorganic mass spectrometry, where uncontrolled secondary reactions
are generally something to be avoided. The limitation restricted the use of hexapole-
based collision cells in ICP-MS to inert gases such as helium or low-reactivity gases
such as hydrogen because of the potential to form undesirable reaction by-products,
which could spectrally interfere with other analytes. Unfortunately, higher-order
multipoles have little control over these secondary reactions because their stability boundaries are very diffuse and not well defined like a quadrupole. As a result,
they do not provide adequate mass separation capabilities to suppress the unwanted
secondary reactions. Thus, the need is to rely mainly on collisional mechanisms and
a process called *kinetic energy discrimination* (KED) to distinguish the interfering
ions and by-product species from the analyte ions. So, what is KED?

KED relies on the principle of separating ions depending on their different ion
energies. As ions enter the interface region, they all have differing kinetic energies
based on the ionization process in the plasma and their mass-to-charge ratio. When
the analyte and plasma/matrix-based interfering ions (sometimes referred to as *precursor ions*) enter the pressurized cell, they will undergo multiple collisions with the
collision gas. Because the collisional cross-sectional area of the precursor ions and
other collision-induced by-product ions are usually larger than the analyte ion, they
will undergo more collisions. This has the effect of lowering the kinetic energy of
these interfering species compared to the analyte ion. If the collision cell rod offset
potential is set slightly more negative than the mass filter potential, the polyatomic

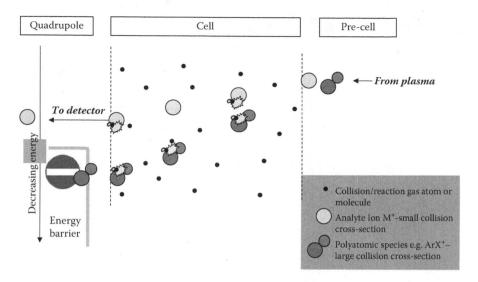

FIGURE 10.2 Principles of kinetic energy discrimination (courtesy of Thermo Scientific).

ions with lower kinetic energy are rejected or discriminated by the potential energy barrier at the cell exit. On the other hand, the analyte ions, which have a higher kinetic energy, are transmitted to the mass analyzer. Figure 10.2 shows the principles of KED.

In addition to the kinetic energy generated by the ionization process, the spread of ion energies will also be dictated by the efficiency of the RF-grounding mechanism (refer to Chapter 5). Therefore, for KED to work properly, the ion energy spread of ions generated in the plasma must be as narrow as possible to ensure that there is very little overlap between the analyte and the polyatomic interfering ion as they enter the mass analyzer. This means that it is absolutely critical for the RF-grounding mechanism to guarantee a low potential at the interface. If this is not the case, and there is a secondary discharge between the plasma and the interface, it will increase the ion energy spread of ions entering the collision cell and make it extremely difficult to separate the polyatomic interfering ion from the analyte of interest based on their kinetic energy difference. The relevance of having a narrow spread of ion energies is shown in Figure 10.3. It can be seen that the ion energy spread of the analyte ion (gray peak) and a polyatomic ion (black peak) is very similar as they enter the collision cell. This allows the collision process and KED system to easily separate the ions as they exit the cell. If the ion energy spread is larger, there would be more of an overlap as the ions enter the mass analyzer, and therefore compromise the detection limit for that analyte.

KED using helium as the collision gas works very well when the interfering polyatomic ion is physically larger than the analyte ion. This is exemplified in Figure 10.4, which shows helium flow optimization plots for six elements in 1:10 diluted seawater.[12] It can be seen that the signal intensity for the analytes—Cr, V, Co, Ni, Cu, and As—are all at a maximum, whereas their respective matrix, argon,

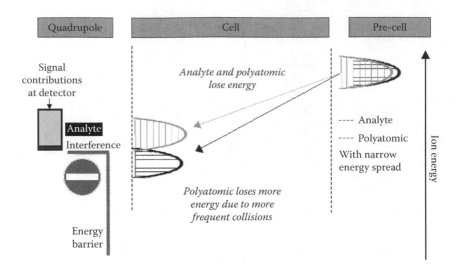

FIGURE 10.3 For optimum separation of analyte ion from the interfering species, it is important to have a narrow spread of ion energies entering the collision cell (courtesy of Thermo Scientific).

and solvent-based polyatomic interferences are all at a minimum at a similar helium flow rate of 5–6 mL/min. All the analytes show very good sensitivity because the KED process has allowed the analytes to be efficiently separated from their respective polyatomic interfering ions. The additional benefit of using helium is that it is inert and even if it is not being used in an interference reduction mode, it can have a beneficial effect on the other elements in a multielement run by increasing sensitivity via the process of collisional focusing. This makes the use of helium and KED very useful for both quantitative and semiquantitative multielement analysis using one set of tuning conditions.

For the KED process to work efficiently, there must be a distinct difference between the kinetic energy of the analyte compared to the interfering ion. In most cases where the polyatomic ion is large, this is not a problem. However, in many cases where the interfering and analyte ions have a similar physical size (cross section), the process requires an extremely large number of collisions, which will have an impact not only on the attenuation of the interfering ion, but also on the analyte ion. This means that for some situations, especially if the requirement is for ultratrace detection capability, the collisional process with KED is not enough to reduce the interference to acceptable levels. That is why most collision cells also have basic reaction mechanism capabilities, which allows low-reactivity gases such as hydrogen and, in some cases, small amounts of highly reactive gases such as ammonia or oxygen mixed with helium to be used.[13] But it should be pointed out that even though this initiates a basic reaction mechanism, rejection of the reaction by-product ions is still handled through the process of KED, which in some applications might not offer the most efficient way of reducing the interfering species. Using the collision cell as a basic reaction cell with low-reactivity gases is described later in this chapter.

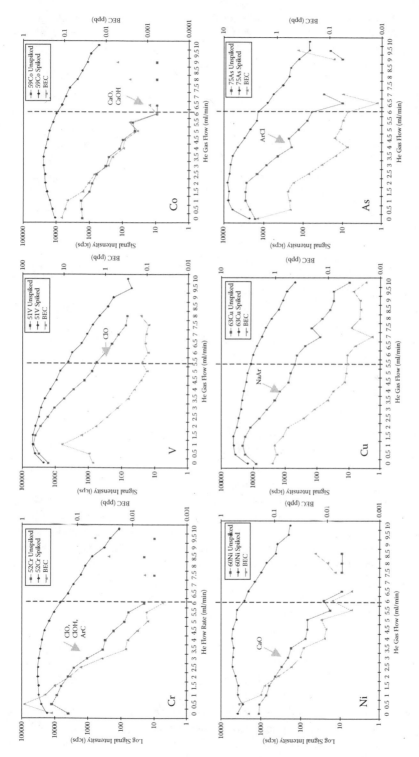

FIGURE 10.4 Helium cell gas flow optimization plots for Cr, V, Co, Ni, Cu, and As in 1:10 diluted seawater, showing that all polyatomic interfering ions are reduced to an acceptable level under one set of cell gas flow conditions (courtesy of Thermo Scientific).

One further point to keep in mind is that a KED-based cell relies on interactions of the interfering ion with an inert or low-reactivity gas, such that it can be separated from the analyte based on their differences in kinetic energy. If the gas contains impurities such as organic compounds or water vapor, the impurity could be the dominant reaction or collision pathway as opposed to the predicted collision/reaction with bulk gas. In addition, other unexpected ion–molecule reactions can readily occur if there are chemical impurities in the gas. This could also pose a secondary problem because of the formation of unexpected cluster ions such as metal oxide and hydroxide species, which have the potential to interfere with other analyte ions. Fortunately, many of these new ions formed in the cell as a result of reactions with the impurities have low energy and are adequately handled by KED. However, depending on the level of the impurity, some of the ions formed have higher energies and are therefore too high to be attenuated by the KED process, which could negatively impact the performance of the interference reduction process. For this reason, it is strongly advised that the highest-purity collision/reaction gases be used. If this is not an option, it is recommended that a gas-purifier (getter) system be placed in the gas line to cleanse the collision/reaction gas of impurities such as H_2O, O_2, CO_2, CO, or hydrocarbons. If you want to learn more about this subject, Yamada and coworkers published a very interesting paper describing the effects of cell–gas impurities and KED in an octapole-based collision cell.[14]

Let us now go on to discuss the other major way of interference rejection in a collision/reaction cell using highly reactive gases and mass (bandpass) filtering discrimination.

REACTION MECHANISMS WITH HIGHLY REACTIVE GASES AND DISCRIMINATION BY SELECTIVE BANDPASS MASS FILTERING

Another way of rejecting polyatomic interfering ions and the products of secondary reactions/collisions is to discriminate them by mass. As mentioned previously, higher-order multipoles cannot be used for efficient mass discrimination because the stability boundaries are diffuse and sequential secondary reactions cannot be easily intercepted. The only way this can be done is to utilize a quadrupole (instead of a hexapole or octapole) inside the reaction/collision cell and use it as a selective bandpass (mass) filter. One such development that uses this approach is called *dynamic reaction cell* (DRC) *technology.*[15] Similar in appearance to the hexapole and octapole collision/reaction cells, the dynamic reaction cell is a pressurized multipole positioned prior to the analyzer quadrupole. However, this is where the similarity ends. In DRC technology, a quadrupole is used instead of a hexapole or octapole. A highly reactive gas such as ammonia, oxygen, or methane is bled into the cell, which is a catalyst for ion molecule chemistry to take place. By a number of different reaction mechanisms, the gaseous molecules react with the interfering ions to convert them into either an innocuous species different from the analyte mass or a harmless neutral species. The analyte mass then emerges from the dynamic reaction cell free of its interference and is steered into the analyzer quadrupole for conventional mass separation.

The advantage of using a quadrupole in the reaction cell is that the stability regions are much better defined than higher-order multipoles, so it is relatively straightforward to operate the quadrupole inside the reaction cell as a mass or bandpass filter

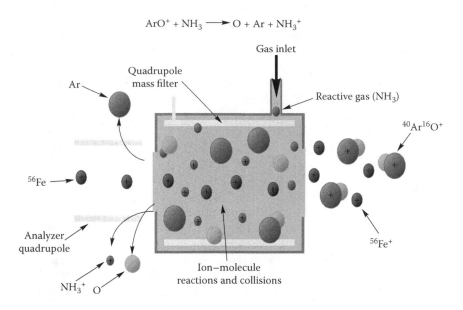

FIGURE 10.5 Elimination of the $^{40}Ar^{16}O^+$ interference with a dynamic reaction cell (copyright © 2003–2007, all rights reserved, PerkinElmer Inc.).

and not just as an ion-focusing guide. Therefore, by careful optimization of the quadrupole electrical fields, unwanted reactions between the gas and the sample matrix or solvent, which could potentially lead to new interferences, are prevented. It means that every time an analyte and interfering ions enter the dynamic reaction cell, the bandpass of the quadrupole can be optimized for that specific problem and then changed on the fly for the next one. This is shown schematically in Figure 10.5, where an analyte ion $^{56}Fe^+$ and an isobaric interference $^{40}Ar^{16}O^+$ enter the dynamic reaction cell. As can be seen, the reaction gas NH_3 picks up a positive charge from the $^{40}Ar^{16}O^+$ ion to form atomic oxygen, argon, and a positive NH_3 ion (this is known as a "charge transfer reaction"). There is no reaction between the $^{56}Fe^+$ and the NH_3, as predicted by thermodynamic reaction kinetics. The quadrupole's electrical field is then set to allow the transmission of the analyte ion $^{56}Fe^+$ to the analyzer quadrupole, free of the problematic isobaric interference, $^{40}Ar^{16}O^+$. In addition, the NH_3^+ is prevented from reacting further to produce a new interfering ion.

The practical benefit of using highly reactive gases is that they increase the number of ion–molecule reactions taking place inside the cell, which results in a faster, more efficient removal of the interfering species. Of course, they will also generate more side reactions which, if not prevented, will lead to new polyatomic ions being formed and could possibly interfere with other analyte masses. However, the quadrupole reaction cell is well characterized by well-defined stability boundaries. So, by careful selection of bandpass parameters, ions outside the mass/charge (m/z) stability boundaries are efficiently and rapidly ejected from the cell. It means that additional reaction chemistries, which could potentially lead to new interferences, are successfully interrupted. In addition, the bandpass of the reaction cell quadrupole can be swept in concert with the bandpass of the quadrupole mass analyzer. This allows a dynamic bandpass to be defined for the reaction cell so that the analyte ion can be efficiently transferred to the analyzer quadrupole. The overall benefit is that within the reaction cell, the most efficient thermodynamic reaction chemistries can be used to minimize the formation of plasma- and matrix-based polyatomic interferences, in addition to simultaneously suppressing the formation of further reaction by-product ions.

The process described can be exemplified by the elimination of $^{40}Ar^+$ by NH_3 gas in the determination of $^{40}Ca^+$. The reaction between NH_3 gas and the $^{40}Ar^+$ interference, which is predominantly charge transfer/exchange, occurs because the ionization potential of NH_3 (10.2 eV) is low compared to that of Ar (15.8 eV). This makes the reaction extremely exothermic and fast. However, as the ionization potential of Ca (6.1 eV) is significantly less than that of NH_3, the reaction, which is endothermic, is not allowed to proceed.[15] This can be seen in greater detail in Figure 10.6.

Of course, other secondary reactions are probably taking place, which you would suspect with such a reactive gas as ammonia, but by careful selection of the cell quadrupole electrical fields, the optimum bandpass only allows the analyte ion to be transported to the analyzer quadrupole, free of the interfering species. This highly efficient reaction mechanism and selection process translates into a dramatic reduction of the spectral background at mass 40, which is shown graphically in Figure 10.7. It can be seen that at the optimum NH_3 flow, a reduction in the $^{40}Ar^+$ background signal of about eight orders of magnitude is achieved, resulting in a detection limit of approximately 0.1 ppt for $^{40}Ca^+$.

One final thing to point out is that when highly reactive gases are used, the purity of the gas is not so critical because the impurity is almost insignificant in determining the ion–molecule reaction mechanism. On the other hand, with collision and low-reactivity gases that contain impurities, such as carbon dioxide, hydrocarbons, or water vapor, the impurity could be the dominant reaction pathway as opposed

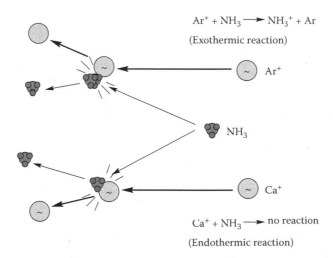

$$Ar^+ + NH_3 \longrightarrow NH_3^+ + Ar$$
(Exothermic reaction)

Ar^+

NH_3

Ca^+

$$Ca^+ + NH_3 \longrightarrow no\ reaction$$
(Endothermic reaction)

FIGURE 10.6   The reaction between NH_3 and Ar^+ is exothermic and fast, whereas there is no reaction between NH_3 and Ca^+ in the dynamic reaction cell (copyright © 2003–2007, all rights reserved, PerkinElmer Inc.).

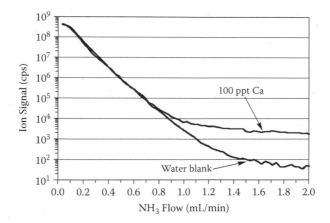

FIGURE 10.7 A reduction of eight orders of magnitude in the $^{40}Ar^+$ background signal is achievable with the dynamic reaction cell, resulting in a 0.1 ppt detection limit for $^{40}Ca^+$ (copyright © 2003–2007, all rights reserved, PerkinElmer Inc.).

to the predicted collision/reaction with the bulk gas. In addition, the formation of unexpected by-product ions or other interfering species, which have the potential to interfere with other analyte ions in a KED-based collision cell, are not such a serious problem with the DRC system because of its ability to intercept and stop these side reactions using the bandpass mass-filtering discrimination process.

These observations were, in fact, made by Hattendorf and Günter, who attempted to quantify the differences between KED and bandpass tuning with regard to suppression of interferences generated in a collision/reaction cell for a group of mainly mono-isotopic element (Sc, Y, La, Th) oxides.[16] They observed that when the collision/reaction gas contains impurities such as water vapor or ammonia, a broad range of additional interferences are produced in the cell. Depending on the relative mass of the precursor ions (ions that are formed in the plasma) compared to the by-product ions formed in the cell, there will be significant differences in the way these interferences are suppressed. They concluded that unless the mass (or energy) differences between the precursor and by-product ions are large, there will be significant overlap of kinetic energy distribution, making it very difficult to separate them, which limits the effectiveness of KED to suppress the cell-generated ions. On the other hand, bandpass tuning can tolerate a much smaller difference in mass between the precursor and by-product interfering ions because of its ability to set the optimum mass/charge cutoff at the point where these interfering ions are rejected. In addition, they found that the bandpass tuning method can use a heavier or denser collision gas if desired, without suffering a loss of sensitivity due to scattering observed with the KED method. The overall conclusion of their study was that "under optimized conditions, the bandpass tuning approach provides superior analytical performance because it retains a significantly higher elemental sensitivity and provides more efficient suppression of cell-generated oxide ions, when compared to kinetic energy discrimination."

Let us now go on to discuss the newest approach to reduce interferences using collision/reaction chemistry—the collision/reaction interface.

THE COLLISION/REACTION INTERFACE

CRI technology was first commercialized in early 2000 but was overlooked to a certain degree because collision/reaction cells, which had been introduced the previous year, offered clear advantages over this design. More recently, another vendor has introduced a CRI system[17] that appears to have solved some of the problems of the earlier designs. Let us take a look at the basic principles of this technology.

Unlike collision and reaction cells, the collision/reaction interface does not use a pressurized multipole-based cell before the mass analyzer. Instead, reaction/collision gases, such as hydrogen or helium, at relatively high flow rates (typically 100–150 mL/min) are injected directly into the plasma at right angles between the sampler and skimmer interface cones. The injection of the collision/reaction gas into this high-density, high-temperature region of the ion beam produces high collision frequency between the argon gas and the injected gas molecules. As a result, most argon-based polyatomic interferences are destroyed or removed before they are extracted into the ion optics. Figure 10.8 shows the basic principles of the CRI.

The limitations of the earlier designs restricted their use for real-world samples because there appeared to be no way to effectively focus the ions, and therefore there was very little control over the collision process. So, even though the addition of a collision/reaction gas helped reduce plasma-based spectral interferences, it did virtually nothing for matrix-induced spectral interferences. In addition, there was no way to carry out KED in the interface region and, as a result, it made it very difficult to take advantage of collisional mechanisms using an inert gas such as helium.

However, the manufacturer of the most recent CRI systems claims that the major improvement in their design is that all the reactions/collisions processes are actually taking place inside the tip of the skimmer and not between the sampler and skimmer cone as with earlier designs. Because of this subtle difference, they say that by using helium as the CRI gas, both argon and matrix-based polyatomic species can receive sufficient energy through vibrational and rotational excitation mechanisms to bring about the dissociation of the interference, whereas the analyte ions simply lose energy as they collide with the gas molecules. Where the collisional impact is not suitable for interference reduction, as in the removal of the argon dimer ($^{40}Ar_2^+$) in the determination of $^{80}Se^+$, low-reactivity gases such as hydrogen can be used to initiate an ion–molecule reaction.[18] They also claim that the effects of KED to reject additional by-product ions are inherent in the design of the interface and the pre-ion optics area.

The CRI is most definitely an interesting concept, which appears to offer a relatively straightforward, non-cell-based solution to minimizing plasma- and matrix-based spectral interferences in ICP-MS. What could be simpler than to inject a collision/reaction gas into the interface region and see your spectral interferences go away? However, CRI is competing against almost 10 years of cell-based knowledge and application material, where the ion–molecule chemistry and collision mechanisms are fairly well understood. So, until there is a substantial body of work in the public domain showing the capabilities of CRI systems, they should be evaluated with some caution. It is worth checking out the vendor application notes, which show the capabilities of the CRI in real world matrices.[19,20]

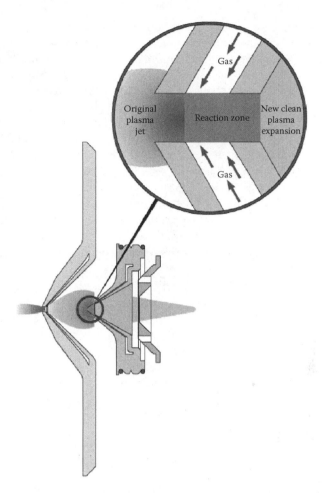

FIGURE 10.8 Principles of the collision/reaction interface (CRI) (courtesy of Varian Instruments Inc.).

USING REACTION MECHANISMS IN A COLLISION CELL

After nearly 10 years of solving real-world application problems, the practical capabilities of both hexapole and octapole collision cells using KED are fairly well understood. It is clear that the majority of applications are being driven by the demand for routine multielement analysis of well-characterized matrices, where a rapid sample turnaround is required. This technique has also been promoted by the vendors as a fast, semiquantitative tool for unknown samples. The fact that it requires very little method development, just one collision gas and one set of tuning conditions, makes it very attractive for these kinds of applications.[21]

However, it is well-recognized that this approach will not work well for many of the more complex interfering species, especially if the analyte is at ultratrace levels. For example, the collision mode using helium is not the best choice for quantifying selenium using its two major isotopes ($^{80}Se^+$, $^{78}Se^+$) because it requires a large number of collisions to separate the argon dimers ($^{40}Ar_2^+$ and $^{40}Ar^{38}Ar^+$) from the analyte ions using kinetic energy discrimination. As a result of the inefficient interference reduction process, the signal intensity for the analyte ion is also suppressed. For this reason, it is well accepted that the best approach is to use a reaction gas in order to initiate some kind of charge-transfer mechanism. This can be seen in Figure 10.9, which shows the calibration for $^{78}Se^+$ using hydrogen as the reaction gas to minimize the impact of the argon dimer, $^{40}Ar^{38}Ar^+$. On the left is the 0, 1.0, 2.5, 5.0, and 10.0 ppb calibration using helium and on the right is the same calibration using hydrogen as the reaction gas. It can be seen very clearly that the signal intensity for the calibration standards using hydrogen gas is approximately five times higher than the calibration standards using helium, producing a selenium detection limit of 15 times lower (1 ppt compared to 15 ppt); this is also seen in the table following the calibration graphs.[22]

Likewise, helium has very little effect on reducing the $^{40}Ar^+$ interference in the determination of $^{40}Ca^+$. So, when using a collision cell with helium, the quantification of Ca must be carried out using the $^{44}Ca^+$ isotope, which is about 50 times less sensitive than $^{40}Ca^+$. For this reason, in order to achieve the lowest detection limits for calcium, a low-reactivity gas such as pure hydrogen is the better option.[22] By initiating an ion–molecule reaction, it allows the most sensitive calcium isotope at 40 amu to be used for quantitation. In fact, even though the use of hydrogen significantly improves the detection limit for calcium, the best interference reduction is achieved using a mixture of ammonia and helium.[13]

Another example of the benefits of using more reactive gases, such as an ammonia and helium mixture over pure helium or hydrogen gas, is in the determination of vanadium in a high-concentration chloride matrix. The collision mode using helium works reasonably well on the reduction of the $^{35}Cl^{16}O^+$ interference at 51 amu. However, when 1% NH_3 in helium is used, the interference is dramatically reduced by the process of charge/electron transfer. This allows the most abundant isotope, $^{51}V^+$, to be used for the quantitation of vanadium in matrices such as seawater or hydrochloric acid. Vanadium detection capability in a chloride matrix is improved by a factor of 50–100 times using the reaction chemistry of NH_3 in helium compared to pure helium in the collision mode.[13,22]

In addition to ammonia–helium mixtures, oxygen is sometimes the best reaction gas to use because it offers the possibility of either moving the analyte ion to a region of the mass spectrum where the interfering ion does not pose a problem or moving the interfering species away from the analyte ion by forming an oxygen-derived polyatomic ion 16 amu higher. An example of changing the mass of the interfering ion is in the determination of $^{114}Cd^+$ in the presence of high concentrations of molybdenum. In the plasma, the molybdenum forms a very stable oxide species $^{98}Mo^{16}O^+$ at 114 amu, which interferes with the major isotope of cadmium, also at mass 114. By using pure oxygen as the reaction gas, the $^{98}Mo^{16}O^+$ interference is converted to the $^{98}Mo^{16}O^{16}O^+$ complex at 16 amu higher than the analyte ion, allowing the $^{114}Cd^+$ isotope to be used for quantitation.[22]

The benefits of using reaction mechanisms for the determination of calcium, vanadium, and cadmium are seen in Figure 10.10, which shows (1) 0–300 ppt cali-

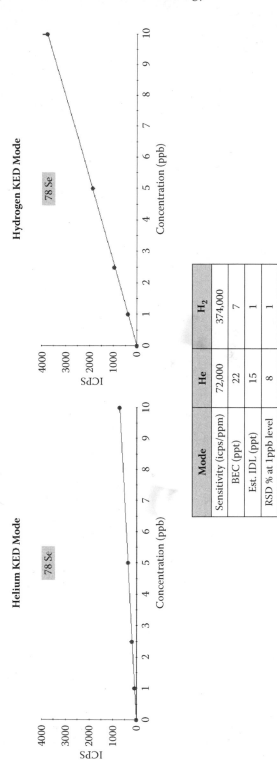

FIGURE 10.9 Comparison of $^{78}Se^+$ calibration plots and detection limits using the collision mode with helium (left) and reaction mode with hydrogen (right) (courtesy of Thermo Scientific).

Mode	He	H₂
Sensitivity (icps/ppm)	72,000	374,000
BEC (ppt)	22	7
Est. IDL (ppt)	15	1
RSD % at 1ppb level	8	1

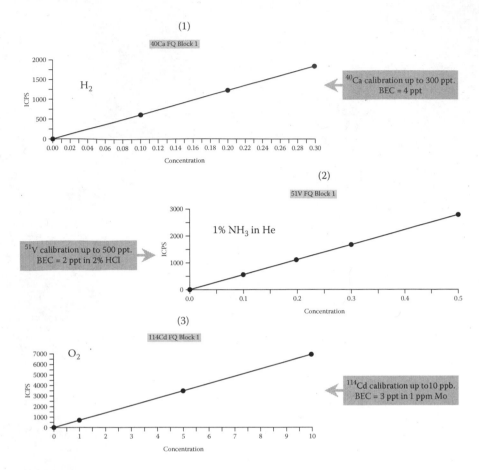

FIGURE 10.10   Calibration plots for (1) $^{40}Ca^+$ using hydrogen, (2) $^{51}V^+$ in 2% HCl using 1% NH_3 in helium, and (3) $^{114}Cd^+$ in high concentrations of molybdenum using oxygen (courtesy of Thermo Scientific).

bration plot for $^{40}Ca^+$ using hydrogen gas, (2) 0–500 ppt calibration plot for $^{51}V^+$ in 2% hydrochloric acid using 1% NH_3 in helium, and (3) 0–10 ppb calibration plot for $^{114}Cd^+$ in a molybdenum matrix using pure oxygen. Background equivalent concentration (BEC) values of 4 ppt, 2 ppt, and 3 ppt were achieved for calcium, vanadium, and cadmium, respectively.[22]

However, it is important to point out that even though a higher-order multipole collision cell with KED can use low-reactivity gases, it usually requires significantly more interactions than a dynamic reaction cell that uses a highly reactive gas.[23] Take for example, the reduction of the $^{40}Ar^+$ interferences in the determination of $^{40}Ca^+$. In a collision cell, even though the kinetic energy of the $^{40}Ar^+$ ion will be reduced by reactive collisions with molecules of hydrogen gas, the $^{40}Ca^+$ will also lose kinetic energy because it, too, will collide with the reaction gas. Now, even though these interactions are basically nonreactive with respect to $^{40}Ca^+$, it will experience the same number of collisions because it has a similar cross-sectional area as the $^{40}Ar^+$ ion. So, in order to achieve many orders of interference reduction with a low-reactivity gas such as hydrogen, a high number of collisions are required. This means that in addition to the interference being suppressed, the analyte will also be affected to a similar extent. As a result, the energy distribution of both the interfering species and the analyte ion at the cell exit will be very close, if not overlapping, resulting in compromised interference reduction capabilities compared to a reaction cell that uses highly reactive gases and discrimination by mass filtering. This compromise translates into a detection limit for $^{40}Ca^+$ with a KED-based collision cell using hydrogen being approximately 50–100 times worse than a dynamic reaction cell using pure ammonia.[24]

DETECTION LIMIT COMPARISON

In general, highly reactive gases are recognized as being more efficient at reducing the interference and generating better signal-to-noise ratio than either an inert gas such as helium, or low-reactivity gases such as hydrogen or mixtures of ammonia and helium. Table 10.1 shows a detection limit comparison (in ppt) between a DRC, an octapole CRC, and a CRI for a group elements across the mass range. These are based on three sigma of 10 replicates of a 1% nitric/deionized water blank. So, besides compensating for the normal argon and water/HNO_3-based spectral features, they do not represent detection limits in a matrix-specific sample, which will vary depending on the analyte and the spectral interference encountered.

It is important to point out that all the detection limits were taken from vendor-generated application notes. For details of instrumental parameters/conditions, please refer to the cited references. With regard to DRC detection limits, data for ammonia was chosen because it is the most widely used reaction gas, which is applicable to the reduction of a wide variety of spectral interferences. However, the elements marked with an (a) were determined in the standard mode, where no reaction was used, whereas for selenium isotopes the optimum detection limits (b) were achieved with methane as the reaction gas. The CRC detection limits were generated with one set of tuning conditions in the helium collision mode because this appears to be the way vendors are promoting its use, particularly for multielement analysis. For that reason, it is important to compare its D/L performance in the way it would be run in a routine laboratory. However, lower detection limits could probably be achieved with the CRC by using different reaction gases and gas mixtures. The CRI detection limit data was generated using hydrogen as the reaction gas, which seems to be the most common gas to use with this technology.

SUMMARY

There is no question that collision/reaction cells have given a new lease on life to quadrupole mass analyzers used in ICP-MS. They have enhanced its performance and flexibility, and most definitely opened up the technique to more demanding applications that were previously beyond its capabilities. However, it must be emphasized that when assessing this technology, it is critical that you fully understand the

TABLE 10.1

Typical Detection Limits in Parts per Trillion (ppt) of a Dynamic Reaction Cell (DRC), an Octapole-Based Collision/Reaction Cell (CRC), and a Collision/Reaction Interface System (CRI)

Element	DRC[24] Using Ammonia DL (ppt)	CRC[21] Using Helium DL (ppt)	CRI[17,18] Using Hydrogen DL (ppt)
^{11}B	1.93[a]	88	n/r
^{23}Na	0.14[a]	490	13
^{24}Mg	0.08[a]	1.6	0.5
^{27}Al	0.05	26	0.8
^{39}K	0.27	400	43
^{40}Ca	0.10	n/r	3
^{44}Ca	n/r	21	n/r
^{48}Ti	0.92	3.7	1
^{51}V	0.12	0.28	0.2
^{52}Cr	0.12	0.53	0.6
^{55}Mn	0.17	0.79	0.4
^{56}Fe	0.12	9.4	2
^{59}Co	0.04	0.5	0.2
^{60}Ni	0.10[a]	1.7	10
^{63}Cu	0.05	2.0	1
^{64}Zn	0.45	3.1	2
^{75}As	0.48[a]	1.4	0.6
^{78}Se	1.2[b]	35	2
^{80}Se	0.7[b]	n/r	9
^{120}Sn	0.12[a]	0.87	n/r
^{121}Sb	0.08[a]	1.0	n/r
^{138}Ba	0.06[a]	0.85	n/r
^{208}Pb	0.07[a]	0.29	0.1

Notes: The DRC data was generated using 1.5 mL/min of pure NH_3 gas, unless stated differently.[24] The CRC used 5 mL/min He gas in the collision mode,[21] and the CRI used 80 mL/min of pure H_2 gas.[17,18]

n/r = not reported.

a = standard mode (no gas).

b = methane gas used.

DL = detection limit.

capabilities of the different approaches, especially how they match up to your application objectives. The KED-based collision cell using an inert gas such as helium is probably better suited to doing multielement analysis in a routine environment. However, you have to be aware that its detection capability is compromised for some elements depending on the type of samples being analyzed. By using ion–molecule reactions as opposed to collisions, detection limits for many of these elements can be improved quite significantly. Of course, if two or even three different gases have to be used, the convenience of using one gas goes away.

On the other hand, using highly reactive gases with discrimination by mass filtering appears to offer the best performance and the most flexibility of all the different commercial approaches. By careful matching of the reaction gas with the analyte ion and polyatomic interference, extremely low detection limits can be achieved by ICP-MS, even for many of the notoriously difficult elements. It should be emphasized that selection of the optimum reaction gas and selection of the best quadrupole bandpass parameters can sometimes translate into quite lengthy method development, especially if there is very little application data available. However, most vendors do a very good job of generating application studies for some of the more routine applications. If you analyze out-of-the-ordinary or complex samples, you might initially need to spend the time to develop an analytical method that is both robust and routine.

So pay attention not only to what the technique can do for your application problem but also to what it cannot do, which is equally important. In other words, make sure you evaluate its capabilities on the basis of all your present and future analytical requirements, such as ease of use, method development, flexibility, sample throughput, and detection capability. When assessing vendor-generated data, make sure the performance is achievable in your laboratory and on your samples.

REFERENCES

1. K. Sakata and K. Kawabata, *Spectrochimica Acta*, **49B**, 1027, 1994.
2. J. T. Rowan and R. S. Houk, *Applied Spectroscopy*, **43**, 976–980, 1989.
3. D. W. Koppenaal, C. J. Barinaga, and M. R. Smith, *Journal of Applied Analytical Chemistry*, **9**, 1053–1058, 1994.
4. G. C. Eiden, C. J. Barinaga, and D. W. Koppenaal, *Journal of Applied Analytical Chemistry*, **11**, 317–322, 1996.
5. P. Turner, T. Merren, J. Speakman, and C. Haines, *Plasma Source Mass Spectrometry: Developments and Applications*, ISBN 0-85404-727-1, 28–34, 1996.
6. D. J. Douglas and J. B. French, *Journal of American Society of Mass Spectrometry*, **3**, 398, 1992.
7. B. A. Thomson, D. J. Douglas, J. J. Corr, J. W. Hager, and C. A. Joliffe, *Analytical Chemistry*, **67**, 1696–1704, 1995.
8. X Series ICP-MS: Enhanced Collision Cell Technology CCT, Thermo Scientific Product Specifications, July 2004, http://www.thermo.com/eThermo/CMA/PDFs/Articles/articlesFile_24138.pdf.
9. E. R. Denoyer, S. D. Tanner, and U. Voellkopf, *Spectroscopy*, **14**, 2, 1999.
10. H. H. Willard, L. L. Merritt, J. A. Dean, and F. A. Settle, *Instrumental Methods of Analysis*, Wadsworth Publishing Co., Belmont, CA, 465–507, 1988.
11. E. De Hoffman, J. Charette, and V. Stroobant, *Mass Spectrometry, Principles and Applications*, John Wiley and Sons, Paris, France, 1996.
12. *Analysis of Ultra-trace Levels of Elements in Seawaters Using 3rd-Generation Collision Cell Technology*, Thermo Scientific Product Application Note: 40718, April 2007, http://www.thermo.com/eThermo/CMA/PDFs/Articles/articlesFile_26161.pdf.

13. J. Takahashi, Determination of Impurities in Semiconductor Grade Hydrochloric Acid Using the Agilent 7500 cs ICP-MS, Agilent Technologies Application Note 5989-4348EN, January 2006, http://www.chem.agilent.com/temp/rad3E0DA/00001358.PDF.

14. N. Yamada, J. Takahashi, and K. Sakata, *Journal of Analytical Atomic Spectrometry*, **17**, 1213–1222, 2002.

15. S. D. Tanner and V. I. Baranov, *Atomic Spectroscopy*, **20**(2), 45–52, 1999.

16. B. Hattendorf and D. Günter, *Journal of Analytical Atomic Spectrometry*, **19**, 600–606, 2004.

17. X. D. Wang, *Typical Detection Limits of the Varian 820-ICP-MS*, Varian Instruments Application Note 33, 20067, http://www.varianinc.com/image/vimage/docs/products/spectr/icpms/atworks/icpms33.pdf.

18. X. D. Wang and I. Kalinitchenko, *Principles and Performance of the Collision Reaction Interface of the Varian 820-MS*, Varian Instruments ICP-MS Advantage Note, 1, October 2005 http://www.varianinc.com/image/vimage/docs/applications/apps/icpms_an1.pdf.

19. A. Tisinger and A. Lynch, *High Matrix Sample Analysis with the Varian 820-MS using the Collision Reaction Interface,* Varian Instruments Application Note 30, 2006, http://www.varianinc.com/image/vimage/docs/products/spectr/icpms/atworks/icpms30.pdf.

20. Y. Abdelnour and J. Murphy, *The Analysis of Whole Blood Samples by Collision Reaction Interface Inductively Coupled Plasma Mass Spectrometry: Varian 820-MS*, Varian Instruments Application Note 28, 2006, http://www.varianinc.com/image/vimage/docs/products/spectr/icpms/atworks/icpms28.pdf.

21. S. Wilbur, *Performance Characteristics of the Agilent 7500cx: Evaluating Helium Collision Mode for Simpler, Faster, More Accurate ICP-MS*, May 2007, http://www.chem.agilent.com/temp/rad68172/00001606.PDF.

22. F. Keenan and W. Spence, *Theory and Applications of Collision/Reaction Cells: How Collision and Reaction Cells Work for Interference Removal in ICP-MS*, Thermo Fisher Scientific Web-based Presentation, 2007, http://breeze.thermo.com/crcs.

23. S. D. Tanner, V. I. Baranov, and D. R. Bandura, *Spectrochimica Acta,* **57B**(9), 1361–1452, 2002.

24. K. Kawabata, Y. Kishi, and R. Thomas, *Spectroscopy*, **18**(1), 16–31, 2003.

11 Ion Detectors

Chapter 11 looks at the detection system—an important area of the mass spectrometer that detects and quantifies the number of ions emerging from the mass analyzer. The detector converts the ions into electrical pulses, which are then counted using its integrated measurement circuitry. The magnitude of the electrical pulses corresponds to the number of analyte ions present in the sample, which is then used for trace element quantitation by comparing the ion signal with known calibration or reference standards.

Since ICP-MS was first introduced in the early 1980s, a number of different ion detection designs have been utilized, the most popular being electron multipliers for low ion count rates and Faraday collectors for high count rates. Today, the majority of ICP-MS systems that are used for ultratrace analysis use detectors that are based on the active film or discrete dynode electron multiplier. They are very sophisticated pieces of equipment and are very efficient at converting ion currents emerging from the mass analyzer into electrical signals. The location of the detector in relation to the mass analyzer is shown in Figure 11.1.

Before I go on to describe discrete dynode detectors in greater detail, it is worth looking at two of the earlier designs—the channel electron multiplier (Channeltron®)[1] and the Faraday cup—to get a basic understanding of how the ICP-MS ion detection process works.

CHANNEL ELECTRON MULTIPLIER

The operating principles of the channel electron multiplier are similar to a photomultiplier tube used in ICP-OES. However, instead of using individual dynodes to convert photons to electrons, the Channeltron is an open glass cone (coated with a semiconductor-type material) to generate electrons from ions that impinge on its surface. For the detection of positive ions, the front of the cone is biased at a negative potential while the far end near the collector is kept at ground. When the ion emerges from the quadrupole mass analyzer, it is attracted to the high negative potential of the cone. When the ion hits this surface, one or more secondary electrons are formed. The potential gradient inside the tube varies based on position, so the secondary electrons move further down the tube. As these electrons strike new areas of the coating, more secondary electrons are emitted. This process is repeated many times. The result is a discrete pulse, which contains many millions of electrons generated from an ion that first hits the cone of the detector.[1] This process is shown simplistically in Figure 11.2.

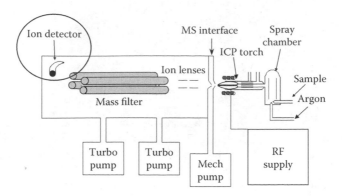

FIGURE 11.1   The location of the detector in relation to the mass analyzer.

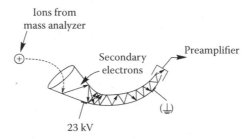

FIGURE 11.2   Basic principles of a channel electron multiplier. (From Channeltron®, *Electron Multiplier Handbook for Mass Spectrometry Applications*, Galileo Electro-Optic Corp., 1991. Channeltron is a registered trademark of Galileo Corp.)

This pulse is then sensed and detected by a very fast preamplifier. The output pulse from the preamplifier then goes to a digital discriminator and counting circuitry that only counts pulses above a certain threshold value. This threshold level needs to be high enough to discriminate against pulses caused by spurious emission inside the tube, from any stray photons from the plasma itself, or photons generated from fast-moving ions striking the quadrupole rods.

It is worth pointing out that the rate at which ions hit the detector is sometimes too high for the measurement circuitry to handle in an efficient manner. This is caused by ions arriving at the detector during the output pulse of the preceding ion and not being detected by the counting system. This "dead time," as it is known, is a fundamental limitation of the multiplier detector and is typically 30–50 s, depending on the detection system. Compensation in the measurement circuitry has to be made for this dead time to count the maximum number of ions hitting the detector.

FARADAY CUP

For some applications, where ultratrace detection limits are not required, the ion beam from the mass analyzer is directed into a simple metal electrode or Faraday cup. With this approach, there is no control over the applied voltage (gain), so they can only be used for high ion currents. Their lower working range is on the order of 10^4 cps, which means that if they are to be used as the only detector, the sensitivity of the ICP mass spectrometer will be severely compromised. For this reason, they are normally used in conjunction with a Channeltron or discrete dynode detector to extend the dynamic range of the instrument. An additional problem with the Faraday cup is that because of the time constant used in the DC amplification process to measure the ion current, they are limited to relatively low scan rates. This limitation makes them unsuitable for the fast scan rates required for traditional pulse counting used in ICP-MS and also limits their ability to handle fast transient peaks.

The Faraday cup was never sensitive enough for quadrupole ICP-MS technology, because it was not suitable for very low ion count rates. An attempt was made in the early 1990s to develop an ICP-MS system using a Faraday cup detector for the environmental market, but its sensitivity was compromised, and as a result was considered more suitable for applications requiring ICP-OES trace level detection capability. However, Faraday cup technology is still utilized in some magnetic sector instruments, particularly where high ion signals are encountered in the determination of high-precision isotope ratios, using a multicollector detection system.

DISCRETE DYNODE ELECTRON MULTIPLIER

These detectors, which are often called *active film multipliers*, work in a similar way to the Channeltron, but utilize discrete dynodes to carry out the electron multiplication.[2] Figure 11.3 illustrates the principles of operation of this device. The detector is positioned off-axis to minimize the background noise from stray radiation and neutral species coming from the ion source. When an ion emerges from the quadrupole, it sweeps through a curved path before it strikes the first dynode. On striking the first dynode, it liberates secondary electrons. The electron-optic design of the dynode produces acceleration of these secondary electrons to the next dynode where they generate more electrons. This process is repeated at each dynode, generating a pulse of electrons that are finally captured by the multiplier collector or anode. Because of the materials used in the discrete dynode detector and the difference in

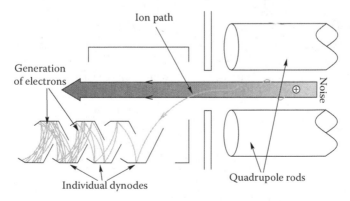

FIGURE 11.3 Schematic of a discrete dynode electron multiplier. (From K. Hunter, *Atomic Spectroscopy,* **15**[1], 17–20, 1994.)

the way electrons are generated, it is typically 50–100% more sensitive than Channeltron technology.

Although most discrete dynode detectors are very similar in the way they work, there are subtle differences in the way the measurement circuitry handles low and high ion count rates. When ICP-MS was first commercialized, it could only handle up to five orders of dynamic range. However, when attempts were made to extend the dynamic range, certain problems were encountered. Before we discuss how modern detectors deal with this issue, let us first look at how it was addressed in earlier instrumentation.

EXTENDING THE DYNAMIC RANGE

Traditionally, ICP-MS using the pulse-counting measurement is capable of about five orders of linear dynamic range. This means that ICP-MS calibration curves, generally speaking, are linear from detection limit up to a few hundred ppb. However, there are a number of ways of extending the dynamic range of ICP-MS another three to four orders of magnitude and working from sub-ppt levels up to a hundred ppm. Here is a brief overview of some of different approaches that have been used.

FILTERING THE ION BEAM

One of the very first approaches to extending the dynamic range in ICP-MS was to filter the ion beam. This was achieved by putting a nonoptimum voltage on one of the ion lens components or the quadrupole itself, to limit the number of ions reaching the detector. This voltage offset, which was set on an individual mass basis, acted as an energy filter to electronically screen the ion beam and reduce the subsequent ion signal to within a range covered by pulse counting ion detection. The main disadvantage with this approach was that the operator had to have prior knowledge of the sample to know what voltage to apply to the high concentration masses.

USING TWO DETECTORS

Another technique that was used on some of the early ICP-MS instrumentation was to utilize two detectors, such as a channel electron multiplier and a Faraday cup, to extend the dynamic range. With this technique, two scans would be made. In the first scan it would measure the high concentration masses using the Faraday cup; in the second scan it would skip over the high concentration masses and carry out pulse counting of the low concentration masses with a channel electron multiplier. This worked reasonably well, but struggled with applications that required rapid switching between the two detectors, because the ion beam had to be physically deflected to select the optimum detector. Not only did this degrade the measurement duty cycle, but detector switching and stabilization times of several seconds also precluded fast transient signal detection.

USING TWO SCANS WITH ONE DETECTOR

The more modern approach is to use just one detector to extend the dynamic range. This has typically been done by using the detector both in pulse and analog mode, so high and low concentrations can be determined in the same sample. There are basically three approaches to using this type of detection system: two of them involve carrying out two scans of the sample, whereas the third only requires one scan.

The first approach uses an electron multiplier operated in both digital and analog mode.[3] Digital counting provides the highest sensitivity, whereas operation in the analog mode (achieved by reducing the high voltage applied to the detector) is used to reduce the sensitivity of the detector, thus extending the concentration range for which ion signals can be measured. The system is implemented by scanning the spectrometer twice for each sample. The first scan, in which the detector is operated in the analog mode, provides signals for elements present at high concentrations. A second scan in which the detector voltage is switched to digital pulse-counting mode provides high-sensitivity detection for elements present at low levels. A major advantage of this technology is that the user does not need to know in advance whether to use analog or digital detection, because the system automatically scans all elements in both modes. However, its major disadvantage is that two independent mass scans are required to gather data across an extended signal range. This not only results in degraded measurement efficiency and slower analyses, but it also means that the system cannot be used for fast transient signal analysis, because mode switching is generally too slow.

An alternative way of extending the dynamic range is similar to the first approach, except that the first scan is used as an investigative tool to examine the sample spectrum before analysis.[4] This first prescan establishes the mass positions at which the analog and pulse modes will be used for subsequently collecting the spectral signal. The second analytical scan is then used for data collection, switching the detector back and forth rapidly between pulse and analog mode at each analytical mass.

Even though these approaches worked very well, their main disadvantage was that two separate scans are required to measure high and low levels. With conventional nebulization, this is not such a major problem except that it can impact sample throughput. However, it does become a concern when it comes to working with transient peaks found in laser sampling (LS), flow injection (FIAS), or electro thermal vaporization (ETV) ICP-MS. Because these transient peaks often only last a few seconds, all the available time must be spent measuring the masses of interest to get the best detection limits. When two scans have to be made, time is wasted collecting data, which is not contributing to the analytical signal.

USING ONE SCAN WITH ONE DETECTOR

The limitation of having to scan the sample twice led to the development of an improved design using a dual stage discrete dynode detector.[5] This technology utilizes measurement circuitry that allows both high and low concentrations to be determined in one scan. This is achieved by measuring the ion signal as an analog signal at the midpoint dynode. When more than a threshold number of ions is detected, the signal is processed through the analog circuitry. When fewer than the threshold number of ions is detected, the signal cascades through the rest of the dynodes and is measured as a pulse signal in the conventional way. This process, which is shown in Figure 11.4, is completely automatic and means that both the analog and the pulse signals are collected simultaneously in one scan.[6]

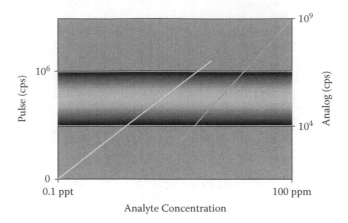

FIGURE 11.4 Dual stage discrete dynode detector measurement circuitry. (From E. R. Denoyer, R. J. Thomas, and L. Cousins, *Spectroscopy,* **12**[2], 56–61, 1997. Covered by U.S. Patent Number 5,463,219.)

The pulse-counting mode is typically linear from zero to about 10^6 cps, whereas the analog circuitry is suitable from 10^4 to 10^9 cps. To normalize both ranges, a cross-calibration is carried out to cover concentration levels, which produces a pulse and an analog signal. This is possible because the analog and pulse outputs can be defined in identical terms of incoming pulse counts per second, based on knowing the voltage at the first analog stage, the output current, and a conversion factor defined by the detection circuitry electronics. By carrying out a cross-calibration across the mass range, a dual mode detector of this type is capable of achieving approximately eight to nine orders of dynamic range in one simultaneous scan. This can be seen in Figures 11.5 and 11.6. Figure 11.5 shows that the pulse counting calibration curve (left-hand line) is linear up to 10^6 cps, whereas the analog calibration curve (right-hand line) is linear from 10^4 to 10^9 cps. Figure 11.6 shows that after cross-calibration, the two curves are normalized, which means the detector is suitable for concentration levels between 0.1 ppt and 100 ppm—typically eight to nine orders of magnitude for most elements.[5]

There are subtle variations of this type of detection system, but its major benefit is that it requires only one scan to determine both high and low concentrations. It, therefore, not only offers the potential to improve sample throughput, but also means that the maximum data can be collected on a transient signal that only lasts a few seconds. This is described in greater detail in Chapter 12, where we discuss different measurement protocols and peak integration routines.

EXTENDING THE DYNAMIC RANGE USING PULSE-ONLY MODE

The most recent development in extending the dynamic range is to use the pulse-only signal. This is achieved by monitoring the ion flux at one of the first few dynodes of the detector (before extensive electron multiplication has taken place) and then attenuating the signal up to 10,000:1 by applying a control voltage. Electron pulses

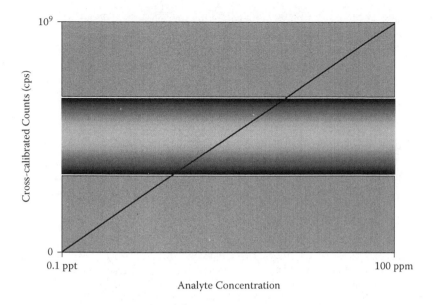

FIGURE 11.5 The pulse counting mode covers up to 10^6 cps, while the analog circuitry is suitable from 10^4 to 10^9 cps, with a dual-mode discrete dynode detector. (From E. R. Denoyer, R. J. Thomas, and L. Cousins, *Spectroscopy,* **12**[2], 56–61, 1997. Covered by U.S. Patent Number 5,463,219.)

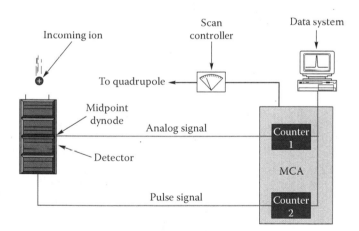

FIGURE 11.6 Using cross-calibration of the pulse and analog modes, quantitation from sub-ppt to high-ppm levels is possible. (From E. R. Denoyer, R. J. Thomas, and L. Cousins, *Spectroscopy,* **12**[2], 56–61, 1997. Covered by U.S. Patent Number 5,463,219.)

passed by the attenuation section are then amplified to yield pulse heights that are typical in normal pulse-counting applications.

There are basically three ways of implementing this technology based on the types of samples being analyzed. It can be run in conventional pulse-only mode for normal low-level work. It can also be run using an operator-selected attenuation factor if the higher level elements being determined are known and similar in concentration. If the samples are complete unknowns and have not been well characterized beforehand, a dynamic attenuation mode of operation is available. In this mode, an additional premeasurement time is built into the quadrupole settling time to determine the optimum detector attenuation for the selected dwell times used.

This novel, pulse-only approach to extending the dynamic range looks to be a very interesting development, which does not have the limitation of having to calibrate where pulse and analog signals cross over. However, it does require a preanalysis attenuation calibration to be carried out on a fairly frequent basis to determine the extent of signal attenuation required. The frequency of this calibration will vary depending on sample workload, but is expected to be on the order of once every 4 weeks.

REFERENCES

1. Channeltron®, *Electron Multiplier Handbook for Mass Spectrometry Applications*, Galileo Electro-Optic Corp., 1991. (Channeltron is a registered trademark of Galileo Corp.)
2. K. Hunter, *Atomic Spectroscopy,* **15**(1), 17–20, 1994.
3. R. C. Hutton, A. N. Eaton, and R. M. Gosland, *Applied Spectroscopy,* **44**(2), 238–242, 1990.
4. Y. Kishi, *Agilent Technologies Application Journal,* August 1997.
5. E. R. Denoyer, R. J. Thomas, and L. Cousins, *Spectroscopy,* **12**(2), 56–61, 1997. Covered by U.S. Patent Number 5,463,219.
6. J. Gray, R. Stresau, and K. Hunter, Ion Counting Beyond 10 GHz, *Poster Presentation Number 890-6P,* Pittsburgh Conference and Exposition, Orlando, FL, 2003.

12 Peak Measurement Protocol

With its multielement capability, superb detection limits, wide dynamic range, and high sample throughput, ICP-MS is proving to be a compelling technique for more and more diverse application areas. However, it is very unlikely that two different application areas have identical analytical requirements. For example, environmental and clinical contract laboratories, although wanting reasonably low detection limits, are not really pushing the technique to its extreme detection capability. Their main requirement usually is high sample throughput, because the number of samples these laboratories can analyze in a day directly impacts their revenue. On the other hand, a semiconductor fabrication plant or a supplier of high-purity chemicals to the electronics industry is interested in the lowest detection limits the technique can offer, because of the contamination problems associated with manufacturing high-performance electronic devices. This chapter looks at the many different measurement protocols associated with identifying and quantifying the analyte peak in ICP-MS and how they impact sample throughput and the quality of the data generated.

To meet such diverse application needs, modern ICP-MS instrumentation has to be very flexible if it is to keep up with the increasing demands of its users. Nowhere is this more important than in the area of peak integration and measurement protocol. The way the analytical signal is managed in ICP-MS has a direct impact on its multielement characteristics, isotopic capability, detection limits, dynamic range, and sample throughput—the five major strengths that attracted the trace element community to the technique almost 25 years ago. To understand signal management in greater detail and its implications on data quality, we will discuss how measurement protocol is optimized based on the application's analytical requirements, and its impact on both continuous signals generated by traditional nebulization devices and transient signals produced by alternative sample introduction techniques such as laser ablation, chromatographic separation, and flow injection.

MEASUREMENT VARIABLES

There are many variables that affect the quality of the analytical signal in ICP-MS. The analytical requirements of the application will often dictate this, but there is no question that instrumental detection and measurement parameters can have a significant impact on the quality of data in ICP-MS. Some of the variables that can potentially impact the quality of the data, particularly when carrying out multielement analysis, are as follows:

- A continuous or transient signal
- The temporal length of the sampling event
- Volume of sample available
- Number of samples being analyzed
- Number of replicates per sample
- Number of elements being determined
- Detection limits required
- Precision/accuracy expected
- Dynamic range needed
- Integration time used
- Peak quantitation routines

Before we go on to discuss these in greater detail and how these parameters affect the data, it is important to remind ourselves how a scanning device like a quadrupole mass analyzer works. Although we will focus on quadrupole technology, the fundamental principles of measurement protocol will be very similar for all types of mass spectrometers that use a sequential approach for multielement peak quantitation.

MEASUREMENT PROTOCOL

The principles of scanning with a quadrupole mass analyzer are shown in Figure 12.1. In this simplified example, the analyte ion in front (black) and four other ions have arrived at the entrance to the four rods of the quadrupole. When a particular RF/DC voltage is applied to each pair of rods, the positive or negative bias on the rods will electrostatically steer the analyte ion of interest down the middle of the four rods to the end, where it will emerge and be converted to an electrical pulse by the detector. The other ions of different mass-to-charge will pass through the spaces between the rods and be ejected from the quadrupole. This scanning process is then repeated for another analyte at a completely different mass-to-charge ratio until all the analytes in a multielement analysis have been measured.

The process for the detection of one particular mass in a multielement run is represented in Figure 12.2. It shows a $^{63}Cu^+$ ion emerging from the quadrupole and being converted to an electrical pulse by the detector. As the optimum RF-to-DC ratio is applied for $^{63}Cu^+$ and repeatedly scanned, the ions as electrical pulses are

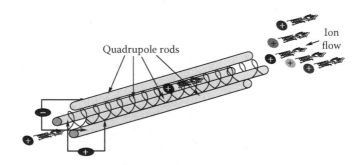

Quadrupole rods

Ion flow

FIGURE 12.1   Principles of mass selection with a quadrupole mass filter.

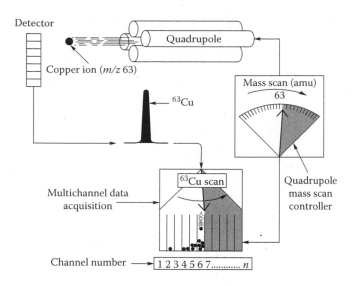

FIGURE 12.2 Detection and measurement protocol using a quadrupole mass analyzer. (From *Integrated MCA Technology in the ELAN ICP-Mass Spectrometer*, Application Note TSMS-25, PerkinElmer Instruments, 1993.)

stored and counted by a multichannel analyzer. This multichannel data acquisition system typically has 20 channels per mass and as the electrical pulses are counted in each channel, a profile of the mass is built up over the 20 channels, corresponding to the spectral peak of $^{63}Cu^+$. In a multielement run, repeated scans are made over the entire suite of analyte masses, as opposed to just one mass represented in this example. The principles of multielement peak acquisition are shown in Figure 12.3. In this example, signal pulses for two masses are continually collected as the quadrupole is swept across the mass spectrum, shown by sweeps 1–3. After a fixed number of sweeps (determined by the user), the total number of signal pulses in each channel is obtained, resulting in the final spectral peak.[1]

When it comes to quantifying an isotopic signal in ICP-MS, there are basically two approaches to consider. One is the multichannel ramp scanning approach, which uses a continuous smooth ramp of 1–n channels (where n is typically 20) per mass across the peak profile. This is shown in Figure 12.4.

Also, there is the peak hopping approach, in which the quadrupole power supply is driven to a discrete position on the peak (normally the maximum point), allowed to settle, and a measurement is taken for a fixed amount of time. This is represented in Figure 12.5.

The multipoint scanning approach is best for accumulating spectral and peak shape information when doing mass scans. It is normally used for doing mass calibration and resolution checks, and as a classical qualitative method development tool to find out what elements are present in the sample and to assess their spectral implications on the masses of interest. Full peak profiling is not normally used for doing rapid quantitative analysis, because valuable analytical time is wasted taking data on the wings and valleys of the peak, where the signal-to-noise ratio is poorest.

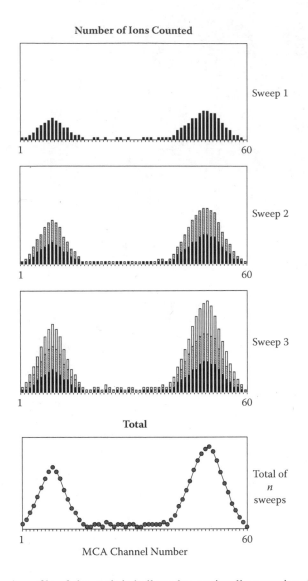

FIGURE 12.3 A profile of the peak is built up by continually sweeping the quadrupole across the mass spectrum. (From *Integrated MCA Technology in the ELAN ICP-Mass Spectrometer*, Application Note TSMS-25, PerkinElmer Instruments, 1993.)

When the best possible detection limits are required, the peak-hopping approach is best. It is important to understand that to get the full benefit of peak hopping, the best detection limits are achieved when single point–peak hopping at the peak maximum is chosen. However, to carry out single point–peak hopping, it is essential that the mass stability be good enough to reproducibly go to the same mass point every time. If good mass stability can be guaranteed (usually by thermostating the quadrupole power supply), measuring the signal at the peak maximum will always give the

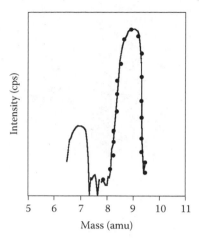

FIGURE 12.4 Multichannel ramp scanning approach using 20 channels per amu.

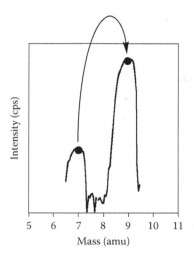

FIGURE 12.5 Peak-hopping approach.

best detection limits for a given integration time. It is well documented that there is no benefit in spreading the chosen integration time over more than one measurement point per mass. If time is a major consideration in the analysis, then using multiple points is wasting valuable time on the wings and valleys of the peak, which contribute less to the analytical signal and more to the background noise. This is shown in Figure 12.6, which demonstrates the degradation in signal-to-background noise of 10 ppb Rh with an increase in the number of points per peak, spread over the same total integration time. Detection limit improvement for a selected group of elements using 1 point/peak compared to 20 points/peak is shown in Figure 12.7.

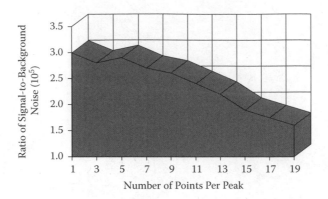

FIGURE 12.6   Signal-to-background noise degrades when more than one point, spread over the same integration time, is used for peak quantitation.

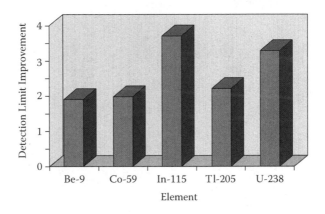

FIGURE 12.7   Detection limit improvement using one point per peak compared to 20 points per peak over the mass range. (From E. R. Denoyer, *Atomic Spectroscopy*, **13**[3], 93–98, 1992.)

OPTIMIZATION OF MEASUREMENT PROTOCOL

Now that the fundamentals of the quadrupole measuring electronics have been described, let us now go into more detail on the impact of optimizing the measurement protocol based on the requirements of the application. When multielement analysis is being carried out by ICP-MS, there are a number of decisions that need to be made. First, we need to know if we are dealing with a continuous signal from a nebulizer or a transient signal from a sampling accessory such as the laser ablation system of a chromatographic device. If it is a transient event, how long will the signal last? Another question that needs to be addressed is, how many elements are going to be determined? With a continuous signal, this is not such a major problem, but could be an issue if we are dealing with a transient signal that lasts only a few seconds. We also need to be aware of the level of detection capability required. This is a major consideration with a single shot laser pulse that lasts 5–10 s, but also with a continuous signal produced by a concentric nebulizer, we might have to accept a compromise of detection limit based on the speed of analysis requirements or amount of sample available. What analytical precision is expected? If isotope ratio/dilution work is being done, how many ions do we have to count to guarantee good precision? Does increasing the integration time of the measurement help the precision? Finally, is there a time constraint on the analysis? A high-throughput laboratory might not be able to afford to use the optimum sampling time to get the ultimate in detection limit. In other words, what compromises need to be made between detection limit, precision, and sample throughput? It is clear that before the measurement protocol can be optimized, the major analytical requirements of the application need to be defined. Let us look at this in greater detail.

MULTIELEMENT DATA QUALITY OBJECTIVES

Because multielement detection capability is probably the major reason why most laboratories invest in ICP-MS, it is important to understand the impact of measurement criteria on detection limits. We know that in a multielement analysis, the quadrupole's RF-to-DC ratio is "driven" or scanned to mass regions, which represent the elements of interest. The electronics are allowed to settle and then "sit" or dwell on the peak and take measurements for a fixed period of time This is usually performed a number of times until the total integration time is fulfilled. For example, if a dwell time of 50 ms is selected for all masses and the total integration time is 1 s, then the quadrupole will carry out 20 complete sweeps per mass, per replicate. It will then repeat the same routine for as many replicates that have been built into the method. This is shown in a simplified manner in Figure 12.8, which displays the scanning protocol of a multielement scan of three different masses.

In this example, the quadrupole is scanned to mass A. The electronics are allowed to settle (settling time), left to dwell for a fixed period of time at one or multiple points

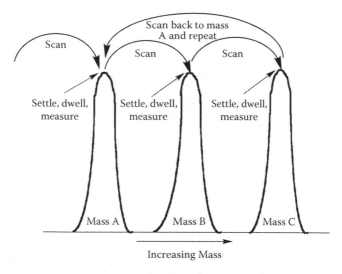

FIGURE 12.8 Multielement scanning and peak measurement protocol used in a quadrupole.

on the peak (dwell time) and intensity measurements taken (based on the dwell time). The quadrupole is then scanned to masses B and C, and the measurement protocol is repeated. The complete multielement measurement cycle (sweep) is repeated as many times as needed to make up the total integration per peak. It should be emphasized that this is a generalization of the measurement routine—management of peak integration by the software will vary slightly based on different instrumentation.

It is clear from this that during a multielement analysis there is a significant amount of time spent scanning and settling the quadrupole, which does not contribute to the quality of the analytical signal. Therefore, if the measurement routine is not optimized carefully, it can have a negative impact on data quality. The dwell time can usually be selected on an individual mass basis, but the scanning and settling times are normally fixed because they are a function of the quadrupole and detector electronics. For this reason, it is essential that the dwell time, which ultimately affects detection limit and precision, dominate the total measurement time, compared to the scanning and settling times. It follows, therefore, that the measurement duty cycle (percentage of actual measuring time compared to total integration time) is maximized when the quadrupole and detector electronics settling times are kept to an absolute minimum. This can be seen in Figure 12.9, which shows a plot of percent measurement duty cycle against dwell time for four different quadrupole settling times—0.2, 1.0, 3.0, and 5.0 ms for one replicate of a multielement scan of five masses, using one point per peak. In this example, the total integration time for each mass was 1 s, with the number of sweeps varying depending on the dwell time used. For this exercise, the % duty cycle is defined by the following equation:

$$\frac{\text{Dwell time} \times \text{\# sweeps} \times \text{\# elements} \times \text{\# replicates}}{\left\{ \left(\text{Dwell time} \times \text{\# sweeps} \times \text{\# elements} \times \text{\# replicates} \right) + \left(\text{Scanning/settling time} \times \text{\# sweeps} \times \text{\# elements} \times \text{\# reps} \right) \right\}} \times 100$$

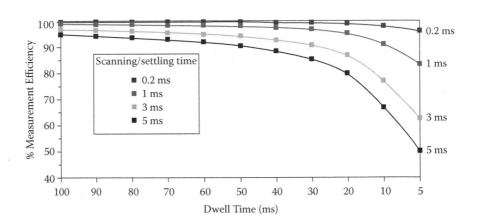

FIGURE 12.9   Measurement duty cycle as a function of dwell time with varying scanning/settling times.

To achieve the highest duty cycle, the nonanalytical time must be kept to an absolute minimum. This leads to more time being spent counting ions and less time scanning, and settling, which do not contribute to the quality of the analytical signal. This becomes critically important when a rapid transient peak is being quantified, because the available measuring time is that much shorter.[3] It is therefore a good rule of thumb, when setting up your measurement protocol in ICP-MS, to avoid using multiple points per peak and long settling times, because it ultimately degrades the quality of the data for a given integration time.

It can also be seen in Figure 12.9 that shorter dwell times translate into a lower duty cycle. For this reason, for normal quantitative analysis work, it is probably desirable to carry out multiple sweeps with longer dwell times (typically, 50 ms) to get the best detection limits. So, if an integration time of 1 s is used for each element, this would translate into 20 sweeps of 50 ms dwell time per mass. Although 1 s is long enough to achieve reasonably good detection limits, longer integration times generally have to be used to reach the lowest possible detection limits. This is shown in Figure 12.10, which shows detection limit improvement as a function of integration time for $^{238}U^+$. As would be expected, there is a fairly predictable improvement in the detection limit as the integration time is increased because more ions are being counted without an increase in the background noise. However, this only holds true up to the point where the pulse-counting detection system becomes saturated and no more ions can be counted. In the case of $^{238}U^+$, it can be seen that this happens at around 25 s, because there is no obvious improvement in D/L at a higher integration time. So from this data, we can say that there appears to be no real benefit in using longer than a 7 s integration time. When deciding the length of the integration time in ICP-MS, you have to weigh up the detection limit improvement against the time taken to achieve that improvement. Is it worth spending 25 s measuring each mass to get 0.02 ppt detection limit, if 0.03 ppt can be achieved using a 7 s integration time?

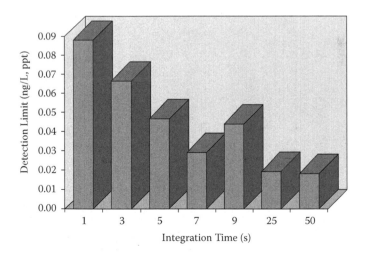

FIGURE 12.10 Plot of detection limit against integration time for $^{238}U^+$ (copyright © 2003–2007, all rights reserved, PerkinElmer Inc.).

Alternatively, is it worth measuring for 7 s when 1 s will only degrade the performance by a factor of 3? It really depends on your data quality objectives.

For some applications like isotope dilution/ratio studies, high precision is also a very important data quality objective.[4] However, to understand what is realistically achievable, we have to be aware of the practical limitations of measuring a signal and counting ions in ICP-MS. Counting statistics tells us that the standard deviation of the ion signal is proportional to the square root of the signal. It follows, therefore, that the RSD or precision should improve with an increase in the number (N) of ions counted as shown by the following equation:

$$\%RSD = \sqrt{\frac{N}{N}} \times 100$$

In practice, this holds up very well as can be seen in Figure 12.11. In this plot of standard deviation as a function of signal intensity for $^{208}Pb^+$, the black dots represent the theoretical relationship as predicted by counting statistics. It can be seen that the measured standard deviation (black bars) follows theory very well up to about 100,000 cps. At that point, additional sources of noise (e.g., sample introduction pulsations/plasma fluctuations) dominate the signal, which lead to poorer standard deviation values.

So, based on counting statistics, it is logical to assume that the more ions are counted, the better the precision will be. To put this in perspective, it means that at least 1 million ions need to be counted to achieve an RSD of 0.1%. In practice, of course, these kinds of precision values are very difficult to achieve with a scanning quadrupole system because of the additional sources of noise. If this information is combined with our knowledge of how the quadrupole is scanned, we begin to understand what is required to get the best precision. This is confirmed by the spectral

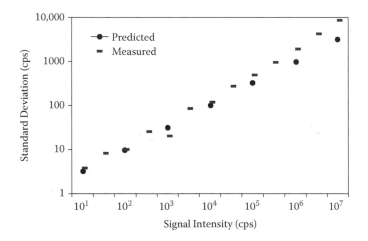

FIGURE 12.11   Comparison of measured standard deviation of a $^{208}Pb^+$ signal against that predicted by counting statistics. (From E. R. Denoyer, *Atomic Spectroscopy*, **13**[3], 93–98, 1992.)

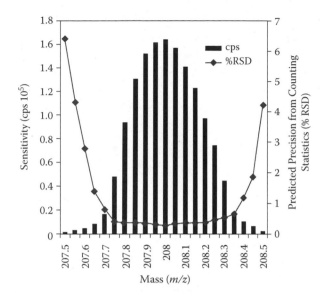

FIGURE 12.12   Comparison of % RSD with signal intensity across the mass profile of a
$^{208}Pb^+$ peak. (From E. R. Denoyer, *Atomic Spectroscopy,* **13**[3], 93–98, 1992.)

scan in Figure 12.12, which shows the predicted precision at all 20 channels of a
5 ppb $^{208}Pb^+$ peak.[2]

This tells us that the best precision is obtained at the channels where the signal
is highest, which as we can see are the ones at or near the center of the peak. For this
reason, if good precision is a fundamental requirement of your data quality objectives, it is best to use single-point peak hopping with integration times on the order
of 5–10 s. On the other hand, if high-precision isotope ratio or isotope dilution work
is being done, where analysts would like to achieve precision values approaching
counting statistics, then much longer measuring times are required. That is why integration times on the order of 5–10 min are commonly used for determining isotope
ratios involving environmental pollutants[5] or clinical metabolism studies.[6] For this
type of analysis, when two or more isotopes are being measured and ratioed to each
other, it follows that the more simultaneous the measurement, the better the precision becomes. Therefore, the ability to make the measurement as simultaneous as
possible is considered more desirable than any other aspect of the measurement. This
is supported by the fact that the best isotope ratio precision data is achieved with
multicollector, magnetic sector ICP-MS technology that carries out many isotopic
measurements at the same time using multiple detectors.[7] Also, TOF technology,
which simultaneously samples all the analyte ions in a slice of the ion beam, offers
excellent precision, particularly when internal standardization measurement is also
carried out in a simultaneous manner.[8] So, the best way to approximate simultaneous measurement with a rapid scanning device such as a quadrupole is to use shorter
dwell times (but not too short that insufficient ions are counted) and keep the scan-
ning/settling times to an absolute minimum, which results in more sweeps for a
given measurement time. This can be seen in Table 12.1, which shows the precision

TABLE 12.1

Precision of Pb Isotope Ratio Measurement as a Function of Dwell Time Using a Total Integration Time of 5.5 s

Dwell time (ms)	% RSD $^{207}Pb^+/^{206}Pb^+$	% RSD $^{208}Pb^+/^{206}Pb^+$
2	0.40	0.36
5	0.38	0.36
10	0.23	0.22
25	0.24	0.25
50	0.38	0.33
100	0.41	0.38

Source: L. Halicz, Y. Erel, and A. Veron, *Atomic Spectroscopy,* **17**(5), 186–189, 1996.

of Pb isotope ratios at different dwell times carried out by researchers at the Geological Survey of Israel.[9] The data are based on nine replicates of a NIST SRM-981 (75 ppb Pb) solution, using 5.5 s integration time per isotope.

From these data, the researchers concluded that a dwell time of 10 or 25 ms offered the best isotope ratio precision measurement (quadrupole settling time was fixed at 0.2 ms). They also found that they could achieve slightly better precision by using a 17.5 s integration time (700 sweeps at 25 ms dwell time), but felt the marginal improvement in precision for nine replicates was not worth spending the approximately three-and-a-half times longer analysis time. This can be seen in Table 12.2.

This work shows the benefit of being able to optimize the dwell time, settling time, and the number of sweeps to get the best isotope ratio precision data. It also helps to be working with relatively healthy ion signals for the three Pb isotopes, ^{206}Pb, ^{207}Pb, and ^{208}Pb (24.1%, 22.1%, and 52.4% abundance, respectively). If the isotopic signals were dramatically different as in the measurement of two of the uranium isotopes, ^{235}U to ^{238}U, which are 0.72% and 99.2745% abundant, respectively, then the ability to optimize the measurement protocol for individual isotopes becomes of even greater importance to guarantee good precision data.

TABLE 12.2

Impact of Integration Time on the Overall Analysis Time for Pb Isotope Ratios

Dwell Time (ms)	No. of Sweeps	Integration Time (s)/mass	%RSD $^{207}Pb^+/^{206}Pb^+$	% RSD $^{207}Pb^+/^{206}Pb^+$	Time for 9 reps (min/s)
25	220	5.5 s	0.24	0.25	2 m 29 s
25	500	12.5 s	0.21	0.19	6 m 12 s
25	700	17.5 s	0.20	0.17	8 m 29 s

Source: L. Halicz, Y. Erel, and A. Veron, *Atomic Spectroscopy,* **17**(5), 186–189, 1996.

It is also worth pointing out that better precision can be achieved with a quadrupole-based instrument if it is fitted with a collision/reaction cell. The collisions actually broaden the residence time of the ions, which means that short-term fluctuations of the ion signal are damped effectively enough to reduce the plasma noise and improve the precision of isotope ratio measurements. Bandura and Tanner achieved external precision values of 0.03–0.06% for lead isotope ratios in NIST 981 SRM using this approach.[10]

It is clear that the analytical demands put on ICP-MS are probably higher than any other trace element technique, because it is continually being asked to solve a wide variety of application problems at increasingly lower levels. However, by optimizing the measurement protocol to fit the analytical requirement, ICP-MS has shown that it has the unique capability to carry out rapid trace element analysis, with superb detection limits and good precision on both continuous and transient signals, and still meet the most stringent data quality objectives.

REFERENCES

1. *Integrated MCA Technology in the ELAN ICP-Mass Spectrometer*, Application Note TSMS-25, PerkinElmer Instruments, 1993.
2. E. R. Denoyer, *Atomic Spectroscopy*, **13**(3), 93–98, 1992.
3. E. R. Denoyer and Q. H. Lu, *Atomic Spectroscopy*, **14**(6), 162–169, 1993.
4. T. Catterick, H. Handley, and S. Merson, *Atomic Spectroscopy*, **16**(10), 229–234, 1995.
5. T. A. Hinners, E. M. Heithmar, T. M. Spittler, and J. M. Henshaw, *Analytical Chemistry*, **59**, 2658–2662, 1987.
6. M. Janghorbani, B. T. G. Ting, and N. E. Lynch, *Microchemica Acta*, **3**, 315–328, 1989.
7. J. Walder, P. A., Freeman, *Journal of Analytical Atomic Spectrometry*, **7**, 571, 1992.
8. F. Vanhaecke, L. Moens, R. Dams, L. Allen, and S. Georgitis, *Analytical Chemistry*, **71**, 3297, 1999.
9. L. Halicz, Y. Erel, and A. Veron, *Atomic Spectroscopy*, **17**(5), 186–189, 1996.
10. D. R. Bandura and S. D. Tanner, *Atomic Spectroscopy*, **20**(2), 69–72, 1999.

13 Methods of Quantitation

There are many different ways to carry out trace element analysis by ICP-MS, depending on your data quality objectives. Such is the flexibility of the technique that it allows detection from sub-ppt up to high-ppm levels using a wide variety of calibration methods from full quantitative and semiquantitative analysis to one of the very powerful isotope ratioing techniques. Chapter 13 looks at the most important quantitation methods available in ICP-MS.

This ability of ICP-MS to carry out isotopic measurements allows the technique to carry out quantitation methods that are not available to any other trace element technique. They include the following:

- Quantitative analysis
- Semiquantitative routines
- Isotope dilution
- Isotope ratio
- Internal standardization

Each of these techniques offers varying degrees of accuracy and precision; so, it is important to understand their strengths and weaknesses to know which one will best meet the data quality objectives of the analysis. Let us look at each of these in greater detail.

QUANTITATIVE ANALYSIS

As in other more mature trace element techniques such as AA and ICP-OES, quantitative analysis in ICP-MS is the fundamental tool used to determine analyte concentrations in unknown samples. In this mode of operation, the instrument is calibrated by measuring the intensity for all elements of interest in a number of known calibration standards that represent a range of concentrations likely to be encountered in your unknown samples. When the full range of calibration standards and blank have been run, the software creates a calibration curve of the measured intensity versus concentration for each element in the standard solutions. Once calibration data are acquired, the unknown samples are analyzed by plotting the intensity of the elements of interest against the respective calibration curves. The software then calculates the concentrations for the analytes in the unknown samples.

This type of calibration is often called external standardization and is usually used when there is very little difference between the matrix components in the standards and the samples. However, when it is difficult to closely match the matrix of the standards with the samples, external standardization can produce erroneous results, because matrix-induced interferences will change analyte sensitivity based on the amount of matrix present in the standards and samples. When this occurs, better accuracy is achieved by using the method of standard addition or a similar approach called addition calibration. Let us look at these three variations of quantitative analysis to see how they differ.

EXTERNAL STANDARDIZATION

As explained earlier, this involves measuring a blank solution followed by a set of standard solutions to create a calibration curve over the anticipated concentration range. Typically, a blank and three standards containing different analyte concentrations are run. Increasing the number of points on the calibration curve by increasing the number of standards may improve accuracy in circumstances where the calibration range is very broad. However, it is seldom necessary to run a calibration with more than five standards. After the standards have been measured, the unknown samples are analyzed and their analyte intensities read against the calibration curve. Over extended analysis times, it is common practice to update the calibration curve, either by recalibrating the instrument with a full set of standards or by running one midpoint standard. The following protocol summarizes a typical calibration using external standardization:

1. Blank >
2. Std. 1 >
3. Std. 2 >
4. Std. 3 >
5. Sample 1 >
6. Sample 2 >
7. Sample...n
8. Recalibrate
9. Sample $n + 1$, etc.

This can be seen more clearly in Figure 13.1, which shows a typical calibration curve using a blank and three standards of 2, 5, and 10 ppb. This calibration curve shows a

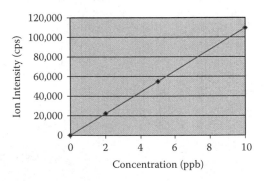

FIGURE 13.1   A simple linear regression calibration curve.

simple *linear regression* but usually other modes of calibration are also available like *weighted linear* to emphasize measurements at the low concentration region of the curve and *linear through zero*, where the linear regression is forced through zero.

It should be emphasized that this graph represents a single-element calibration. However, because ICP-MS is usually used for multielement analysis, multielement standards are typically used to generate calibration data. For that reason, it is absolutely essential to use multielement standards that have been manufactured specifically for ICP-MS. Single-element AA standards are not suitable, because they usually have only been certified for the analyte element and not for any others. The purity of the standard cannot be guaranteed for any other element and as a result cannot be used to make up multielement standards for use with ICP-MS. For the same reason, ICP-OES multielement standards are not advisable either, because they are only certified for a group of elements and could contain other elements at higher levels, which will affect the ICP-MS multielement calibration.

STANDARD ADDITIONS

This mode of calibration provides an effective way to minimize sample-specific matrix effects by spiking samples with known concentrations of analytes.[1,2] In standard addition calibration, the intensity of a blank solution is first measured. Next, the sample solution is "spiked" with known concentrations of each element to be determined. The instrument measures the response for the spiked samples and creates a calibration curve for each element for which a spike has been added. The calibration curve is a plot of the blank subtracted intensity of each spiked element against its concentration value. After creating the calibration curve, the unspiked sample solutions are then analyzed and compared to the calibration curve. Based on the slope of the calibration curve and where it intercepts the x-axis, the instrument software determines the unspiked concentration of the analytes in the unknown samples. This can be seen in Figure 13.2, which shows a calibration of the sample intensity and the

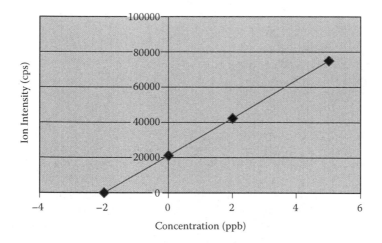

FIGURE 13.2 A typical "method of additions" calibration curve.

sample spiked with 2 and 5 ppb of the analyte. The concentration of sample is where the calibration line intercepts the negative side of the x-axis.

The following protocol summarizes a typical calibration using the method of standard additions:

1. Blank >
2. Spiked Sample 1 (Spike Conc. 1) >
3. Spiked Sample 1 (Spike Conc. 2) >
4. Unspiked Sample 1 >
5. Blank >
6. Spiked Sample 2 (Spike Conc. 1) >
7. Spiked Sample 2 (Spike Conc. 2) >
8. Unspiked Sample 2 >
9. Blank >
10. Etc.

ADDITION CALIBRATION

Unfortunately, with the method of standard additions, each and every sample has to be spiked with all the analytes of interest, which becomes extremely labor intensive when many samples have to be analyzed. For this reason, a variation of standard additions called *addition calibration* is more widely used in ICP-MS. However, this method can only be used when all the samples have a similar matrix. It uses the same principle as standard additions, but only the first (or representative) sample is spiked with known concentrations of analytes and then analyzes the rest of the sample batch against the calibration, assuming all samples have a similar matrix to the first one. The following protocol summarizes a typical calibration using the method of addition calibration:

1. Blank >
2. Spiked Sample 1 (Spike Conc. 1) >
3. Spiked Sample 1 (Spike Conc. 2) >
4. Unspiked Sample 1 >
5. Unspiked Sample 2 >
6. Unspiked Sample 3 >
7. Etc.

SEMIQUANTITATIVE ANALYSIS

If your data quality objectives for accuracy and precision are less stringent, ICP-MS offers a very rapid semiquantitative mode of analysis. This technique enables you to automatically determine the concentrations of up to 75 elements in an unknown sample, without the need for calibration standards.[3,4] There are slight variations in the way different instruments approach semiquantitative analysis, but the general principle is to measure the entire mass spectrum, without specifying individual elements or masses. It relies on the principle that each element's natural isotopic abundance is fixed. By measuring the intensity of all their isotopes, correcting for common spectral interferences, including molecular, polyatomic, and isobaric species and applying heuristic, knowledge-driven routines in combination with numerical calculations, a positive or negative confirmation can be made for each element present in the sample. Then, by comparing the corrected intensities against a stored isotopic response table, a good semiquantitative approximation of the sample components can be made.

Semiquant, as it is often called, is an excellent approach to rapidly characterizing unknown samples. Once the sample has been characterized, you can choose to either update the response table with your own standard solutions to improve analytical accuracy, or switch to the quantitative analysis mode to focus on specific elements and determine their concentrations with even greater accuracy and precision. Whereas a semiquantitative determination can be performed without using a series of standards, the use of a small number of standards is highly recommended for improved accuracy across the full mass range. Unlike traditional quantitative analysis, in which you analyze standards for all the elements you want to determine, semiquant calibration is achieved using just a few elements distributed across the mass range. This calibration process, shown more clearly in Figure 13.3, is used to update the reference response curve data that correlates measured ion intensities to the concentrations of elements in a solution. During calibration, this response data is adjusted to account for changes in the instrument's sensitivity due to variations in the sample matrix.

This process is often called semiquantitative analysis using external calibration, and like traditional quantitative analysis using external standardization, works extremely well for samples that have a similar matrix. However, if you are analyzing samples containing widely different concentrations of matrix components, external calibration does not work very well because of matrix-induced suppression effects on the analyte signal. If this is the case, semiquant using a variation of standard addition calibration should be used. Similar to standard addition calibration used in quantitative analysis, this procedure involves adding known quantities of specific elements to every unknown sample before measurement. The major difference with semiquant is that the elements you add must not already be present in significant quantities in

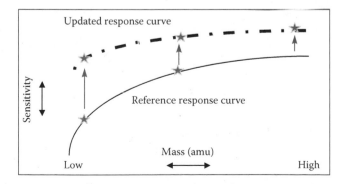

FIGURE 13.3 In semiquantitative analysis, a small group of elements are used to update the reference response curve to improve the accuracy as the sample matrix changes.

the unknown samples, because they are being used to update the stored reference response curve. As with external calibration, the semiquant software then adjusts the stored response data for all remaining analytes relative to the calibration elements. This procedure works very well, but tends to be very labor intensive because the calibration standards have to be added to every unknown sample.

The other thing to be wary of with semiquantitative analysis is the spectral complexity of unknown samples. If you have a spectrally rich sample and are not making any compensations for spectral overlaps close to the analyte peaks, it could possibly give you a false-positive for that element. Therefore, you have to be very cautious when reporting semiquantitative results on completely unknown samples. They should be characterized first, especially with respect to the types of spectral interferences generated by the plasma gas, the matrix, and the solvents/acids/chemicals used for sample preparation. Collision/reaction cells/interfaces can help in the reduction of some of these interferences, but extreme care should be taken, as these devices are known to have no effect on some polyatomic interferences, and in some cases can increase the spectral complexity by generating other interfering complexes.

ISOTOPE DILUTION

Although quantitative and semiquantitative analysis methods are suitable for the majority of applications, there are other calibration methods available, depending on your analytical requirements. For example, if your application requires even greater accuracy and precision, the "isotope dilution" technique may offer some benefits. Isotope dilution is an absolute means of quantitation based on altering the natural abundance of two isotopes of an element by adding a known amount of one of the isotopes, and is considered one of the most accurate and precise approaches to elemental analysis.[5,6,7,8]

For this reason, a prerequisite of isotope dilution is that the element must have at least two stable isotopes. The principle works by spiking a known weight of an enriched stable isotope into your sample solution. By knowing the natural abundance of the two isotopes being measured, the abundance of the spiked enriched isotopes, the weight of the spike, and the weight of the sample, the original trace element concentration can be determined by using the following equation:

$$C = \frac{\left[\text{Aspike} - \left(R \times \text{Bspike}\right)\right] \times \text{Wspike}}{\left[R \times \left(\text{Bsample} - \text{Asample}\right)\right] \times \text{Wsample}}$$

where
$\qquad C$ = Concentration of trace element
$\quad$ Aspike = % of higher abundance isotope in spiked enriched isotope
$\quad$ Bspike = % of lower abundance isotope in spiked enriched isotope
$\quad$ Wspike = Weight of spiked enriched isotope
$\qquad R$ = Ratio of the % of higher abundance isotope to lower abundance isotope in the spiked sample
$\quad$ Bsample = % of higher natural abundance isotope in sample

Asample = % of lower natural abundance isotope in sample
Wsample = Weight of sample

This might sound complicated, but in practice, it is relatively straightforward. This is illustrated in Figure 13.4, which shows an isotope dilution method for the determination of copper in a 250 mg sample of orchard leaves, using the two copper isotopes ^{63}Cu and ^{65}Cu.

In the bar graph on top it can be seen that the natural abundance of the two isotopes are 69.09% and 30. 91%, respectively, for ^{63}Cu and ^{65}Cu. In the middle graph it shows that 4 μg of an enriched isotope of 100% ^{65}Cu (and 0 % ^{63}Cu) is spiked into the sample, which now produces a spiked sample containing 71.4% of ^{65}Cu and 28.6% of ^{63}Cu, as seen in the bottom plot.[9] If we plug these data into the preceding equation, we get:

$$C = \frac{\left[100 - \left(71.4/28.6 \times 0\right)\right] \times 4 \, \mu g}{\left[\left(71.4/28.6 \times 69.09\right) - 30.91\right] \times 0.25 \, g}$$

$$C = 400/35.45 = 11.3 \, \mu g/g$$

The major benefit of the isotope dilution technique is that it provides measurements that are extremely accurate because you are measuring the concentration of the isotopes in the same solution as your unknown sample, and not in a separate external calibration solution. In addition, because it is a ratioing technique, loss of solution during the sample preparation stage has no influence on the accuracy of the result. The technique is also extremely precise, because using a simultaneous detection system such as a magnetic sector multicollector or a simultaneous ion sampling device such as a TOF ICP-MS, the results are based on measuring the two isotopes solution at the same instant in time, which compensates for imprecision of the signal due to sources of sample introduction related noise, such as plasma instability, peristaltic pump pulsations, and nebulization fluctuations. Even when using a scanning mass analyzer such as a quadrupole, the measurement protocol can be optimized to scan very rapidly between the two isotopes and achieve very good precision. However, isotope dilution has some limitations, which makes it suitable only for certain applications. These limitations include the following:

- The element you are determining must have more than one isotope, because calculations are based on the ratio of one isotope to another isotope of the same element; this makes it unsuitable for approximately 15 elements that can be determined by ICP-MS.
- It requires certified enriched isotopic standards, which can be very expensive, especially those that are significantly different from the normal isotopic abundance of the element.
- It compensates for interferences due to signal enhancement or suppression, but does not compensate for spectral interferences. For this reason, an external blank solution must always be run.

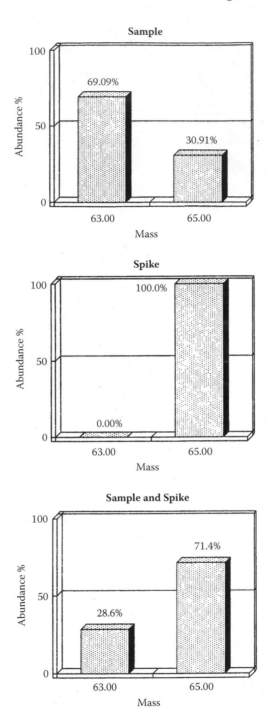

FIGURE 13.4 Quantitation of trace levels of copper in a sample of SRM orchard leaves using isotope dilution methodology. (From *Multi-elemental Isotope Dilution Using the Elan* ICP-MS *Elemental Analyzer,* ICP-MS *Technical Summary TSMS-1,* PerkinElmer Instruments, 1985.)

ISOTOPE RATIOS

The ability of ICP-MS to determine individual isotopes also makes it suitable for another isotopic measurement technique called "isotope ratio" analysis. The ratio of two or more isotopes in a sample can be used to generate very useful information, including an indication of the age of a geological formation, a better understanding of animal metabolism, and it can also help to identify sources of environmental contamination.[10,11,12,13,14] Similar to isotope dilution, isotope ratio analysis uses the principle of measuring the exact ratio of two isotopes of an element in the sample. With this approach, the isotope of interest is typically compared to a reference isotope of the same element. For example, you might want to compare the concentration of ^{204}Pb to that of ^{206}Pb. Alternatively, the requirement might be to compare one isotope to all remaining reference isotopes of an element, e.g., the ratio of ^{204}Pb to ^{206}Pb, ^{207}Pb, and ^{208}Pb. The ratio is then expressed in the following manner:

$$\text{Isotope ratio} = \frac{\text{Intensity of isotope of interest}}{\text{Intensity of reference isotope}}$$

As this ratio can be calculated from within a single sample measurement, classic external calibration is not normally required. However, if there is a large difference between the concentrations of the two isotopes, it is recommended to run a standard of known isotopic composition. This is done to verify that the higher concentration isotope is not suppressing the signal of the lower concentration isotope and biasing the results. This effect, called mass discrimination, is less of a problem if the isotopes are relatively close in concentration, e.g., ^{107}Ag to ^{109}Ag, which are 51.839% and 48.161% abundant, respectively. However, it can be an issue if there is a significant difference in their concentration values, e.g., ^{235}U to ^{238}U, which are 0.72% and 99.275% abundant, respectively. Mass discrimination effects can be reduced by running an external reference standard of known isotopic concentration, comparing the isotope ratio with the theoretical value, and then mathematically compensating for the difference. The principles of isotope ratio analysis and how to achieve optimum precision values are explained in greater detail in Chapter 12.

INTERNAL STANDARDIZATION

Another method of standardization commonly employed in ICP-MS is called *internal standardization*. It is not considered an absolute calibration technique, but instead is used to correct for changes in analyte sensitivity caused by variations in the concentration and type of matrix components found in the sample. An internal standard is a nonanalyte isotope that is added to the blank solution, standards, and samples before analysis. It is typical to add three or four internal standard elements to the samples to cover the analyte elements of interest. The software adjusts the analyte concentration in the unknown samples by comparing the intensity values of the internal standard intensities in the unknown sample to those in the calibration standards.

The implementation of internal standardization varies according to the analytical technique that is being used. For quantitative analysis, the internal standard elements are selected on the basis of the similarity of their ionization characteristics to the analyte elements. Each internal standard is bracketed with a group of analytes. The software then assumes that the intensities of all elements within a group are affected in a similar manner by the matrix. Changes in the ratios of the internal standard intensities are then used to correct the analyte concentrations in the unknown samples.

For semiquantitative analysis that uses a stored response table, the purpose of the internal standard is similar, but a little different in implementation to quantitative analysis. A semiquant internal standard is used to continuously compensate for instrument drift or matrix-induced suppression over a defined mass range. If a single internal standard is used, all the masses selected for the determination are updated by the same amount based on the intensity of the internal standard. If more than one internal standard is used, which is recommended for measurements over a wide mass range, the software interpolates the intensity values based on the distance in mass between the analyte and the nearest internal standard element.

It is worth emphasizing that if you do not want to compare your intensity values to a calibration graph, most instruments allow you to report raw data. This enables you to analyze your data using external data-processing routines, to selectively apply a minimum set of ICP-MS data processing methods, or to just view the raw data file before reprocessing it. The availability of raw data is primarily intended for use in nonroutine applications such as chromatography separation techniques and laser sampling devices that produce a time-resolved transient peak or by users whose sample set requires data processing using algorithms other than those supplied by the instrument software.

REFERENCES

1. D. Beauchemin, J. W. McLaren, A. P. Mykytiuk, and S. S. Berman, *Analytical Chemistry*, **59**, 778, 1987.
2. E. Pruszkowski, K. Neubauer, and R. Thomas, *Atomic Spectroscopy,* **19**(4), 111–115, 1998.
3. M. Broadhead, R. Broadhead, and J. W. Hager, *Atomic Spectroscopy,* **11**(6), 205–209, 1990.
4. E. Denoyer, *Journal of Analytical Atomic Spectrometry,* **7**, 1187, 1992.
5. J. W. McLaren, D. Beauchemin, and S. S. Berman, *Analytical Chemistry,* **59**, 610, 1987.
6. H. Longerich, *Atomic Spectroscopy,* **10**(4), 112–115, 1989.
7. A. Stroh, *Atomic Spectroscopy,* **14**(5), 141–143, 1993.
8. T. Catterick, H. Handley, and S. Merson, *Atomic Spectroscopy,* **16**(10), 229–234, 1995.
9. *Multi-elemental Isotope Dilution Using the Elan ICP-MS Elemental Analyzer, ICP-MS Technical Summary TSMS-1*, PerkinElmer Instruments, 1985.
10. B. T. G. Ting and M. Janghorbani, *Analytical Chemistry,* **58**, 1334, 1986.
11. M. Janghorbani, B. T. G. Ting, and N. E. Lynch, *Microchemica Acta,* **3**, 315–328, 1989.
12. T. A. Hinners, E. M. Heithmar, T. M. Spittler, and J. M. Henshaw, *Analytical Chemistry,* **59**, 2658–2662, 1987.
13. L. Halicz, Y. Erel, and A. Veron, *Atomic Spectroscopy,* **17**(5), 186–189, 1996.
14. M. Chaudhary-Webb, D. C. Paschal, W. C. Elliott, H. P. Hopkins, A. M. Ghazi, B. C. Ting, and I. Romieu, *Atomic Spectroscopy,* **19**(5), 156, 1998.

14 Review of Interferences

Now that we have covered the fundamental principles of ICP-MS and its measurement and calibration routines, let us turn our attention to the technique's most common interferences and the methods that are used to compensate for them. Although interferences are reasonably well understood in ICP-MS, it can often be difficult and time consuming to compensate for them, particularly in complex sample matrices. Prior knowledge of the interferences associated with a particular set of samples will often dictate the sample preparation steps and the instrumental methodology used to analyze them.

Interferences in ICP-MS are generally classified into three major groups: spectral, matrix, and physical. Each of them has the potential to be problematic in its own right, but modern instrumentation and good software combined with optimized analytical methodologies has minimized their negative impact on trace element determinations by ICP-MS. Let us look at these interferences in greater detail and describe the different approaches used to compensate for them.

SPECTRAL INTERFERENCES

Spectral overlaps are probably the most serious types of interferences seen in ICP-MS. The most common are known as a polyatomic or molecular spectral interference and are produced by the combination of two or more atomic ions. They are caused by a variety of factors, but are usually associated with either the plasma/nebulizer gas used, matrix components in the solvent/sample, other elements in the sample, or entrained oxygen/nitrogen from the surrounding air. For example, in the argon plasma, spectral overlaps caused by argon ions and combinations of argon ions with other species are very common. The most abundant isotope of argon is at mass 40, which dramatically interferes with the most abundant isotope of calcium at mass 40, whereas the combination of argon and oxygen in an aqueous sample generates the $^{40}Ar^{16}O^+$ interference, which has a significant impact on the major isotope of Fe at mass 56. The complexity of these kinds of spectral problems can be seen in Figure 14.1, which shows a mass spectrum of deionized water from mass 40 to mass 90.

In addition, argon can form polyatomic interferences with elements found in the acids used to dissolve the sample. For example, in a hydrochloric acid medium, $^{40}Ar^+$ combines with the most abundant chlorine isotope at 35 amu to form $^{40}Ar^{35}Cl^+$, which interferes with the only isotope of arsenic at mass 75, whereas in an organic solvent matrix, argon and carbon combine to form $^{40}Ar^{12}C^+$, which interferes with $^{52}Cr^+$, the most abundant isotope of chromium. Sometimes, matrix or solvent ions combine to form spectral interferences of their own. A good example is in a sample that contains sulfuric acid. The dominant sulfur isotope, $^{32}S^+$, combines with two

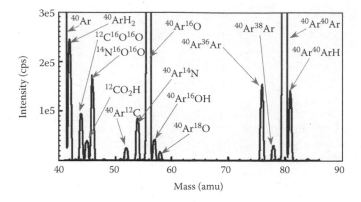

FIGURE 14.1 ICP mass spectrum of deionized water from mass 40 to mass 90.

oxygen ions to form a $^{32}S^{16}O^{16}O^+$ molecular ion, which interferes with the major isotope of Zn at mass 64. In the analysis of samples containing high concentrations of sodium, such as seawater, the most abundant isotope of Cu at mass 63 cannot be used because of interference from the $^{40}Ar^{23}Na^+$ molecular ion. There are many more examples of these kinds of polyatomic and molecular interferences, which have been comprehensively reviewed.[1] Table 14.1 represents some of the most common matrix–solvent spectral interferences seen in ICP-MS.

TABLE 14.1
Some Common Plasma/Matrix/Solvent-Related Polyatomic Spectral Interferences Seen in ICP-MS

Element/Isotope	Matrix/Solvent	Interference
$^{39}K^+$	H_2O	$^{38}ArH^+$
$^{40}Ca^+$	H_2O	$^{40}Ar^+$
$^{56}Fe^+$	H_2O	$^{40}Ar^{16}O^+$
$^{80}Se^+$	H_2O	$^{40}Ar^{40}Ar^+$
$^{51}V^+$	HCl	$^{35}Cl^{16}O^+$
$^{75}As^+$	HCl	$^{40}Ar^{35}Cl^+$
$^{28}Si^+$	HNO_3	$^{14}N^{14}N^+$
$^{44}Ca^+$	HNO_3	$^{14}N^{14}N\ ^{16}O^+$
$^{55}Mn^+$	HNO_3	$^{40}Ar^{15}N^+$
$^{48}Ti^+$	H_2SO_4	$^{32}S^{16}O^+$
$^{52}C^+r$	H_2SO_4	$^{34}S^{18}O^+$
$^{64}Zn^+$	H_2SO_4	$^{32}S^{16}O^{16}O^+$
$^{63}Cu^+$	H_3PO_4	$^{31}P^{16}O^{16}O^+$
$^{24}Mg^+$	Organics	$^{12}C^{12}C^+$
$^{52}Cr^+$	Organics	$^{40}Ar^{12}C^+$
$^{65}Cu^+$	Minerals	$^{48}Ca^{16}OH^+$
$^{64}Zn^+$	Minerals	$^{48}Ca^{16}O^+$
$^{63}Cu^+$	Seawater	$^{40}Ar^{23}Na^+$

OXIDES, HYDROXIDES, HYDRIDES, AND DOUBLY CHARGED SPECIES

Another type of spectral interference is produced by elements in the sample combining with H^+, $^{16}O^+$, or $^{16}OH^+$ (either from water or air) to form molecular hydrides (+ H^+), oxides (+ $^{16}O^+$), and hydroxides (+ $^{16}OH^+$), which occur at 1, 16, and 17 mass units, respectively, higher than the element's mass.[2] These interferences are typically produced in the cooler zones of the plasma, immediately before the interface region. They are usually more serious when rare earth or refractory-type elements are present in the sample, because many of them readily form molecular species (particularly oxides), which create spectral overlap problems on other elements in the same group. If the oxide species is mainly derived from entrained air around the plasma, it can be reduced by either using an elongated outer tube to the torch or a metal shield between the plasma and the RF coil.

Associated with oxide-based spectral overlaps are doubly charged spectral interferences. These are species that are formed when an ion is generated with a double positive charge as opposed to a normal single charge and produces an isotopic peak at half its mass. Similar to the formation of oxides, the level of doubly charged species is related to the ionization conditions in the plasma and can usually be minimized by careful optimization of the nebulizer gas flow, RF power, and sampling position within the plasma. It can also be impacted by the severity of the secondary discharge present at the interface,[3] which was described in greater detail in Chapter 5. Table 14.2 shows a selected group of elements that readily form oxides, hydroxides, hydrides, and doubly charged species, together with the analytes affected by them.

TABLE 14.2
Some Elements That Readily Form Oxides, Hydroxides, Hydrides, and Doubly Charged Species in the Plasma, Together with the Analytes Affected by the Interference

Oxide, Hydroxide, Hydride, Doubly Charged Species	Analyte Affected by Interference
$^{40}Ca^{16}O^+$	$^{56}Fe^+$
$^{48}Ti^{16}O^+$	$^{64}Zn^+$
$^{98}Mo^{16}O^+$	$^{114}Cd^+$
$^{138}Ba^{16}O^+$	$^{154}Sm^+$, $^{154}Gd^+$
$^{139}La^{16}O^+$	$^{155}Gd^+$
$^{140}Ce^{16}O^+$	$^{156}Gd^+$, $^{156}Dy^+$
$^{40}Ca^{16}OH^+$	$^{57}Fe^+$
$^{31}P^{18}O^{16}OH^+$	$^{66}Zn^+$
$^{79}BrH^+$	$^{80}Se^+$
$^{31}P^{16}O_2H^+$	$^{64}Zn^+$
$^{138}Ba^{2+}$	$^{69}Ga^+$
$^{139}La^{2+}$	$^{69}Ga^+$
$^{140}Ce^{2+}$	$^{70}Ge^+$, $^{70}Zn^+$

ISOBARIC INTERFERENCES

The final classification of spectral interferences is called isobaric overlaps, produced mainly by different isotopes of other elements in the sample creating spectral interferences at the same mass as the analyte. For example, vanadium has two isotopes at 50 and 51 amu. However, mass 50 is the only practical isotope to use in the presence of a chloride matrix because of the large contribution from the $^{16}O^{35}Cl^+$ interference at mass 51. Unfortunately, mass 50 amu, which is only 0.25% abundant, also coincides with isotopes of titanium and chromium, which are 5.4% and 4.3% abundant, respectively. This makes the determination of vanadium in the presence of titanium and chromium very difficult unless mathematical corrections are made. Figure 14.2 shows all the possible naturally occurring isobaric spectral overlaps in ICP-MS.[4]

WAYS TO COMPENSATE FOR SPECTRAL INTERFERENCES

Let us now look at the different approaches used to compensate for spectral interferences. One of the very first ways used to get around severe matrix-derived spectral interferences was to remove the matrix somehow. In the early days, this involved precipitating the matrix with a complexing agent and then filtering off the precipitate. However, more recently, this has been carried out by automated matrix removal/analyte preconcentration techniques using chromatography-type equipment. In fact, this is the preferred method for carrying out trace metal determinations in seawater, because of the matrix and spectral problems associated with such high concentrations of sodium and chloride ions.[5]

Mathematical Correction Equations

Another method that has been successfully used to compensate for isobaric interferences and some less severe polyatomic overlaps (when no alternative isotopes are available for quantitation) is to use mathematical interference correction equations. Similar to interelement corrections (IECs) in ICP-OES, this method works on the principle of measuring the intensity of the interfering isotope or interfering species at another mass, which is ideally free of any interferences. A correction is then applied by knowing the ratio of the intensity of the interfering species at the analyte mass to its intensity at the alternate mass. Let us look at a "real-world" example to exemplify this type of correction. The most sensitive isotope for cadmium is at mass 114. However, there is also a minor isotope of tin at mass 114. This means that if there is any tin in the sample, quantitation using $^{114}Cd^+$ can only be carried out if a correction is made for $^{114}Sn^+$. Fortunately, Sn has a total of 10 isotopes, which means that probably at least one of them is going to be free of a spectral interference. Therefore, by measuring the intensity of Sn at one of its most abundant isotopes (typically, $^{118}Sn^+$) and ratioing it to $^{114}Sn^+$, a correction is made in the method software in the following manner:

$$\text{Total counts at mass } 114 = {}^{114}Cd^+ + {}^{114}Sn^+$$

$$\text{Therefore, } {}^{114}Cd^+ = \text{Total counts at mass } 114 - {}^{114}Sn^+$$

Relative Abundance of the Natural Isotopes

Isotope	%	%	%	Isotope	%	%	%	Isotope	%	%	%	Isotope	%	%	%
1	H 99.985			61			Ni 1.140	121			Sb 57.36	181	Ta 99.988		
2	H 0.015			62			Ni 3.634	122	Sn 4.63	Te 2.603		182		W 26.3	
3		He 0.000137		63	Cu 69.17			123		Te 0.908	Sb 42.64	183		W 14.3	
4		He 99.999863		64		Zn 48.6	Ni 0.926	124	Sn 5.79	Te 4.816	Xe 0.10	184	Os 0.02	W 30.67	
5				65	Cu 30.83			125		Te 7.139		185			Re 37.40
6			Li 7.5	66		Zn 27.9		126		Te 18.95	Xe 0.09	186	Os 1.58	W 28.6	
7			Li 92.5	67		Zn 4.1		127	I 100			187	Os 1.6		Re 62.60
8				68		Zn 18.8		128		Te 31.69	Xe 1.91	188	Os 13.3		
9	Be 100			69			Ga 60.108	129			Xe 26.4	189	Os 16.1		
10		B 19.9		70	Ge 21.23	Zn 0.6		130	Ba 0.106	Te 33.80	Xe 4.1	190	Os 26.4		Pt 0.01
11		B 80.1		71			Ga 39.892	131			Xe 21.2	191		Ir 37.3	
12			C 98.90	72	Ge 27.66			132	Ba 0.101		Xe 26.9	192	Os 41.0		Pt 0.79
13			C 1.10	73	Ge 7.73			133		Cs 100		193		Ir 62.7	
14	N 99.643			74	Ge 35.94	Se 0.89		134	Ba 2.417		Xe 10.4	194			Pt 32.9
15	N 0.366			75			As 100	135	Ba 6.592			195			Pt 33.8
16		O 99.762		76	Ge 7.44	Se 9.36		136	Ba 7.854	Ce 0.19	Xe 8.9	196	Hg 0.15		Pt 25.3
17		O 0.038		77		Se 7.63		137	Ba 11.23			197		Au 100	
18		O 0.200		78	Kr 0.35	Se 23.78		138	Ba 71.70	Ce 0.25	La 0.0902	198	Hg 9.97		Pt 7.2
19			F 100	79			Br 50.69	139			La 99.9098	199	Hg 16.87		
20	Ne 90.48			80	Kr 2.25	Se 49.61		140		Ce 88.48		200	Hg 23.10		
21	Ne 0.27			81			Br 49.31	141			Pr 100	201	Hg 13.18		
22	Ne 9.25			82	Kr 11.6	Se 8.73		142	Nd 27.13	Ce 11.08		202	Hg 29.86		
23		Na 100		83	Kr 11.5			143	Nd 12.18			203			Tl 29.524
24			Mg 78.99	84	Kr 57.0	Sr 0.56		144	Nd 23.80	Sm 3.1		204	Hg 6.87	Pb 1.4	
25			Mg 10.00	85			Rb 72.165	145	Nd 8.30			205			Tl 70.476
26			Mg 11.01	86	Kr 17.3	Sr 9.86		146	Nd 17.19			206		Pb 24.1	
27	Al 100			87		Sr 7.00	Rb 27.835	147		Sm 15.0		207		Pb 22.1	
28		Si 92.23		88		Sr 82.58		148	Nd 5.76	Sm 11.3		208		Pb 52.4	
29		Si 4.67		89			Y 100	149		Sm 13.8		209	Bi 100		
30		Si 3.10		90	Zr 51.45			150	Nd 5.64	Sm 7.4		210			
31			P 100	91	Zr 11.22			151			Eu 47.8	211			
32	S 95.02			92	Zr 17.15	Mo 14.84		152	Gd 0.20	Sm 26.7		212			
33	S 0.75			93			Nb 100	153			Eu 52.2	213			
34	S 4.21			94	Zr 17.38	Mo 9.25		154	Gd 2.18	Sm 22.7		214			
35		Cl 75.77		95		Mo 15.92		155	Gd 14.80			215			
36	S 0.02		Ar 0.337	96	Zr 2.80	Mo 16.68	Ru 5.52	156	Gd 20.47	Dy 0.06		216			
37		Cl 24.23		97		Mo 9.55		157	Gd 15.65			217			
38			Ar 0.063	98		Mo 24.13	Ru 1.88	158	Gd 24.84	Dy 0.10		218			
39	K 93.2581			99			Ru 12.7	159			Tb 100	219			
40	K 0.0117	Ca 96.941	Ar 99.600	100		Mo 9.63	Ru 12.6	160	Gd 21.86	Dy 2.34		220			
41	K 6.7302			101			Ru 17.0	161		Dy 18.9		221			
42		Ca 0.647		102	Pd 1.02		Ru 31.6	162	Er 0.14	Dy 25.5		222			
43		Ca 0.135		103		Rh 100		163		Dy 24.9		223			
44		Ca 2.086		104	Pd 11.14		Ru 18.7	164	Er 1.61	Dy 28.2		224			
45			Sc 100	105	Pd 22.33			165			Ho 100	225			
46	Ti 8.0	Ca 0.004		106	Pd 27.33	Cd 1.25		166	Er 33.6			226			
47	Ti 7.3			107			Ag 51.839	167	Er 22.95			227			
48	Ti 73.8	Ca 0.187		108	Pd 26.46	Cd 0.89		168	Er 26.8	Yb 0.13		228			
49	Ti 5.5			109			Ag 48.161	169			Tm 100	229			
50	Ti 5.4	V 0.250	Cr 4.345	110	Pd 11.72	Cd 12.49		170	Er 14.9	Yb 3.05		230			
51		V 99.750		111		Cd 12.80		171		Yb 14.3		231	Pa 100		
52			Cr 83.789	112	Sn 0.97	Cd 24.13		172		Yb 21.9		232	Th 100		
53			Cr 9.501	113		Cd 12.22	In 4.3	173		Yb 16.12		233			
54	Fe 5.8		Cr 2.365	114	Sn 0.65	Cd 28.73		174		Yb 31.8	Hf 0.162	234	U 0.0055		
55		Mn 100		115	Sn 0.34		In 95.7	175	Lu 97.41			235	U 0.7200		
56	Fe 91.72			116	Sn 14.53	Cd 7.49		176	Lu 2.59	Yb 12.7	Hf 5.206	236			
57	Fe 2.2			117	Sn 7.68			177			Hf 18.606	237			
58	Fe 0.28		Ni 68.077	118	Sn 24.23			178			Hf 27.297	238	U 99.2745		
59		Co 100		119	Sn 8.59			179			Hf 13.629				
60			Ni 26.223	120	Sn 32.59	Te 0.096		180	Ta 0.012	W 0.13	Hf 35.100				

"Isotopic Compositions of the Elements 1989" Pure Appl. Chem., Vol. 63, No. 7, pp. 991-1002, 1991. © 1991 IUPAC

FIGURE 14.2 Relative isotopic abundances of the naturally occurring elements, showing all the potential isobaric interferences. (From UIPAC Isotopic Composition of the Elements, *Pure and Applied Chemistry* **75**[6], 683–799, 2003.)

To find out the contribution from $^{114}Sn^+$, it is measured at the interference-free isotope of $^{118}Sn^+$ and a correction of the ratio of $^{114}Sn^+/^{118}Sn^+$ is applied, which means $^{114}Cd^+$ = Counts at mass 114 − ($^{114}Sn^+/^{118}Sn^+$) × ($^{118}Sn^+$).

Now, the ratio ($^{114}Sn^+/^{118}Sn^+$) is the ratio of the natural abundances of these two isotopes (065%/24.23%) and is always constant.

$$\text{Therefore, } ^{114}Cd^+ = \text{mass } 114 - (0.65\%/24.23\%) \times (^{118}Sn^+)$$

$$\text{or } ^{114}Cd^+ = \text{mass } 114 - (0.0268) \times (^{118}Sn^+)$$

An interference correction for $^{114}Cd^+$ would then be entered in the software as

$$- (0.0268) \times (^{118}Sn^+)$$

This is a relatively simple example, but explains the basic principles of the process. In practice, especially in spectrally complex samples, corrections often have to be made to the isotope being used for the correction in addition to the analyte mass, which makes the mathematical equation far more complex.

This approach can also be used for some less severe polyatomic-type spectral interferences. For example, in the determination of V at mass 51 in diluted brine (typically 1000 ppm NaCl), there is a substantial spectral interference from $^{35}C^{16}lO^+$ at mass 51. By measuring the intensity of the $^{37}C^{16}lO^+$ at mass 53, which is free of any interference, a correction can be applied in a similar way to the previous example.

Cool/Cold Plasma Technology

If the intensity of the interference is large, and analyte intensity is extremely low, mathematical equations are not ideally suited as a correction method. For that reason, alternative approaches have to be considered to compensate for the interference. One such approach, which has helped to reduce some of the severe polyatomic overlaps, is to use cold/cool plasma conditions. This technology, which was reported in the literature in the late 1980s, uses a low-temperature plasma to minimize the formation of certain argon-based polyatomic species.[6] Under normal plasma conditions (typically 1000–1400 W RF power and 0.8–1.0 L/min of nebulizer gas flow), argon ions combine with matrix and solvent components to generate problematic spectral interferences such as $^{38}ArH^+$, $^{40}Ar^+$, and $^{40}Ar^{16}O^+$, which impact the detection limits of a small number of elements including K, Ca, and Fe. By using cool plasma conditions (500–800 W RF power and 1.5–1.8 L/min nebulizer gas flow), the ionization conditions in the plasma are changed so that many of these interferences are dramatically reduced. The result is that detection limits for this group of elements, are significantly enhanced.[7] An example of this improvement is shown in Figure 14.3. It shows a spectral scan of 100 ppt of $^{56}Fe^+$ (its most sensitive isotope) using cool plasma conditions. It can be clearly seen that there is virtually no contribution from $^{40}Ar^{16}O^+$, as indicated by the extremely low background for deionized water, resulting in single figure ppt detection limits for iron. Under normal plasma conditions, the $^{40}Ar^{16}O^+$ intensity is so large that it would completely overlap the $^{56}Fe^+$ peak.

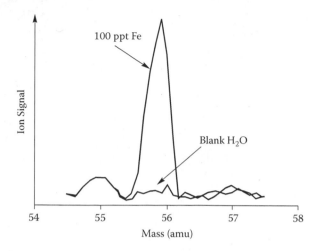

FIGURE 14.3 Spectral scan of 100 ppt ^{56}Fe and deionized water using cool plasma conditions. (From S. D. Tanner, M. Paul, S. A. Beres, and E. R. Denoyer, *Atomic Spectroscopy*, **16**[1], 16, 1995.)

Unfortunately, even though the use of cool plasma conditions is recognized as being a very useful tool for the determination of a small group of elements, its limitations are well documented.[9] A summary of the limitations of cool plasma technology includes the following:

- As a result of less energy being available in a cool plasma, elements that form a strong bond with one of the matrix/solvent ions cannot be easily decomposed and, as a result, their detection limits are compromised.
- Elements with high ionization potentials cannot be ionized, because there is much less energy compared to a normal, high-temperature plasma.
- Elemental sensitivity is severely affected by the sample matrix; so, cool plasma often requires the use of standard additions or matrix matching to achieve satisfactory results.
- When carrying out multielement analysis, normal plasma conditions must also be used. This necessitates the need for stabilization times on the order of 3 min to change from a normal to a cool plasma, which degrades productivity and results in higher sample consumption.

For this reason, it is not ideally suited for the analysis of complex samples. However, it does offer real detection limit improvement for elements with low ionization potential, such as sodium and lithium, which benefit from the ionization conditions of the cooler plasma.

Collision/Reaction Cells

The limitations of cool plasma technology have led to the development of collision/reaction cells and interfaces, which utilize ion–molecule collisions and reactions to cleanse the ion beam of harmful polyatomic and molecular interferences before

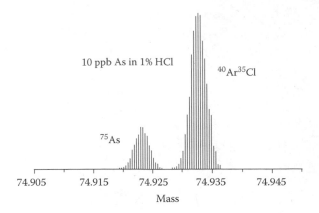

10 ppb As in 1% HCl

$^{40}Ar^{35}Cl$

^{75}As

74.905   74.915   74.925   74.935   74.945

Mass

FIGURE 14.4 Separation of $^{75}As^+$ from $^{40}Ar^{35}Cl^+$ using high resolving power (5,000) of a double focusing magnetic sector instrument. (From W. Tittes, N. Jakubowski, and D. Stuewer, Poster Presentation at Winter Conference on Plasma Spectrochemistry, San Diego, 1994.)

they enter the mass analyzer. Quadrupole mass analyzers fitted with these devices are showing enormous potential to eliminate many spectral interferences, allowing the use of the most sensitive elemental isotopes that were previously unavailable for quantitation. A full description and review of collision/reaction cell and interface technology and how they handle spectral interferences is given in Chapter 10.

High-Resolution Mass Analyzers

The best and probably most efficient way to remove spectral overlaps is to resolve them using a high-resolution mass spectrometer.[10] Over the past 10 years this approach, particularly double-focusing magnetic sector mass analyzers, has proved to be invaluable for separating many of the problematic polyatomic and molecular interferences seen in ICP-MS, without the need to use cool plasma conditions or collision/reaction cells. This can be seen in Figure 14.4, which shows a spectral peak for 10 ppb of $^{75}As^+$ resolved from the $^{40}Ar^{35}Cl^+$ interference in a 1% hydrochloric acid matrix, using a resolution setting of 5000.[11]

Although their resolving capability is far more powerful than quadrupole-based instruments, there is a sacrifice in sensitivity if extremely high resolution is used, which can often translate into a degradation in detection capability for some elements, compared to other spectral interference correction approaches. A full review of magnetic sector technology for ICP-MS is given in Chapter 8.

MATRIX INTERFERENCES

Let us now look at the other class of interference in ICP-MS—suppression of the signal by the matrix itself. There are basically three types of matrix-induced interferences. The first and simplest to overcome is often called a *sample transport effect* and is a physical suppression of the analyte signal, brought on by the level of dissolved solids or acid concentration in the sample. It is caused by the sample's impact on droplet formation in the nebulizer or droplet size selection in the spray chamber.

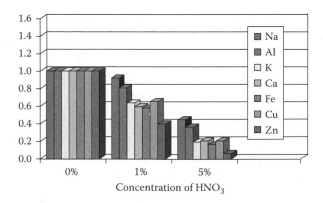

FIGURE 14.5 Matrix suppression caused by increasing concentrations of HNO_3 using cool plasma conditions (RF power: 800 W, nebulizer gas: 1.5 L/min). (From J. M. Collard, K. Kawabata, Y. Kishi, and R. Thomas, *Micro*, January 2002.)

In the case of organic matrices, it is usually caused by variations in the pumping rate of solvents with different viscosities. The second type of matrix suppression is caused when the sample affects the ionization conditions of the plasma discharge. This results in the signal being suppressed by varying amounts, depending on the concentration of the matrix components. This type of interference is exemplified when different concentrations of acids are aspirated into a cool plasma. The ionization conditions in the plasma are so fragile that higher concentrations of acid result in severe suppression of the analyte signal. This can be seen very clearly in Figure 14.5, which shows sensitivity for a selected group of elements in varying concentrations of nitric acid in a cool plasma.[9]

COMPENSATION USING INTERNAL STANDARDIZATION

The classic way to compensate for a physical interference is to use internal standardization (IS). With this method of correction, a small group of elements (usually at the ppb level) are spiked into the samples, calibration standards, and blank to correct for any variations in the response of the elements caused by the matrix. As the intensities of the internal standards change, the element responses are updated every time a sample is analyzed. The following criteria are typically used for selecting an internal standard:

- It is not present in the sample.
- The sample matrix or analyte elements do not spectrally interfere with it.
- It does not spectrally interfere with the analyte masses.
- It should not be an element that is considered an environmental contaminant.
- It is usually grouped with analyte elements of a similar mass range. For example, a low-mass internal standard is grouped with the low-mass analyte elements, and so on, up the mass range.
- It should be of a similar ionization potential to the group of analyte elements, so it behaves in a similar manner in the plasma.

Some of the most common elements/masses reported to be good candidates for internal standards include ^{9}Be, ^{45}Sc, ^{59}Co, ^{74}Ge, ^{89}Y, ^{103}Rh, ^{115}In, ^{169}Tm, ^{175}Lu, ^{187}Re, and ^{232}Th. An internal standard is also used to compensate for long-term signal drift as a result of matrix components slowly blocking the sampler and skimmer cone orifices. Even though total dissolved solids are usually kept below 0.2% in ICP-MS, this can still produce instability of the analyte signal over time with some sample matrices. It should also be emphasized that the difference in intensities of the internal standard elements across the mass range will indicate the flatness of the mass response curve. The flatter the mass response curve (i.e., less mass discrimination), the easier it is to compensate for matrix-based suppression effects using IS.

SPACE-CHARGE-INDUCED MATRIX INTERFERENCES

Many of the early researchers reported that the magnitude of signal suppression in ICP-MS increased with decreasing atomic mass of the analyte ion.[12] More recently, it has been suggested that the major cause of this kind of suppression is the result of poor transmission of ions through the ion optics due to matrix-induced space charge effects.[13] This has the effect of defocusing the ion beam, which leads to poor sensitivity and detection limits, especially when trace levels of low-mass elements are being determined in the presence of large concentrations of high-mass matrices. Unless any compensation is made, the high-mass matrix element will dominate the ion beam, pushing the lighter elements out of the way.[14] This can be seen in Figure 14.6, which shows the classic space charge effects of a uranium (major isotope ^{238}U$^+$) matrix on the determination of ^{7}Li$^+$, ^{9}Be$^+$, ^{24}Mg$^+$, ^{55}Mn$^+$, ^{85}Rb$^+$, ^{115}In$^+$, ^{133}Cs$^+$, ^{205}Tl$^+$, and ^{208}Pb$^+$. It can clearly be seen that the suppression of the low-mass elements such as Li and Be is significantly higher than with the high-mass elements such as Tl and Pb in the presence of 1000 ppm uranium.

There are a number of ways to compensate for space charge matrix suppression in ICP-MS. IS has been used, but unfortunately does not address the fundamental cause of the problem. The most common approach used to alleviate or at least reduce space charge effects is to apply voltages to individual lens components of the ion optics. This is achieved in a number of different ways, but irrespective of the design of the ion-focusing system, its main function is to reduce matrix-based suppression

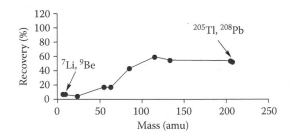

FIGURE 14.6 Space charge matrix suppression caused by 1000 ppm uranium is significantly higher on low-mass elements such as Li and Be than it is with the high-mass elements such as Tl and Pb. (From S. D. Tanner, *Journal of Analytical Atomic Spectrometry*, **10**, 905, 1995.)

effects by steering as many of the analyte ions through to the mass analyzer while rejecting the maximum number of matrix ions. For more details on space charge effects and different designs of ion optics, refer to Chapter 6.

REFERENCES

1. M. A. Vaughan and G. Horlick, *Applied Spectroscopy*, **41**(4), 523, 1987.
2. S. N. Tan and G. Horlick, *Applied Spectroscopy*, **40**(4), 445, 1986.
3. D. J. Douglas and J. B. French, *Spectrochimica Acta*, **41B**(3), 197, 1986.
4. UIPAC Isotopic Composition of the Elements, *Pure and Applied Chemistry* **75**(6), 683–799, 2003.
5. S. N. Willie, Y. Iida, and J. W. McLaren, *Atomic Spectroscopy,* **19**(3), 67, 1998.
6. S. J. Jiang, R. S. Houk, and M. A. Stevens, *Analytical Chemistry,* **60**, 1217, 1988.
7. K. Sakata and K. Kawabata, *Spectrochimica Acta*, **49B**, 1027, 1994.
8. S. D. Tanner, M. Paul, S. A. Beres, and E. R. Denoyer, *Atomic Spectroscopy*, **16**(1), 16, 1995.
9. J. M. Collard, K. Kawabata, Y. Kishi, and R. Thomas, *Micro*, January 2002.
10. R. Hutton, A. Walsh, D. Milton, and J. Cantle, *ChemSA*, **17**, 213–215, 1992.
11. W. Tittes, N. Jakubowski, and D. Stuewer, Poster Presentation at *Winter Conference on Plasma Spectrochemistry*, San Diego, 1994.
12. J. A. Olivares and R. S Houk, *Analytical Chemistry,* **58**, 20, 1986.
13. S. D. Tanner, D. J. Douglas, and J. B. French, *Applied Spectroscopy*, **48**, 1373, 1994.
14. S. D. Tanner, *Journal of Analytical Atomic Spectrometry*, **10**, 905, 1995.

15 Contamination Issues Associated with Sample Preparation

Serious consideration must be given to contamination issues in ICP-MS, particularly in the area of sample preparation. If you have been using flame AA or ICP-OES, you will probably have to rethink your sample preparation procedures for ICP-MS. Chapter 15 takes a closer look at the major causes of contamination and analyte loss in ICP-MS and how they affect both the analysis and the method development process.

There are many factors that influence the ability to get the correct result with any trace element technique. Unfortunately with ICP-MS, the problem is magnified even more because of its extremely high sensitivity. So, in order to ensure that the data reported is an accurate reflection of the sample in its natural state, the analyst must not only be aware of all the potential sources of contamination, but also the many reasons why analyte loss is a problem in ICP-MS. Figure 15.1 shows the major factors than can impact the analytical result in ICP-MS.

COLLECTING THE SAMPLE

Collecting the sample and maintaining its integrity is a science all by itself and is beyond the scope of this book. However, it is worth discussing briefly, in order to understand its importance in the overall scheme of collecting, preparing, and analyzing the sample. The object of sampling is to collect a portion of the material that is small enough in size to be conveniently transported and handled, and at the same time accurately represents the bulk material being sampled. Depending on the sampling requirements and the type of sample, there are basically three main types of sampling procedures. They are as follows:

- **Random sampling** is the most basic type of sampling and only represents the composition of the bulk material at the time and place it was sampled. If the composition of the material is known to vary with time, individual samples collected at suitable intervals and analyzed separately can reflect the extent, frequency, and duration of these variations.
- **Composite sampling** is when a number of samples are collected at the same point, but at different times and mixed together before being analyzed.
- **Integrated sampling** is achieved by mixing together a number of samples that have been collected simultaneously from different points.

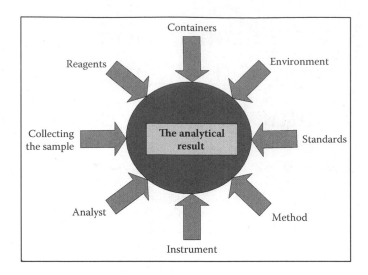

FIGURE 15.1   Major factors that can influence the analytical result in ICP-MS.

We will not go into which type of sampling is the most effective, but it must be emphasized that unless the correct sampling or subsampling procedure is used, the analytical data generated by the ICP-MS instrumentation may be seriously flawed because it may not represent the original bulk material. If the sample is a liquid, it is also important to collect the sample in clean containers (as shown later) that have been thoroughly washed. In addition, if the sample is to be kept for a long period of time before analysis, it is essential that the analytes stay in solution in a preservative such as a dilute acid (this will also help stop the analytes from being absorbed into the walls of the container). It is also important to keep the samples as cool as possible to avoid losses through evaporation. Kratochvil and Taylor give an excellent review of the importance of sampling for chemical analysis.[1]

PREPARING THE SAMPLE

As mentioned previously, ICP-MS was originally developed for the analysis of liquid samples, even though it can be modified to analyze solid materials (see Chapter 17). Thus, if the sample is not in a liquid form, some kind of sample preparation has to be carried out in order to make it so. There is no question that collecting a solid sample, preparing it, and getting it into solution probably represent the most crucial steps in the overall ICP-MS analytical methodology because of the potential sources of contamination from grinding, sieving, weighing, dissolving, and diluting the sample. Let us take a look at these steps in greater detail and, in particular, focus on their importance when being used for ICP-MS.

GRINDING THE SAMPLE

Some fine-powder solid samples are ready to be dissolved without grinding; for them, mere passage through a fine-mesh sieve (mesh is typically 0.1–0.2 mm^2) is enough.

Other types of coarser solid samples, such as soils, need to be passed through a coarse-mesh sieve (typically 2 mm² mesh) to make it ready for dissolution.[2] However, if the solid sample is not in a convenient form to be dissolved, it has to be ground to a smaller particle size, mainly to improve the homogeneity of the original sample taken and make it more representative when taking a subsample. The ideal particle size will vary depending on the sample, but the sample is typically ground to pass through a fine-mesh sieve (0.1 mm² mesh). This uniform particle size ensures that the particles in the test portion are the same size as the particles in the rest of the ground sample. Another reason for grinding the sample into small uniform particles is that they are easier to dissolve.

The process of grinding a sample with mortar and pestle or ball mill and passing it through a metallic sieve can be a major cause of contamination. This can occur from the remains of a previous sample that had been prepared earlier or from materials used in the manufacture of the grinding or sieving equipment. For example, if tungsten carbide equipment is used to grind the sample, major elements like tungsten and carbon as well additive elements like cobalt and titanium can also be a problem. Additionally, sieves, which are made from stainless steel, bronze, or nickel, can also introduce metallic contamination into the sample. In order to minimize some of these problems, plastic sieves are often used. However, still remaining is the problem of contamination from the grinding equipment. For this reason, it is usual to discard the first portion of the sample or even to use different grinding and sieving equipment for different kinds of samples.

SAMPLE DISSOLUTION METHODS

Unfortunately, there is no single dissolution procedure that can be used for all types of solid samples. There are many different approaches to getting solid samples into solution. For some samples, this is fairly straightforward and fast, whereas for others it can be very complex and time consuming. However, all the successful sample dissolution procedures used in ICP-MS typically have a number of things in common:

- Complete dissolution is a usual requirement.
- Ultrapure reagents should be used.
- The reagents should not contaminate or interfere with the analysis.
- Equipment should show no chemical attack or corrosion.
- There should be no loss of analyte.
- In speciation studies, the integrity of elemental form/valency state/species should be maintained.
- Ideally, dissolution should be fast.
- Safety is paramount.

Even though the contamination issues are exaggerated with ICP-MS, the most common approaches to getting samples into solution are very similar to the ones used for other trace element techniques. The most common dissolution techniques include the following:

- Hot plate, pressure bombs,[3] or microwave digestion[4] using concentrated acids/oxidizing agents—such as nitric acid, perchloric acid, hydrofluoric acid, aqua regia, hydrogen peroxide, or various mixtures of these—are among the most common approaches to dissolution and are typically used for metals, soils or sediments,[5] minerals,[6] and biological samples.[7]
- Dissolution with strong bases such as caustic or trimethyl ammonium hydroxide (TMAH)—typically used for biological samples.[8]
- Heating with fusion mixtures or fluxes such as lithium metaborate, sodium carbonate, or sodium peroxide in a metal crucible (e.g., platinum, silver, or nickel) and redissolving in a dilute mineral acid—typically used for ceramics, stubborn minerals, ores, rocks, and slags.[9,10]
- Dry ashing using a flame, heat lamp, or a heated muffle furnace and redissolving the residue in a dilute mineral acid—typically used for organic or biological matrices.[11]
- Wet ashing using concentrated acids (usually with some kind of heat)—typically used for organic, petrochemical, or biomedical samples.[12]
- Dissolution with organic solvents—typically used for organic or oil-type samples.[13]

The choice of which dissolution technique to use is often very complicated and depends on criteria such as the size of the sample, the matrix components in the sample, the elements to be analyzed, the concentration of elements being determined, the types of interferences anticipated, the type of ICP-MS equipment being used, the time available for analysis, safety concerns, and the expertise of the analyst. However, with ICP-MS, contamination issues are probably the greatest concern. For that reason, the most common approach to sample preparation is to keep the process as simple as possible, because the more steps that are involved, the more chance there is of contaminating the sample. This means that, ideally, if the sample is already a liquid, a simple acidification might be all that is needed. If the sample is a solid, a straightforward acid dissolution is preferred over the more complex and time-consuming fusion and ashing procedures. An excellent handbook of decomposition methods used for analytical chemistry was written by Bock in 1979.[14]

It is also important to emphasize that many acids that are used for AA and ICP-OES are not ideal for ICP-MS because of the polyatomic spectral interferences they produce. Although this is not strictly a contamination problem, it can significantly impact your data if not taken into consideration. For example, if vanadium or arsenic is being determined, it is advisable not to use hydrochloric acid (HCl) or perchloric acid ($HClO_4$) because they generate polyatomic ions such as $^{35}Cl^{16}O^+$ and $^{40}Ar^{35}Cl^+$, which interfere with the isotopes $^{51}V^+$ and $^{75}As^+$, respectively. Sulfuric acid (H_2SO_4) and phosphoric acid (H_3PO_4) are also acids that should be avoided if possible because they generate sulfur- and phosphorus-based polyatomic ions. Therefore, if there is a choice of which acid to use for dissolution, nitric acid (HNO_3) is the preferred one to use. Even though it can generate interferences of its own, they are generally less severe than those of the other acids.[15] Table 15.1 shows the kinds of polyatomic spectral interferences generated by the most common mineral acids and dissolution chemicals.

TABLE 15.1

Typical Polyatomic Spectral Interferences Generated by Common Mineral Acids and Dissolution Chemicals

Acid/Solvent/Fusion Mixture	Interference	Element/Isotope
HCl	$^{35}Cl^{16}O^+$	$^{51}V^+$
HCl	$^{40}Ar^{35}Cl^+$	$^{75}As^+$
HNO_3	$^{14}N^{14}N^+$	$^{28}Si^+$
HNO_3	$^{14}N^{14}N^{16}O^+$	$^{44}Ca^+$
HNO_3	$^{40}Ar^{15}N^+$	$^{55}Mn^+$
H_2SO_4	$^{32}S^{16}O^+$	$^{48}Ti^+$
H_2SO_4	$^{34}S^{18}O^+$	$^{52}Cr^+$
H_2SO_4	$^{32}S^{16}O^{16}O^+$	$^{64}Zn^+$
H_3PO_4	$^{31}P^{16}O^{16}O^+$	$^{63}Cu^+$
Any organic solvent	$^{12}C^{12}C^+$	$^{24}Mg^+$
Any organic solvent	$^{40}Ar^{12}C^+$	$^{52}Cr^+$
Lithium-based fusion mixtures	$^{40}Ar^7Li^+$	$^{47}Ti^+$
Boron-based fusion mixtures	$^{40}Ar^{11}B^+$	$^{51}V^+$
Sodium-based fusion mixtures	$^{40}Ar^{23}Na^+$	$^{63}Cu^+$

In addition, fusion mixtures present unique problems for ICP-MS. Not only because the major elements form polyatomic spectral interferences with the argon gas, but the high levels of dissolved solids in the sample can cause blockage of the interface cones, which over time can lead to signal drift. An additional problem with a fusion procedure is the risk of losing volatile analytes due to the high temperature of the muffle furnace or flame used to heat the crucible.

CHOICE OF REAGENTS AND STANDARDS

Careful consideration must be given to the choice and purity of reagents, especially if sub-ppt concentration levels are expected. General laboratory- or reagent-grade chemicals used for AA or ICP-OES sample preparation are not usually pure enough. For that reason, most manufacturers of laboratory chemicals now offer ultra-high-purity grades of chemicals, acids, and fusion mixtures specifically for use with ICP-MS. It is therefore absolutely essential that the highest-grade chemicals and water be used in the preparation and dilution of the sample. In fact, the grade of deionized water used for dilution and the cleaning of vessels and containers is very important in ICP-MS.

Less pure water such as single-distilled or deionized water is fine for flame AA or ICP-OES, but is not suitable for use with ICP-MS because it could possibly contain contaminants such as dissolved inorganic or organic matter, suspended dust or scale particles, and microorganisms. All these contaminants can affect reagent blank levels and negatively impact instrument and method detection limits. This necessitates the use of the most chemically pure water for ICP-MS work. There are a number of water purification systems on the market, which use combinations of filters, ion exchange cartridges, and/or reverse osmosis systems to remove the particulates, organic matter, and trace metal contaminants. These ultra-high-purity water systems (similar to the ones used to for semiconductor processing) typically produce water with a resistance of better than 18 mΩ.[16]

Another area of concern with regard to contamination is in the selection of calibration standards. Because ICP-MS is a technique capable of quantifying up to 75 different elements, it will be detrimental to the analysis to use calibration standards that are developed for a single-element technique such as atomic absorption. These single-element standards are usually certified only for the analyte element and not for any others, although they are often quoted on the certificate. It is therefore absolutely critical to use calibration standards that have been specifically made for a multielement technique such as ICP-MS. It does not matter whether they are single or multielement standards, as long as the certificate contains information on the suite of analyte elements you are interested in as well as any other potential interferents.

It is also desirable that the certified values be confirmed by both a classical wet technique and an instrumental technique, all of which are traceable to National Institute of Standards and Technology (NIST) reference materials. It is also important to fully understand the uncertainty or error associated with a certified value, so that you know how it impacts the data you report.[17] Figure 15.2 is a certificate for a 1000 mg/L erbium certified reference standard used in ICP-MS, showing values for over 30 trace metal contaminants. Make sure that all your calibration standards come with similar certification, so you have the confidence that your reported data can be scrutinized to the highest standards.

The same case applies if a calibration standard is being made from a high-purity salt of the metal. The salt has to be certified for not only the element of interest, but also for the full suite of analyte elements as well as other elements that could be potential interferents. It is also important to understand the shelf life of these standards and chemicals and how long-term storage affects the concentration of the analyte elements, especially at low levels.

VESSELS, CONTAINERS, AND SAMPLE PREPARATION EQUIPMENT

The containers used for preparation, dilution, storage, and introduction of the sample can have a huge impact on your data in ICP-MS. Traditional glassware such as beakers, volumetric flasks, and autosampler tubes, which are fine for AA and ICP-OES work, are not ideally suited to ICP-MS. The major problem is potential contamination from the major elemental components of the glassware. For example, glass made from soda lime contains percentage concentrations of silicon, sodium, calcium, magnesium, and aluminum; also, borosilicate glass contains high levels of boron. Besides these major elements, glass might also contain minor concentrations of Zr, Li, Ba, Fe, K, and Mn. Unfortunately, if the sample solution is highly acidic, there is a strong possibility that these elements can be leached out of the glassware.

In addition to the contamination issues, analytes can be absorbed into the walls of volumetric flasks and beakers made of glass. This can be a serious problem if the sample or standard is to be stored for extended periods of time, especially if the

FIGURE 15.2 Certificate for a 1000 mg/L erbium certified reference standard used in ICP-MS, showing values for over 30 trace metal contaminants (courtesy of SPEX Certiprep).

analyte concentrations are extremely low. If using glassware is unavoidable, it is a good idea to clean the glassware on a regular basis using chromic acid or some kind of commercial glass detergent such as Decon™ or Citranox™. If long-term storage is a necessity, either avoid using glassware or minimize the analyte loss by keeping the solutions acidified (~pH 2), so there is very little chance of absorption into the walls of the glassware.[18]

Glassware is such a universal material used for sample preparation that it is very difficult to completely avoid it. However, serious consideration should be given to looking for alternative materials in as many of the ICP-MS sample preparation steps as possible. Today, the most common materials used to manufacture beakers, volumetric containers, and autosampler tubes for ultratrace element techniques such as GFAA and ICP-MS, are mainly plastic based. Over the past 10–15 years, the demand for these kinds of materials has increased significantly because of the contamination issues associated with glassware.

Some plastics are more inert and more pure than others, so thought should be given as to which one is optimal for your samples. Selection should be made based on the suite of elements being analyzed, analyte concentration levels, matrix components, or whether it is an aqueous-, acid-, or organic-based solution. Some of the most common plastic materials used in the manufacture of sample preparation vessels and sample introduction components include polypropylene (PP), polyethylene (PE), polysulfide (PS), polycarbonate (PC), polyvinylchloride (PVC), polyvinylfluoride (PVF), perfluoroalkoxy (PFA), and polytetrafluoroethylene (PTFE).

It is generally felt that PTFE or PFA probably represent the cleanest materials, and even though they are the most expensive, they are considered the most suitable for ultratrace ICP-MS work. However, even though these types of plastics are generally much cleaner than glass, they still contain some trace elements. For example, certain plastics might contain phosphorus from the mold-releasing agent, and some plastic tube caps and covers are manufactured with barium compounds to enhance their color. These are all potential sources of contamination that can cause serious problems in ICP-MS, especially if heat is involved in sample preparation. This is particularly true if microwave dissolution is used to prepare the sample because of the potential for high-temperature breakdown of the polymer material over time. Table 15.2, which was taken from a publication from about 30 years ago, gives trace element contamination levels of some common plastics used in the manufacture of laboratory beakers, volumetric ware, and autosampler tubes.[19] It should be strongly emphasized that these data might not be representative of current-day products, but may only be an approximation for comparison purposes.

TABLE 15.2

Typical Trace Element Contamination Levels of Some Common Plastic Materials Used in the Manufacture of Laboratory Beakers, Volumetric Ware, and Autosampler Tubes

Material	Na (ppm)	Al (ppm)	K (ppm)	Sb (ppm)	Zn (ppm)
Polyethylene (CPE)	1.3	0.5	5	0.005	—
Polyethylene (LPE)	15	30	0.6	0.2	520
Polypropylene (PP)	4.8	55	—	0.6	—
Polysulfide (PS)	2.2	0.5	—	—	—
Polycarbonate (PC)	2.7	3.0	—	—	—
Polyvinylchloride (PVC)	20	—	—	—	—
Polytetrafluoroethylene (PTFE)	0.16	0.23	90	—	—

Source: J. R. Moody and R. M. Lindstrom, *Analytical Chemistry,* **49**, 14, 2264–2267, 1977.

Even though microwave dissolution is rapidly becoming the sample dissolution method of choice over conventional hot plate digestion methods, it will not be discussed in great detail in this chapter. Such is the maturity of this approach nowadays that there are a multitude of books and reference papers in the public domain covering just about every type of sample being analyzed by ICP-MS, including geological materials,[20] soils,[21] sediments,[22] waters,[23] biological materials,[24] and foodstuffs.[25]

Consideration should also be given to the selection of other equipment and materials used in sample preparation as they can impact the analysis; some areas of concern include the following:

- The filtering materials, if the sample needs to be filtered—whether to use conventional filter papers or ones made from cellulose or acetate glass, or the method of vacuum filtration using sintered disks.
- If blood is to be drawn for analysis, the cleanliness of the syringe and the material it is made from can contribute to contamination of the sample.
- Paper towels used for many different reasons in a laboratory. These are generally high in zinc and also contain trace levels of transition metals such as Fe, Cr, and Co, so avoid using them in and around your sample preparation areas.
- Pipettes, pipette tips, and suction bulbs can all contribute to trace metal contamination levels, so the disposable variety is recommended.

It is important to emphasize that whatever containers, vessels, beakers, volumetric ware, or equipment is used to prepare the sample for ICP-MS analysis, it is absolutely critical that when not in use, they be soaked and washed in a dilute acid (1–2% HNO_3 is typical). In addition, if they are not being used for extended periods, they should be stored with dilute acid in them. Wherever possible, disposable equipment such as autosampler tubes and pipette tips should be used and then thrown away after use, to cut down on contamination.

THE ENVIRONMENT

The environment in this case refers to the cleanliness of the surrounding area where the instrumentation is installed, where sample preparation is carried out, and any other area the sample comes in contact with. It is advisable that the sample preparation area be as close to the instrument as possible, without actually being in the same room, so that the sample is not exposed to any additional sources of contamination. It is recommended that dissolution be carried out in clean, metal-free fume extraction hoods and, if possible, in an area separate from where samples are to be prepared for less sensitive techniques such as flame AA or ICP-OES. In addition to having a clean area for dissolution, it is also important to carry out other sample preparation tasks such as weighing, filtering, pipetting and diluting, etc., in a clean environment.

These kinds of environmental contamination problems are everyday occurrences in the semiconductor industry because of the strict cleanliness demands required for the fabrication of silicon wafers and production of semiconductor devices. The purity of silicon wafers has a direct effect on the yield of devices, so it is crucial that trace element contamination levels are kept to a minimum in order to reduce defects. This means that any analytical methodology used to determine purity levels on the surface of silicon wafers, or in the high-purity chemicals used to manufacture the devices, must use spotlessly clean instruments. These unique demands of the semiconductor industry have led to the development of special air filtration systems that continually pump air through ultraclean HEPA filters to remove the majority of airborne particulates.

The efficiency of particulate removal will depend on the analytical requirements, but for the semiconductor industry it is typical to work in environments that contain 1 or 10 particles (<0.2 μ) per cubic foot of air (class 1 and 10 clean rooms, respectively). These kinds of precautions are absolutely necessary to maintain low instrument background levels for the analysis of semiconductor-related samples, but might not be required for other types of applications. So, even though contamination-free analysis is important, it might be sufficient to work in a class 100, 1000, or 10,000 clean room and still meet your cleanliness objectives.[26]

These clean rooms tend to be very expensive to build, so if your budget does not stretch to a "full-blown" clean room, it might be worth investing in special HEPA filter enclosures just for your instrument and sample preparation area. These are typically either mobile units that can be wheeled around the laboratory and placed around different equipment or hood-based enclosures that are placed over a particular instrument. Whatever system is used, their objective is to ensure that the area around the equipment is free of airborne contamination and the instrument background levels are as low as possible.

THE ANALYST

The expertise of the analyst who actually prepares the samples and carries out the analysis can be a major factor in getting the right result by ICP-MS. Even if all precautions have been taken to cut down on contamination, if the analyst is not experienced in working with ICP-MS and does not understand all the potential pitfalls, the analysis could be doomed. For example, analysts have to be aware of all the potential contaminants that are generated by their own bodies or the clothes or jewelry they are wearing. Table 15.3 shows some common trace elements found on the human body. It is by no means an exhaustive list, but at least it gives you an idea of the problem.

These kinds of contamination problems are the reason you often see operators of equipment used in the semiconductor industry wearing "bunny suits." These are white suits that cover the entire body of the operator, including head, hands, and feet, to stop any human-based contamination

TABLE 15.3

Some Common Trace Elements Contaminants Found on and around the Human Body

Source of Contamination	Trace Metal Contaminant
Hair	Zn, Cu, Fe, Pb, Mn
Skin	Zn, Cu
Nails	Ca, Si
Jewelry	Au, Ag, Cu, Fe, Ni, Cr
Cigarette smoke	Cd, As, K, Fe, B
Cosmetics	Zn, Bi
Deodorants	Al

from getting into the equipment or instrumentation. They are not so important for higher levels of quantitation, but are absolutely necessary for the kind of ultratrace contamination levels found in the electronics industry.

INSTRUMENT AND METHODOLOGY

The instrument and the methodology itself can also be potential sources of error. It is therefore important to be aware of this and to understand what is required when developing a method to carry out the determination of ultratrace levels by ICP-MS. As mentioned previously, the choice of sample preparation methodology can impact the analysis by either causing corrosion problems for some of the instrument components, producing spectral interferences on the analyte, or creating matrix-induced signal drift problems. However, in addition to optimizing sample preparation, a great deal of thought must also go into the choice of instrumental components and to understand how they impact the method development process. Some of the criteria that should be under consideration when deciding on the analytical methodology include the following:

1. The acid concentration in the final solution being presented to the instrument should ideally be 2–3% maximum because of the sample transport interferences associated with high concentrations of mineral acids.
2. If highly corrosive acids such as hydrofluoric acid are being used, appropriate corrosion-resistant sample introduction components, such as plastic spray chamber and nebulizer, sapphire sample injector and platinum interface cones, should be used.
3. Hydrochloric, sulfuric, and phosphoric acids should be avoided because of the spectral problems created by the high concentration of chlorine, sulfur, and phosphorus ions in the matrix.
4. The choice of fusion mixture should be given serious consideration because of the potential for the lithium-, sodium-, or potassium-based salts to deposit themselves around the sampler or skimmer cone orifice, which over time can lead to serious drift problems.
5. The sample weight might have to be compromised if a fusion mixture is required, because 0.2% is the maximum level of dissolved solids that can be aspirated into the ICP mass spectrometer.
6. There are many grades of argon gas available for spectrochemical analysis. For ultratrace determinations by ICP-MS, the highest grade should always be used (usually ultra-high-purity-grade argon is 99.99999% pure).
7. The use of high-purity collision gases is absolutely critical when collision cells are being used, because of the potential to create additional interfering species.
8. Petrochemical-type samples usually require the addition of oxygen to the nebulizer gas flow in order to burn off the organic matrix, so the highest quality of oxygen should be used.

9. The choice of pneumatic tubing should be compatible with the sample solution. For example, when analyzing organic samples, suitable pump tubing and sample capillary should be used that are resistant to the organic solvent.

10. There are many different kinds of pump tubing and capillary. If a polyvinylchloride-based tubing is used, chlorine could potentially be leached out and may cause spectral interferences.

11. Peristaltic pump speed, washout times, read delays, and stability times should be optimized based on the sample matrix and suite of elements, because of memory effects in the sample introduction/interface areas—which may facilitate contamination from the previously analyzed sample.

12. What are the expected analyte concentrations and matrix levels? This will decide whether the sample can be diluted or whether the analytes need to be preconcentrated or the matrix components removed.

13. If the samples are completely unknown, it is a good strategy to dilute the sample 1:100 and get an approximation of the analyte concentrations using the instrument's semiquant routine. This can also give you an insight into understanding the potential interferences from the other elements in the sample.

These are generally considered some of the most important criteria for deciding on an analytical methodology to analyze a set of samples by conventional solution nebulization. However, it should be emphasized that the strategy might also include the use of sampling accessories, such as laser ablation or flow injection. For example, the ability to analyze a solid directly by laser ablation eliminates most of the contamination issues with the preparation, dilution, and aspiration of liquid samples. Even though this might sound attractive, solid sampling has unique problems of its own. So, before this approach is chosen, it is important to also understand all its limitations, especially for a particular set of samples. On the other hand, if solution nebulization is the preferred approach, it should be determined whether there will be any benefit to using segmented flow analysis to reduce the amount of matrix entering the mass spectrometer. Clearly, for some matrices it is advantageous, but for others it might not be worth the effort. It is therefore important to understand these issues before a decision is made (refer to Chapter 17).

It is also worth explaining point number 7 in greater detail because it is not a classical contamination issue. In a collision cell that uses KED to suppress the polyatomic interference, the cell relies on interactions of the interfering ion with an inert or low-reactivity gas, such that it can be separated from the analyte based on their differences in kinetic energy. If the gas contains impurities such as H_2O, O_2, CO_2, CO, or hydrocarbons, the impurity could be the dominant reaction pathway as opposed to the predicted collision/reaction with bulk gas, resulting in the formation of additional and unexpected spectral interferences. For this reason, it is strongly advised that the highest-purity collision/reaction gases be used or a gas-purifier system be placed in the gas line to cleanse the collision/reaction gas of any impurities (refer to Chapter 10).

Whatever analytical methodology approach is used, the issue of contamination must always be at the forefront of the decision.[27] ICP-MS is such a sensitive technique that to take advantage of its unparallel detection capability and sample throughput capabilities, analytical cleanliness and optimized method development is of the utmost importance. If attention is paid to these areas, there is no question that data of the highest quality can be obtained, even at the ultratrace level. This chapter is not intended to be an exhaustive look at contamination or analyte loss issues, but just to make the reader aware that in order to get the right result in ICP-MS, it is important to examine all aspects of analysis from first collection of the sample, all the way through to quantitation by the instrument.

REFERENCES

1. B. Kratochvil and J. K. Taylor, *Analytical Chemistry,* **53**(8), 925–938A, 1981.
2. Environmental Protection Agency (EPA) ICP-MS Method 200.8 for soils and sediments.
3. B. Bernas, *Analytical Chemistry,* **40**(11), 1682–1586, 1986.
4. *Introduction to Microwave Sample Preparation—Theory and Practice* H. M. Kingston and L. B. Jassie, Eds., American Chemical Society, 1988.
5. A. Hewitt and C. M. Reynolds, *Atomic Spectroscopy,* **11**(5), 187–192, 1990.
6. R. A. Nadkarni, *Analytical Chemistry,* **56**, 2233–2237, 1984.
7. A. Abu-Samra, *Analytical Chemistry,* **47**(8), 1475–1477, 1975.
8. E. Pruszkowski, K. Neubauer, and R. Thomas, *Atomic Spectroscopy,* **19**(4), 111–115, 1998.
9. C. O. Ingamells, *Analytica Chimica Acta,* **52**, 323–334, 1970.
10. C. B. Belcher, *Talanta,* **10**, 75–81, 1963.
11. The Dry Ashing Method for Preparing Sewage Sludges, *American Laboratory,* August 1982.
12. S. Bajo and U. Suter, *Analytical Chemistry,* **54**(1), 49–51, 1982.
13. F. McElroy, A. Mennito, E. Debrah, and R. Thomas, *Spectroscopy,* **13**(2), 42–53, 1998.
14. R. Bock, *A Handbook for Decomposition Methods in Analytical Chemistry,* International Textbook Company, Ltd., 1979.
15. S. Tan, G. Horlick, *Applied Spectroscopy,* **40**, 445, 1986.
16. Suggested Guidelines for Pure Water, *Book of SEMI Standards (BOSS),* Semiconductor Equipment and Materials International, San Jose, CA, 2002.
17. N. Kocherlakota, R. Obernauf, and R. Thomas, *Spectroscopy,* **17**(7), 2002.
18. D. E. Robertson, *Analytical Chemistry,* **40**(7), 1067–1072, 1968.
19. J. R. Moody and R. M. Lindstrom, *Analytical Chemistry,* **49**, 14, 2264–2267, 1977.
20. M. Totland, I. Jarvis, and K. E. Jarvis, *Chemical Geology,* **95**, 35-62, 1992.
21. V. L. Verma and T. M. McKee, *Paper Presented at the Seventh Annual Waste Testing and Quality Assurance Symposium* (EnvirACS), Washington, DC, July 10, 1991.
22. ASTM Method Number D5258-92, *Standard Practice for Acid Extraction of Elements from Sediments Using Closed Vessel Microwave Heating,* American Society for Testing and Materials, Annual Book of ASTM Standards, 1992.
23. ASTM Method Number D4309-91, *Standard Practice for Sample Digestion Using Closed Vessel Microwave Heating Technique for the Determination of Total Recoverable Metals in Water,* Annual Book of ASTM Standards, American Society for Testing and Materials, 1991.
24. H. T. McCarthy and P. C. Ellis, *Journal of Analytical Chemistry,* **74**(3), 566–569, 1991.
25. D. Sears, Jr. and Z. Grosser, Food Testing and Analysis, June/July 1997.
26. T. Talasek, *Solid State Technology,* 44-46, December 1993.
27. M. Zief and J. W. Mitchel, *Contamination Control in Trace Metal Analysis,* John Wiley and Sons, New York, 1976.

16 Routine Maintenance

The components of an ICP mass spectrometer are generally more complex than other atomic spectroscopic techniques, and as a result, more time is required to carry out routine maintenance to ensure that the instrument is performing to the best of its ability. Some tasks involve a simple visual inspection of a part, whereas others involve cleaning or changing components on a regular basis. However, routine maintenance is such a critical part of owning an ICP-MS system that it can impact both the performance and the lifetime of the instrument.

The fundamental principle of inductively coupled plasma mass spectrometry (ICP-MS), which gives the technique its unequalled isotopic selectivity and sensitivity, also unfortunately contributes to some of its weaknesses. The fact that the sample "flows into" the spectrometer and is not "passed by it" at right angles, such as flame AA and inductively coupled plasma optical emission spectroscopy (ICP-OES), means that the potential for thermal problems, corrosion, chemical attack, blockage, matrix deposits, and drift is much higher than with the other atomic spectrometry (AS) techniques. However, being fully aware of this fact and carrying out regular inspection of instrumental components can reduce and sometimes eliminate many of these potential problem areas. There is no question that a laboratory which initiates a routine maintenance schedule stands a much better chance of having an instrument ready and available for analysis whenever it is needed, compared to a laboratory that basically ignores these issues and assumes the instrument will look after itself.

Let us now look at the areas of the instrument that a user needs to pay attention to. I will not go into great detail but just give a brief overview of what is important, so you can compare it with maintenance procedures of trace element techniques you are more familiar with. These areas should be very similar with all commercial ICP-MS systems, but depending on the design of the instrument and the types of samples being analyzed, the regularity of changing or cleaning components might be slightly different (particularly if the instrument is being used for laser ablation work). The main areas that require inspection and maintenance on a routine or semiroutine basis include the following:

- Sample introduction system
- Plasma torch
- Interface region
- Ion optics
- Roughing pumps
- Air/water filters

Other areas of the instrument require less attention, but nevertheless the user should also be aware of maintenance procedures required to maximize their lifetime. They will be discussed at the end of this section.

SAMPLE INTRODUCTION SYSTEM

The sample introduction system, comprising the peristaltic pump, nebulizer, spray chamber, and drain system, takes the initial abuse from the sample matrix, and as a result, is an area of the ICP mass spectrometer that requires a great deal of attention. The principles of the sample introduction area have been described in great detail in Chapter 3, so let us now examine what kind of routine maintenance it requires.

PERISTALTIC PUMP TUBING

If the instrument uses a peristaltic pump, the sample is pumped at about 1 mL/min into the nebulizer. The constant motion and pressure of the pump rollers on the pump tubing, which is typically made from a polymer-based material, ensure a continuous flow of liquid to the nebulizer. However, over time, this constant pressure of the rollers on the pump tubing has the tendency to stretch it, which changes its internal diameter, and therefore, the amount of sample being delivered to the nebulizer. The impact could be a change in the analyte intensity, and therefore, a degradation in short-term stability.

Therefore, the condition of the pump tubing should be examined every few days, particularly if your laboratory has a high sample workload or if extremely corrosive solutions are being analyzed. The peristaltic pump tubing is probably one of the most neglected areas, so it is absolutely essential that it be a part of your routine maintenance schedule. Here are some suggested tips to reduce pump tubing–based problems:

- Manually stretch the new tubing before use.
- Maintain the proper tension on tubing.
- Ensure tubing is placed correctly in channel of the peristaltic pump.
- Periodically check flow of sample delivery, and throw away tubing if in doubt.
- Replace tubing if there is any sign of wear; do not wait until it breaks.
- With high sample workload, change tubing every day or every other day.
- Release pressure on pump tubing when instrument is not in use.
- Pump and capillary tubing can be a source of contamination.
- Pump tubing is a consumable item—keep a large supply of it on hand.

NEBULIZERS

The frequency of nebulizer maintenance will primarily depend on the types of samples being analyzed and the design of nebulizer being used. For example, in a cross-flow nebulizer, the argon gas is directed at right angles to the sample capillary tip, in contrast to the concentric nebulizer, where the gas flow is parallel to the capillary. This can be seen in Figures 16.1 and 16.2, which show schematics of a concentric and cross-flow nebulizer, respectively.

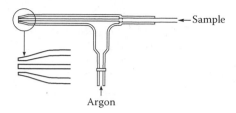

FIGURE 16.1 Schematic of a concentric nebulizer (courtesy of Meinhard Glass Products).

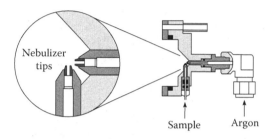

FIGURE 16.2 Schematic of a cross-flow nebulizer (copyright © 2003–2007, all rights reserved, PerkinElmer Inc.).

The larger diameter of the liquid capillary and longer distance between the liquid and gas tips of the cross-flow design make it far more tolerant to dissolved solids and suspended particles in the sample than the concentric design. On the other hand, aerosol generation of a cross-flow nebulizer is far less efficient than a concentric nebulizer, and therefore it produces droplets of less optimum size than that required for the ionization process. As a result, concentric nebulizers generally produce higher sensitivity and slightly better precision than the cross-flow design, but are more prone to clogging.

So, the choice of which nebulizer to use is usually based on the types of samples being aspirated and the data quality objectives of the analysis. However, whichever type is being used, attention should be paid to the tip of the nebulizer to ensure it is not getting blocked. Sometimes, microscopic particles can build up on the tip of the nebulizer without the operator noticing, which, over time, can cause a loss of sensitivity, imprecision, and poor long-term stability. In addition, O-rings and the sample capillary can be affected by the corrosive solutions being aspirated, which can also degrade performance. For these reasons, the nebulizer should always be a part of the regular maintenance schedule. Some of the most common things to check include the following:

- Visually check the nebulizer aerosol by aspirating water—a blocked nebulizer will usually result in an erratic spray pattern with lots of large droplets.
- Remove blockage by either using backpressure from argon line or dissolving the material by immersing the nebulizer in an appropriate acid or solvent—an ultrasonic bath can sometimes be used to aid dissolution, but

check with manufacturer first in case it is not recommended. (Note: Never stick any wires down the end of the nebulizer, because it could do permanent damage.)

- Ensure that the nebulizer is securely seated in spray chamber end cap.
- Check all O-rings for damage or wear.
- Ensure the sample capillary is inserted correctly into the sample line of the nebulizer.
- Nebulizer should be inspected every 1–2 weeks, depending on the workload.

SPRAY CHAMBER

By far the most common design of spray chamber used in commercial ICP-MS instrumentation is the double-pass design, which selects the small droplets by directing the aerosol into a central tube. The larger droplets emerge from the tube, and by gravity, exit the spray chamber via a drain tube. The liquid in the drain tube is kept at positive pressure (usually by way of a loop), which forces the small droplets back between the outer wall and the central tube; they emerge from the spray chamber into the sample injector of the plasma torch. Scott double-pass spray chambers come in a variety of shapes, sizes, and materials, but are generally considered the most rugged design for routine use. Figure 16.3 shows a double-pass spray chamber (made of a polymer material) coupled to a cross-flow nebulizer.

The most important maintenance with regard to the spray chamber is to make sure that the drain is functioning properly. A malfunctioning or leaking drain can produce a change in the spray chamber backpressure, producing fluctuations in the analyte signal, resulting in erratic and imprecise data. Less frequent problems can result from degradation of O-rings between the spray chamber and sample injector of the plasma torch. Typical maintenance procedures regarding the spray chamber include the following:

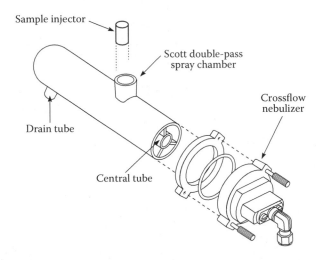

FIGURE 16.3 A double-pass spray chamber coupled to a cross-flow nebulizer (copyright © 2003–2007, all rights reserved, PerkinElmer Inc.).

- Make sure the drain tube fits tightly and there are no leaks.
- Ensure the waste solution is being pumped from the spray chamber into the drain properly.
- If a drain loop is being used, make sure the level of liquid in the drain tube is constant.
- Check O-ring or ball joint between spray chamber exit tube and torch sample injector—ensure the connection is snug.
- The spray chamber can be a source of contamination with some matrices/analytes, so flush thoroughly between samples.
- Empty spray chamber of liquid when instrument is not in use.
- Spray chamber and drain maintenance should be inspected every 1–2 weeks, depending on workload.

PLASMA TORCH

Not only are the plasma torch and sample injector exposed to the sample matrix and solvent, but they also have to sustain the analytical plasma at approximately 10,000 K. This combination makes for a very hostile environment and therefore is an area of the system that requires regular inspection and maintenance. A plasma torch positioned in the RF coil is shown in Figure 16.4.

As a result, one of the main problems is staining and discoloration of the outer tube of the quartz torch because of heat and the corrosiveness of the liquid sample. If the problem is serious enough, it has the potential to cause electrical arcing. Another potential problem area is blockage of the sample injector due to matrix components in the sample. As the aerosol exits the sample injector, desolvation takes place, and the sample changes from small liquid droplets to minute solid particles prior to entering the base of the plasma. Unfortunately, with some sample matrices, these particles can deposit themselves on the tip of the sample injector over time, leading to possible clogging and drift. In fact, this can be a potentially serious problem when aspirating organic solvents, because carbon deposits can rapidly build up on the sample injector and cones unless a small amount of oxygen is added to the nebulizer gas flow. Some torches also use metal plates or shields to reduce the secondary discharge between the plasma and the interface. These are consumable items, because of the intense heat and the effect of the RF field on the shield. A shield in poor condition can affect

FIGURE 16.4   A plasma torch mounted in the torch box (courtesy of Varian Inc.).

instrument performance, so the user should always be aware of this and replace it when necessary.

Some useful maintenance tips with regard to the torch area include the following:

- Look for discoloration or deposits on the outer tube of the quartz torch. Remove material by soaking the torch in appropriate acid or solvent if required.
- Check torch for thermal deformation. A nonconcentric torch can cause loss of signal.
- Check sample injector for blockages. If the injector is demountable, remove the material by immersing it in an appropriate acid or solvent if required (if the torch is one piece, soak the entire torch in the acid).
- Ensure that the torch is positioned in the center of the load coil and at the correct distance from the interface cone when replacing the torch assembly.
- If the coil has been removed for any reason, make sure the gap between the turns is correct as per recommendations in the operator's manual.
- Inspect any O-rings or ball joints for wear or corrosion. Replace if necessary.
- If a shield or plate is used to ground the coil, ensure it is always in good condition; otherwise, replace when necessary.
- The torch should be inspected every 1–2 weeks, depending on workload.

INTERFACE REGION

As the name suggests, the interface is the region of the ICP mass spectrometer where the plasma discharge at atmospheric pressure is "coupled" to the mass spectrometer at 10^{-6} torr by way of two interface cones—a sampler and skimmer. This coupling of a high-temperature ionization source such as an ICP to the metallic interface of the mass spectrometer imposes demands on this region of the instrument that are unique to this AS technique. When this is combined with matrix, solvent, and analyte ions together with particulates and neutral species being directed at high velocity at the interface cones, an extremely harsh environment is the result. The most common types of problems associated with the interface are blocking or corrosion of the sampler cone and, to a lesser extent, the skimmer cone. A schematic of the interface cones showing potential areas of blockage are shown in Figure 16.5.

A blockage is not always obvious, because often the buildup of material on the cone or corrosion around the orifice can take a long time to reveal itself. For that reason, the sampler and skimmer interface cones have to be inspected and cleaned on a regular basis. The frequency will often depend on the types of samples being analyzed and also the design of the ICP mass spectrometer. For example, it is well documented that a secondary discharge at the interface can prematurely discolor and degrade the sampler cone, especially when complex matrices are being analyzed or if the instrument is being used for high sample throughput.

Besides the cones, the metal interface housing itself is also exposed to the high-temperature plasma. Therefore, it needs to be cooled by a recirculating water system, usually containing some kind of antifreeze or corrosion inhibitor or by a continuous supply of mains water. Recirculating systems are probably more widely used because the temperature of the interface can be controlled much better. There is no

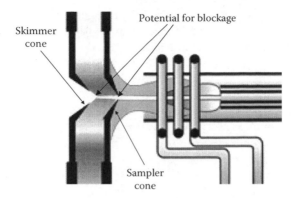

FIGURE 16.5   A schematic of the interface cones showing potential areas of blockage.

real routine maintenance involved with the interface housing, except maybe to check the quality of the coolant from time to time, to make sure there is no corrosion of the interface cooling system. If for any reason the interface gets too hot, there are usually built-in safety interlocks that will turn the plasma off. Some useful hints to prolong the lifetime of the interface and cones include the following:

- Check that both sampler and skimmer cone are clean and free of sample deposits. The typical frequency is weekly, but will depend on sample type and workload.
- If necessary, remove and clean cones using the manufacturer's recommendations. Typical approaches include immersion in a beaker of weak acid or detergent placed in a hot water or ultrasonic bath. Abrasion with fine wire wool or a coarse polishing compound has also been used.
- Never stick any wire into the orifice; it could do permanent damage.
- Nickel cones will degrade rapidly with harsh sample matrices. Use platinum cones for highly corrosive solutions and organic solvents.
- Periodically check cone orifice diameter and shape with a magnifying glass (10–20× magnification). An irregular-shaped orifice will affect instrument performance.
- Thoroughly dry cones before installing them back in the instrument because water/solvent could be pulled back into the mass spectrometer.
- Check coolant in recirculating system for signs of interface corrosion such as copper or aluminum salts (or predominant metal of interface).

ION OPTICS

The ion optic system is usually positioned just behind or close to the skimmer cone to take advantage of the maximum number of ions entering the mass spectrometer. There are many different commercial designs and layouts, but they all have one attribute in common, and that is to transport the maximum number of analyte ions while allowing the minimum number of matrix ions through to the mass analyzer.

The ion-focusing system is not traditionally thought of as a component that needs frequent inspection, but because of its proximity to the interface region, it can accumulate minute particulates and neutral species that over time can dislodge, find their way into the mass analyzer, and affect instrument performance. Signs of a dirty or contaminated ion optic system are poor stability or a need to gradually increase lens voltages over time. For that reason, no matter what design of ion optics is used, inspection and cleaning every 3–6 months (depending on workload and sample type) should be an integral part of a preventative maintenance plan. Some useful maintenance tips for the ion optics to ensure maximum ion transmission and good stability include the following:

- Look for sensitivity loss over time, especially in complex matrices.
- If sensitivity is still low after cleaning the sample introduction system, torch, and interface cones, it could indicate that the ion lens system is becoming dirty.
- Try retuning or reoptimizing the lens voltages.
- If voltages are significantly different (usually higher than previous settings), it probably means lens components are getting dirty.
- When the lens voltages become unacceptably high, the ion lens system will probably need replacing or cleaning. Use recommended procedures outlined in the operator's manual.
- Depending on the design of the ion optics, some single-lens systems are considered consumables and are discarded after a period of time, whereas multicomponent lens systems are usually cleaned using abrasive papers or polishing compounds, and rinsed with water and an organic solvent.
- If cleaning ion optics, make sure they are thoroughly dry because water or solvent could be sucked back into the mass spectrometer.
- Gloves are usually recommended when reinstalling an ion optic system because of the possibility of contamination.
- Do not forget to inspect or replace O-rings or seals when replacing ion optics.
- Depending on instrument workload, you should expect to see some deterioration in the performance of the ion lens system after 3–4 months of use. This is a good approximation of when it should be inspected and cleaned or replaced if necessary.
- With some instruments, you will need to break the vacuum to get to the ion optic region. Even though vacuum can be re-established very quickly, this should be a consideration when carrying out your own ion lens cleaning procedures.

ROUGHING PUMPS

Typically, two roughing pumps are used in commercial instruments. One pump is used on the interface region, and the other is used as a backup to the turbomolecular pumps on the main vacuum chamber. They are usually oil-based rotary or diffusion pumps, where the oil needs to be changed on a regular basis, depending on the instrument usage. The oil in the interface pump will need changing more often than the one on the main vacuum chamber because it is pumping for a longer period. A good indication of when the oil needs to be changed is the color in the "viewing glass." If it appears dark brown, there is a good chance that heat has degraded its lubricating properties, and it needs to be changed. With the roughing pump on the interface, the oil should be changed every 1–2 months, and with the main vacuum chamber pump, it should be changed every 3–6 months. These times are only approximations and will vary depending on the sample workload and the time the instrument is actually running. Some important tips when changing the roughing pump oil:

- Do not forget to turn the instrument and the vacuum off. If the oil is being changed from "cold," it might be useful to run the instrument for 10–15 min beforehand to get the oil to flow better.
- Drain the oil into a suitable vessel; caution, the oil might be very hot if the instrument has been running all day.
- Fill the oil to the required level in the "viewing glass."
- Check for any loose hose connections.
- Replace oil filter if necessary.
- Turn the instrument back on. Check for any oil leaks around filling cap, and tighten if necessary.

AIR FILTERS

Most of the electronic components, especially the ones in the RF generator, are air-cooled. Therefore, the air filters should be checked, cleaned, or replaced on a fairly regular basis. Although this is not carried out as routinely as the sample introduction system, a typical time frame to inspect the air filters is every 3–6 months, depending on the workload and instrument usage.

OTHER COMPONENTS TO BE PERIODICALLY CHECKED

It is also important to emphasize that other components of the ICP mass spectrometer have a finite lifetime, and will need to be replaced or at least inspected from time to time. These components are not considered a part of the routine maintenance schedule, and usually require a service engineer (or at least an experienced user) to clean or to change them. These areas to be cleaned are described in the following text.

THE DETECTOR

Depending on the usage and levels of ion signals measured on a routine basis, the electron multiplier should last about 12 months. A sign of a failing detector is a rapid decrease in the "gain" setting despite attempts to increase the detector voltage. The lifetime of a detector can be increased by avoiding measurements at masses that produce extremely high ion signals, such as those associated with the argon gas, solvent or acid used to dissolve the sample (e.g., hydrogen, oxygen, and nitrogen), or any mass associated with the matrix itself. It is important to emphasize that the detector should be replaced by an experienced person wearing gloves, to reduce the possibility of contamination from grease or organic/water vapor from the operator's hands. It is advisable that a spare detector be purchased with the instrument.

TURBOMOLECULAR PUMPS

Most of the instruments running today use two turbomolecular pumps to create the operating vacuum for the main mass analyzer/detector chamber and the ion optic region. However, some of the newer instruments use a single, twin-throated turbo pump. The lifetime of turbo pumps, in general, is dependant on a number of factors, including the pumping capacity of the pump (usually expressed as L/s), the size (or volume) of the vacuum chamber to be pumped, the orifice diameter of the interface cones (in mm), and the time the instrument is running. Although some instruments still use the same turbo pumps after 5–10 years of operation, the normal lifetime of a pump in an instrument that has a reasonably high sample workload is on the order of 3–4 years. This is an approximation and will obviously vary depending on the make and design of the pump (especially the type of bearings used). As the turbomolecular pump is one of the most expensive components of an ICP-MS system, this should be factored into the overall running costs of the instrument over its operating lifetime.

It is worth pointing out that although the turbo pump is not generally included in routine maintenance, most instruments use a "Penning" (or similar) gauge to monitor the vacuum in the main chamber. Unfortunately, this gauge can become dirty over time and lose its ability to measure the correct pressure. The frequency of this is almost impossible to predict but is closely related to types and numbers of samples analyzed. A sudden drop in pressure or fluctuations in the signal are two of the most common indications of a dirty Penning gauge. When this happens, the gauge must be removed and cleaned. This should be performed by an experienced operator or service engineer because removing the gauge, cleaning it, maintaining the correct electrode geometry, and reinstalling it correctly into the instrument is a fairly complicated procedure. It is further complicated by the fact that a Penning gauge is operated at high voltage.

MASS ANALYZER

Under normal circumstances, there is no need for the operator to be concerned about routine maintenance of the mass analyzer. With modern turbomolecular pumping systems, it is highly unlikely there will be any pump- or sample-related contamination problems associated with the quadrupole, magnetic sector, or TOF mass analyzer. This certainly was not the case with some of the early instruments that used oil-based diffusion pumps, because many researchers found that the quadrupole and prefilters were contaminated by oil vapors from the pumps. Today, it is fairly common for turbomolecular-based mass analyzers to require no maintenance of the quadrupole rods over the lifetime of the instrument, other than an inspection carried out by a service engineer on an annual basis. However, in extreme cases, particularly with older instruments, removal and cleaning of the quadrupole assembly might be required to get acceptable peak resolution and abundance sensitivity performance.

The overriding message I would like to leave you with on this subject is that routine maintenance cannot be overemphasized in ICP-MS. Even though it might be considered a mundane and time-consuming chore, it can have a significant impact on the uptime of your instrument. Read the routine maintenance section of the operator's manual and understand what is required. It is essential that time be scheduled on a weekly, monthly, and quarterly basis for preventative maintenance on your instrument. In addition, you should budget for an annual preventative maintenance contract under which the service engineer checks out all the important instrumental components and systems on a regular basis to make sure they are all working correctly. This might not be as critical if you work in an academic environment, where the instrument might be down for extended periods, but in my opinion, it is absolutely critical if you work in commercial laboratory, which is using the instrument to generate revenue. There is no question that spending the time to keep your ICP mass spectrometer in good working order can mean the difference between owning an instrument whose performance could be slowly degrading without your knowledge or one that is always working in "peak" condition.

17 Alternative Sample Introduction Techniques

Conventional sample introduction systems using a spray chamber and nebulizer account for the majority of ICP-MS applications being carried out today. However, nonstandard sampling accessories such as laser ablation systems, flow injection analyzers, electrothermal vaporizers, cooled spray chambers, desolvation equipment, direct injection nebulizers, and automated sample delivery systems and dilutors are considered critical to enhancing the practical capabilities of the technique. Initially regarded as novel sampling devices, they have since proved themselves to be invaluable for solving real-world application problems. Chapter 17 describes the basic principles of these accessories and gives an overview of their practical capabilities.

It is recognized that standard ICP-MS instrumentation using a traditional sample introduction system comprising a spray chamber and nebulizer has certain limitations, particularly when it comes to the analysis of complex samples. Some of these known limitations include the following:

- Inability to analyze solids.
- Contamination issues with samples requiring multiple sample preparation steps.
- Liquid aerosol can impact ionization process.
- Total dissolved solids must be kept below 0.2%.
- If matrix has to be removed, it has to be done offline.
- Long washout times required for samples with a heavy matrix.
- Dilutions and addition of internal standards can be labor intensive and time consuming.
- Matrix components can generate severe spectral overlaps on many analytes.
- The analysis of slurries is very difficult.
- Matrix suppression can be quite severe with some samples.
- Spectral interferences generated by solvent-induced species can limit detection capability.
- Organic solvents can present unique problems.
- Sample throughput is limited by the sample introduction process.
- Not suitable for the determination of elemental species or oxidation states.

Such were the demands of real-world users to overcome these kinds of problem areas that instrument manufacturers devised different strategies based on the type of samples being analyzed. Some of these strategies involved parameter optimization or modification of instrument components, but it was clear that this approach alone was not going to solve every conceivable problem. For this reason, they turned

163

their attention to the development of sampling accessories, which were optimized for a particular application problem or sample type. Over the past 10–15 years, this demand has led to the commercialization of specialized sample introduction tools, not only by the instrument manufacturers themselves, but also by third-party vendors specializing in these kinds of accessories. The most common ones used today include the following:

- Laser ablation systems (LA)
- Flow injection analyzers (FIA)
- Electrothermal vaporizors (ETV)
- Chilled spray chambers and desolvation systems
- Direct injection nebulizers (DIN)
- Fast automated sampling procedures
- Chromatographic separation equipment

Let us now take a closer look at some of these techniques to understand their basic principles and what benefits they bring to ICP-MS.

LASER ABLATION

The limitation of ICP-MS in analyzing solid materials (without the need for wet chemical dissolution/digestion methods) led to the development of laser ablation. The principle behind this approach is the use of a high-powered laser to ablate the surface of a solid and sweep the sample aerosol into the ICP mass spectrometer for analysis in the conventional way.[1]

Before I go on to describe some typical applications suited to laser ablation ICP-MS, let us first take a brief look at the history of analytical lasers and how they eventually became such a useful sampling tool. The use of lasers as vaporization devices was first investigated in the early 1960s. When light energy with an extremely high-power density interacts with a solid material, the photon-induced energy is converted into thermal energy, resulting in vaporization and removal of the material from the surface of the solid.[2] Some of the early researchers used ruby lasers to induce a plasma discharge on the surface of the sample and measure the emitted light with an atomic emission spectrometer.[3] Although this proved useful for certain applications, the technique suffered from low sensitivity, poor precision, and severe matrix effects caused by nonreproducible excitation characteristics. Over the years, various improvements were made to this basic design with very little success,[4] because the sampling process and the ionization/excitation process (both under vacuum) were still intimately connected and interacted strongly with each other.

This limitation led to the development of laser ablation as a sampling device for atomic spectroscopy instrumentation, where the sampling step was completely separated from the excitation or ionization step. The major benefit is that each step can be independently controlled and optimized. These early devices used a high-energy laser to ablate the surface of a solid sample, and the resulting aerosol was swept into some kind of atomic spectrometer for analysis. Although initially used with atomic absorption[5,6] and plasma-based emission techniques,[7,8] it was not until the mid-1980s, when lasers were coupled with ICP-MS, that the analytical community sat up and took notice.[9] For the first time, researchers were coming up with evidence that virtually any type of solid could be vaporized, irrespective of electrical characteristics, surface topography, size, or shape, and be transported into the ICP for analysis by atomic emission or mass spectrometry. This was an exciting breakthrough for ICP-MS, because it meant the technique could be used for the bulk sampling of solids, or if required, for the analysis of small spots or microinclusions, in addition to being used for the analysis of solutions.

COMMERCIAL SYSTEMS FOR ICP-MS

The first laser ablation systems developed for ICP instrumentation were based on solid-state ruby lasers, operating at 694 nm. These were developed in the early 1980s, but did not prove to be successful for a number of reasons, including poor stability, low power density, low repetition rate, and large beam diameter, which made them limited in their scope and flexibility as a sample introduction device for trace element analysis. It was at least another 5 years before any commercial instrumentation became available. These early commercial laser ablation systems, which were specifically developed for ICP-MS, used the Nd:YAG (neodymium doped yttrium aluminum garnet) design, operated at the primary wavelength of 1064 nm—in the infrared.[10] They initially showed a great deal of promise because analysts were finally able to determine trace levels directly in the solid without sample dissolution. However, it soon became apparent that they did not meet the expectations of the analytical community, for many reasons, including complex ablation characteristics, poor precision, nonoptimization for microanalysis and, because of poor laser coupling, they were unsuitable for many types of solids. By the early 1990s, most of the LAs purchased were viewed as novel and interesting, but not suited to solving real-world application problems.

These basic limitations in IR laser technology led researchers to investigate the benefits of shorter wavelengths. Systems were developed that were based on Nd:YAG technology at the 1064 nm primary wavelength, but utilizing optical components to double (532 nm), quadruple (266 nm), and quintuple (213 nm) the frequency. Innovations in lasing materials and electronic design together with better thermal characteristics produced higher energy with higher pulse-to-pulse stability. These more advanced UV lasers showed significant improvements, particularly in the area of coupling efficiency, making them more suitable for a wider array of sample types. In addition, the use of higher-quality optics allowed for a more homogeneous laser beam profile, which provided the optimum energy density to couple with the sample matrix. This resulted in the ability to make spots much smaller and with more controlled ablations irrespective of sample material, which were critical for the analysis of surface defects, spots, and microinclusions. Figure 17.1 shows the optical layout of a commercially available frequency-quintupled 213 nm Nd:YAG laser ablation system.

EXCIMER LASERS

The successful trend toward shorter wavelengths and the improvements in the quality of optical components also drove the development of UV gas-filled lasers, such as XeCl (308 nm), KrF (248 nm), and ArF (193 nm) excimer lasers. These showed

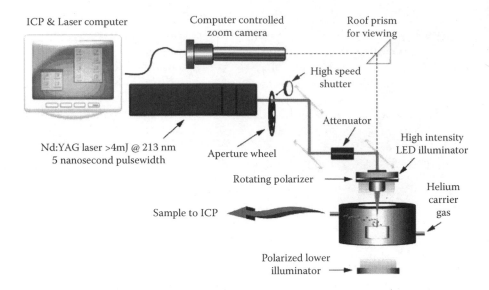

FIGURE 17.1 Schematic of a commercially available frequency-quintupled 213 nm Nd:YAG laser ablation system (courtesy of Cetac Technologies).

great promise, especially the ones that operated at shorter wavelengths and were specifically designed for ICP-MS. Unfortunately, they necessitated a more sophisticated beam delivery system, which tended to make them more expensive. In addition, the complex nature of the optics and the fact that gases had to be changed on a routine basis made them a little more difficult to use and maintain, and as a result required a more skilled operator to run them. However, their complexity was far outweighed by their better absorption capabilities for UV-transparent materials such as calcites, fluorites, and silicates, smaller particle size, and higher flow of ablated material. There was also evidence to suggest that the shorter-wavelength excimer lasers exhibit better elemental fractionation characteristics (typically defined as the intensity of certain elements varying with time, relative to the dry aerosol volume) than the longer-wavelength Nd:YAG design, because they produce smaller particles that are easier to volatilize.

BENEFITS OF LASER ABLATION FOR ICP-MS

Today there are a number of commercial laser ablation systems on the market designed specifically for ICP-MS, including 266 nm and 213 nm Nd:YAG and 193 nm ArF excimer lasers. They all have varying output energy, power density, and beam profiles, and even though each one has different ablation characteristics, they all work extremely well, depending on the types of samples being analyzed and the data quality requirements. Laser ablation is now considered a very reliable sampling technique for ICP-MS, which is capable of producing data of the very highest quality directly on solid samples and powders. Some of the many benefits offered by this technique include the following:

- Direct analysis of solids without dissolution.
- Ability to analyze virtually any kind of solid material, including rocks, minerals, metals, ceramics, polymers, plastics, plant material, and biological specimens.
- Ability to analyze a wide variety of powders by pelletizing with a binding agent.
- No requirement for sample to be electrically conductive.
- Sensitivity in the ppb to ppt range, directly in the solid.
- Labor-intensive sample preparation steps are eliminated, especially for samples such as plastics and ceramics that are extremely difficult to get into solution.
- Contamination is minimized because there are no digestion/dilution steps.
- Reduced polyatomic spectral interferences compared to solution nebulization.
- Examination of small spots, inclusions, defects, or microfeatures on surface of sample.
- Elemental mapping across the surface of a mineral.
- Depth profiling to characterize thin films or coatings.

Let us now take a closer look at the strengths and weaknesses of the different laser designs with respect to application requirements.

OPTIMUM LASER DESIGN BASED ON APPLICATION REQUIREMENTS

The commercial success of laser ablation was initially driven by its ability to directly analyze solid materials such as rocks, minerals, ceramics, plastics, and metals, without going through a sample dissolution stage. Table 17.1 represents some typical multielement detection limits in NIST 612 glass generated with a 266 nm Nd:YAG design coupled to an ICP-MS system.[11] It can be seen that sub-ppb detection limits in the solid material are achievable for most of the elements. This kind of performance is typically obtained using larger spot sizes on the order of 100–1000 μm in diameter, which is ideally suited to 266 nm laser technology.

However, the desire for ultratrace analysis of optically challenging materials, such as calcite, quartz, glass, and fluorite, combined with the capability to characterize small spots and microinclusions, proved very challenging for the 266 nm design. The major reason is that the ablation process is not very controlled and precise, and as a result it is difficult to ablate a minute area without removing some of the surrounding material. In addition, erratic ablating of the sample initially generates larger particles (>1 μm size), which are not efficiently ionized in the plasma and therefore contribute to poor precision.[12] Even though modifications helped improve ablation behavior, it was not totally successful because of the basic limitation of the 266 nm to couple efficiently to UV-transparent materials. The drawbacks in 266 nm technology eventually led to the development of 213 nm lasers[13] because of the recognized superiority of shorter wavelengths to exhibit a higher degree of absorbance in transparent materials.[14]

Analytical chemists, particularly in the geochemical community, welcomed 213 nm UV lasers with great enthusiasm, because they now had a sampling tool

TABLE 17.1

Typical Detection Limits Achievable in NIST 612 SRM Glass Using a 266 nm Nd:YAG Laser Ablation System Coupled to an ICP Mass Spectrometer

Element	3σ DLs (ppb)	Element	3σ DLs (ppb)
B	3.0	Ce	0.05
Sc	3.4	Pr	0.05
Ti	9.1	Nd	0.5
V	0.4	Sm	0.1
Fe	13.6	Eu	0.1
Co	0.05	Gd	1.5
Ni	0.7	Dy	0.5
Ga	0.2	Ho	0.01
Rb	0.1	Er	0.2
Sr	0.07	Yb	0.4
Y	0.04	Lu	0.04
Zr	0.2	Hf	0.4
Nb	0.5	Ta	0.1
Cs	0.2	Th	0.02
Ba	0.04	U	0.02
La	0.05		

Source: Courtesy of Cetac Technologies.

that offered much better control of the ablation process, even for easily fractured minerals. This is demonstrated in Figures 17.2 and 17.3, which show the ablation differences between 266 and 213 nm, respectively, for NIST 612 glass SRM. It can be seen that the 200 μm ablation crater produced with the 266 nm laser is irregular and shows redeposited ablated material around the edges of the crater, whereas the crater with the 213 nm system is very clean and symmetrical, with no ablated material around the edges. The absence of any redeposited material with the 213 nm laser means that a higher proportion of ablated material actually makes it to the plasma. Both craters are shown at 10× magnification.

This significant difference in crater geometry between the two systems is predominantly a result of the effective absorption of laser energy and the difference in the delivered power per unit area—also known as laser irradiance (or fluence per laser pulse width).[15] The result is a difference in depth penetration and the size/volume of particles reaching the plasma. With the 266 nm laser system, a high-volume burst of material is initially observed, whereas with the 213 nm laser, the signal gradually increases and levels off quickly, indicating a more consistent stream of small particles being delivered to the plasma and the mass spectrometer. Therefore, when analyzing this type of mineral with the 266 nm design, it is typical that the first 100 to 200 shots of the ablation process are filtered out to ensure that no data are taken during

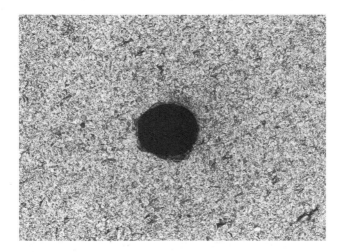

FIGURE 17.2 A 200 μm crater produced by the ablation of a NIST 612 glass SRM, using a 266 nm laser ablation system, showing excess ablated material around the edges of the crater (courtesy of Cetac Technologies).

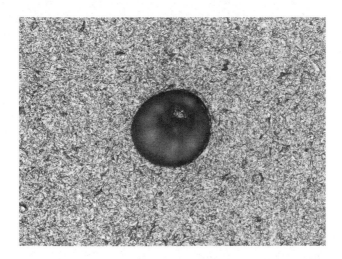

FIGURE 17.3 A 200 μm crater produced by the ablation of a NIST 612 glass SRM using a 213 nm laser system, showing a symmetrical, well-defined crater (courtesy of Cetac Technologies).

the initial burst of material. This can be somewhat problematic when analyzing small spots or inclusions, because of the limited amount of sample being ablated.

The benefits of 213 nm lasers emphasize that matrix independence, high spatial resolution, and the ability to couple with UV-transparent materials without fracturing (particularly for small spots or depth analysis studies) was very important for geochemical-type applications. These findings led researchers to study even shorter wavelengths, in particular, 193 nm ArF excimer technology. Besides their accepted

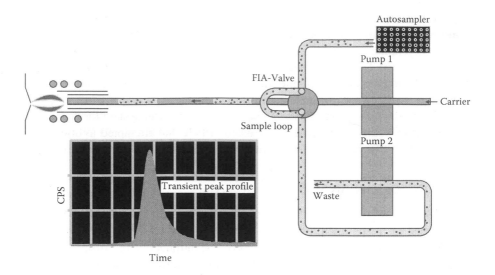

FIGURE 17.5 Schematic of a flow injection system used for the process of microsampling.

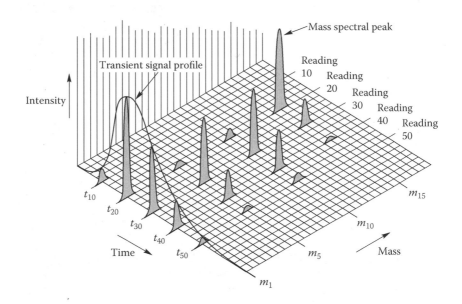

FIGURE 17.6 3D plot of intensity versus mass in the time domain for the determination of a group of elements in a transient peak (copyright © 2003–2007, all rights reserved, Perkin-Elmer Inc.).

detail in Figure 17.6, which shows a 3D transient plot of intensity versus mass in the time domain for the determination of a group of elements.

Some of the many on-line procedures that are applicable to FI-ICP-MS include the following:

- Microsampling for improved stability with heavy matrices[26]
- Automatic dilution of samples/standards[27]
- Standards addition[28]
- Cold vapor and hydride generation for enhanced detection capability for elements such as Hg, As, Sb, Bi, Te, and Se[29]
- Matrix separation and analyte preconcentration using ion exchange procedures[30]
- Elemental speciation[31]

Flow injection coupled to ICP-MS has shown itself to be very diverse and flexible in meeting the demands presented by complex samples, as indicated in the foregoing references. However, one of the most interesting areas of research is in the direct analysis of seawater by flow injection ICP-MS. Traditionally, seawater is very difficult to analyze by ICP-MS because of two major problems. First, the high NaCl content will block the sampler cone orifice over time, unless a 10–20-fold dilution is made of the sample. This is not such a major problem with coastal waters, because the levels are high enough. However, if the sample is open-ocean seawater, this is not an option, because the trace metals are at a much lower level. The other difficulty associated with the analysis of seawater is that ions from the water, chloride matrix, and the plasma gas can combine to generate polyatomic spectral interferences, which are a problem, particularly for the first-row transition metals.

Attempts have been made over the years to remove the NaCl matrix and preconcentrate the analytes using various types of chromatography and ion exchange column technology. One such early approach was to use an HPLC system coupled to an ICP mass spectrometer utilizing a column packed with silica-immobilized 8-hydroxyquinoline.[32] This worked reasonably well, but was not considered a routine method, because silica-immobilized 8-hydroxyquinoline was not commercially available, and also spectral interferences produced by HCl and HNO_3 (used to elute the analytes) precluded determination of a number of the elements, such as Cu, As, and V. More recently, chelating agents based on the iminodiacetate acid functionality group have gained wider success, but are still not considered truly routine for a number of reasons, including the necessity for calibration using standard additions, the requirement of large volumes of buffer to wash the column after loading the sample, and the need for conditioning between samples because some ion exchange resins swell with changes in pH.[33–35]

However, a research group at the NRC in Canada has developed a very practical on-line approach, using a flow injection sampling system coupled to an ICP mass spectrometer.[30] Using a special formulation of a commercially available, iminodiacetate ion exchange resin (with a macroporous methacrylate backbone), trace elements can be separated from the high concentrations of matrix components in the seawater, with a pH 5.2 buffered solution. The trace metals are subsequently eluted into the plasma with 1M HNO_3, after the column has been washed out with deionized water. The column material has sufficient selectivity and capacity to allow accurate determinations at ppt levels using simple aqueous standards, even for elements such as V and Cu, which are notoriously difficult in a chloride matrix. This can be seen in Figure 17.7, which shows spectral scans for a selected group of elements in a certified

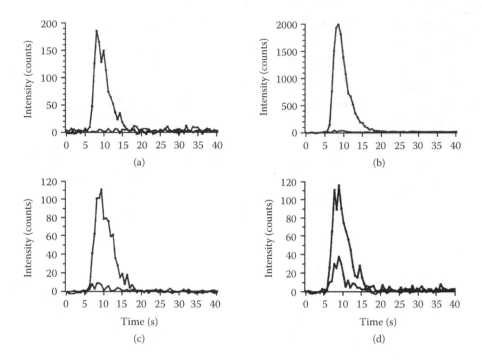

FIGURE 17.7   Analyte and blank spectral scans of (a) Co, (b) Cu, (c) Cd, and (d) Pb in NASS-4 open-ocean seawater certified reference material, using flow injection coupled to ICP-MS. (From S. N. Willie, Y. Iida and J. W. McLaren, *Atomic Spectroscopy*, **19**[3], 67, 1998.)

reference material open-ocean seawater sample (NASS-4), and Table 17.2, which compares the results for the this methodology with the certified values, together with the limits of detection (LOD). Using this on-line method, the turnaround time is less than 4 min per sample, which is considerably faster than other high-pressure chelation techniques reported in the literature.

ELECTROTHERMAL VAPORIZATION

ETA for use with AA has proved to be a very sensitive technique for trace element analysis over the last three decades. However, the possibility of using the atomization/heating device for ETV sample introduction into an ICP mass spectrometer was identified in the late 1980s.[36] The ETV sampling process relies on the basic principle that a carbon furnace or metal filament can be used to thermally separate the analytes from the matrix components and then sweep them into the ICP mass spectrometer for analysis. This is achieved by injecting a small amount of the sample (usually 20–50 μL via an autosampler) into a graphite tube or onto a metal filament. After the sample is introduced, drying, charring, and vaporization are achieved by slowly heating the graphite tube/metal filament. The sample material is vaporized into a flowing stream of carrier gas, which passes through the furnace or over the filament

TABLE 17.2

Analytical Results for NASS-4 Open-Ocean Seawater Certified Reference Material, Using Flow Injection ICP-MS Methodology

		NASS-4 (ppb)	
Isotope	LOD (ppt)	Determined	Certified
$^{51}V^+$	4.3	1.20 ± 0.04	Not certified
$^{63}Cu^+$	1.2	0.210 ± 0.008	0.228 ± 0.011
$^{60}Ni^+$	5	0.227 ± 0.027	0.228 ± 0.009
$^{66}Zn^+$	9	0.139 ± 0.017	0.115 ± 0.018
$^{55}Mn^+$	Not reported	0.338 ± 0.023	0.380 ± 0.023
$^{59}Co^+$	0.5	0.0086 ± 0.0011	0.009 ± 0.001
$^{208}Pb^+$	1.2	0.0090 ± 0.0014	0.013 ± 0.005
$^{114}Cd^+$	0.7	0.0149 ± 0.0014	0.016 ± 0.003

Source: From S. N. Willie, Y. Iida and J. W. McLaren, *Atomic Spectroscopy*, **19**(3) 67, 1998.

during the heating cycle. The analyte vapor recondenses in the carrier gas and is then swept into the plasma for ionization.

One of the attractive characteristics of ETV for ICP-MS is that the vaporization and ionization steps are carried out separately, which allows for the optimization of each process. This is particularly true when a heated graphite tube is used as the vaporization device, because the analyst typically has more control of the heating process and as a result can modify the sample by means of a very precise thermal program before it is introduced to the ICP for ionization. By boiling off and sweeping the solvent and volatile matrix components out of the graphite tube, spectral interferences arising from the sample matrix can be reduced or eliminated. The ETV sampling process consists of six discrete stages: sample introduction, drying, charring (matrix removal), vaporization, condensation, and transport. Once the sample has been introduced, the graphite tube is slowly heated to drive off the solvent. Opposed gas flows, entering from each end of the graphite tube, and then purges the sample cell by forcing the evolving vapors out the dosing hole. As the temperature increases, volatile matrix components are vented during the charring steps. Just prior to vaporization, the gas flows within the sample cell are changed. The central channel (nebulizer) gas then enters from one end of the furnace, passes through the tube, and exits out of the other end. The sample-dosing hole is then automatically closed, usually by means of a graphite tip, to ensure no analyte vapors escape. After this gas flow pattern has been established, the temperature of the graphite tube is ramped up very quickly, vaporizing the residual components of the sample. The vaporized analytes either recondense in the rapidly moving gas stream or remain in the vapor phase. These particulates and vapors are then transported to the ICP in the carrier gas, where they are ionized by the ICP for analysis in the mass spectrometer.

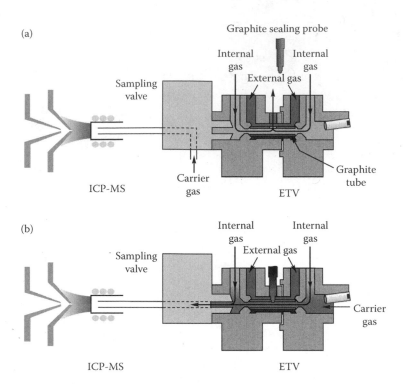

FIGURE 17.8 A graphite furnace ETV sampling device for ICP-MS, showing the two distinct steps of sample pretreatment (a) and vaporization (b) into the plasma (copyright © 2003–2007, all rights reserved, PerkinElmer Inc.).

Another benefit of decoupling the sampling and ionization processes is the opportunity for chemical modification of the sample. The graphite furnace itself can serve as a high-temperature reaction vessel where the chemical nature of compounds within it can be altered. In a manner similar to that used in atomic absorption, chemical modifiers can change the volatility of species to enhance matrix removal and increase elemental sensitivity.[37] An alternative gas such as oxygen may also be introduced into the sample cell to aid in the charring of the carbon in organic matrices such as biological or petrochemical samples. Here, the organically bound carbon reacts with the oxygen gas to produce CO_2, which is then vented from the system. A typical ETV sampling device, showing the two major steps of sample pretreatment (drying and ashing) and vaporization into the plasma, is seen schematically in Figure 17.8.

Over the past 15 years, ETV sampling for ICP-MS has mainly been used for the analysis of complex matrices including geological materials,[38] biological fluids,[39] seawater,[40] and coal slurries,[41] which have proved difficult or impossible by conventional nebulization. By removal of the matrix components, the potential for severe spectral and matrix-induced interferences is dramatically reduced. Even though ETV-ICP-MS was initially applied to the analysis of very small sample volumes, the advent of low-flow nebulizers has limited its use for this type of work.

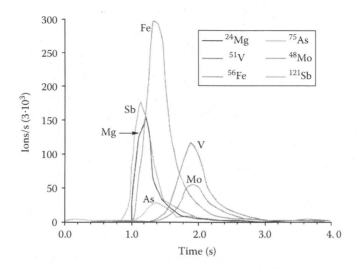

FIGURE 17.9 A temporal display of 50 pg of Mg, Sb, As, Fe, V, and Mo in 37% hydrochloric acid by ETV-ICP-MS. (From S. A. Beres, E. R. Denoyer, R. Thomas, P. Bruckner, *Spectroscopy*, **9**(1), 20–26, 1994.)

An example of the benefit of ETV sampling is in the analysis of samples containing high concentrations of mineral acids such as HCl, HNO_3, and H_2SO_4. Besides physically suppressing analyte signals, these acids generate massive polyatomic spectral overlaps, which interfere with many analytes, including As, V, Fe, K, Si, Zn, and Ti. By carefully removing the matrix components with the ETV device, the determination of these elements becomes relatively straightforward. This is illustrated in Figure 17.9, which shows a spectral display in the time domain for 50 pg spikes of a selected group of elements in concentrated hydrochloric acid (37% w/w), using a graphite-furnace-based ETV-ICP-MS.[42] It can be seen in particular that good sensitivity is obtained for $^{51}V^+$, $^{56}Fe^+$, and $^{75}As^+$, which would have been virtually impossible by direct aspiration because of spectral overlaps from $^{39}ArH^+$, $^{35}Cl^{16}O^+$, $^{40}Ar^{16}O^+$, and $^{40}Ar^{35}Cl^+$, respectively. The removal of the chloride and water from the matrix translates into ppt detection limits directly in 37% HCl, as shown in Table 17.3.

It can also be seen in Figure 17.9 that the elements are vaporized off the graphite tube in order of their boiling points. In other words, magnesium, which is the most volatile, is driven off first, whereas V and Mo, which are the most refractory, come off last. However, even though they emerge at different times, the complete transient event lasts less than 3 s. This physical time limitation, imposed by the duration of the transient signal, makes it imperative that all isotopes of interest be

TABLE 17.3

Detection Limits for V, Fe, and As in 37% Hydrochloric Acid by ETV-ICP-MS

Element	DL (ppt)
$^{51}V^+$	50
$^{56}Fe^+$	20
$^{75}As^+$	40

Source: S. A. Beres, E. R. Denoyer, R. Thomas, P. Bruckner; *Spectroscopy*, **9**(1), 20–26, 1994.

measured under the highest signal-to-noise conditions throughout the entire event. The rapid nature of the transient has also limited the usefulness of ETV sampling for routine multielement analysis, because realistically only a small number of elements can be quantified with good accuracy and precision in less than 3 s. In addition, the development of low-flow nebulizers, desolvation devices, cool plasma technology, and collision/reaction cells and interfaces has meant that multielement analysis can now be carried out on difficult matrices without the need for ETV sample introduction.

CHILLED SPRAY CHAMBERS AND DESOLVATION DEVICES

Chilled/cooled spray chambers and desolvation devices are becoming more and more common in ICP-MS, primarily to cut down the amount of liquid entering the plasma in order to reduce the severity of the solvent-induced spectral interferences such as oxides, hydrides, hydroxides, and argon/solvent-based polyatomic interferences. They are very useful for aqueous-type samples, but probably more important for volatile organic solvents, because there is a strong possibility that the sample aerosol would extinguish the plasma unless modifications are made to the sampling procedure. The most common chilled spray chambers and desolvation systems being used today include the following:

- Water-cooled spray chambers
- Peltier-cooled spray chambers
- Ultrasonic nebulizers (USN)
- Ultrasonic nebulizers (USN) coupled with membrane desolvation
- Specialized microflow nebulizers coupled with desolvation techniques

Let us take a closer look at these devices.

WATER-COOLED AND PELTIER-COOLED SPRAY CHAMBERS

Water-cooled spray chambers have been used in ICP-MS for many years and are standard on a number of today's commercial instrumentation to reduce the amount of water or solvent entering the plasma. However, the trend today is to cool the sample using a thermoelectric device called a Peltier cooler. Thermoelectric cooling (or heating) uses the principle of generating a hot or cold environment by creating a temperature gradient between two different materials. It uses electrical energy via a solid-state heat pump to transfer heat from a material on one side of the device to a different material on the other side, thus producing a temperature gradient across the device (similar to a household air conditioning system). Peltier cooling devices, which are typically air-cooled (but water cooling is an option), can be used with any kind of spray chamber and nebulizer, but commercial products for use with ICP-MS are normally equipped with a cyclonic spray chamber and a low-flow pneumatic nebulizer.

    The main purpose of cooling the sample aerosol is to reduce the amount of water or solvent entering the plasma by lowering the temperature of the spray chamber. This can be a few degrees below ambient or as low −20°C, depending on the type of samples being analyzed. This can have a three-fold effect: First, it helps to minimize solvent-based spectral interferences, such as oxides and hydroxides formed in the plasma, and second, because very little plasma energy is needed to vaporize the solvent, it allows more energy to be available to excite and ionize the analyte ions. There is also evidence to suggest that cooling the spray chamber will help minimize signal drift due to external environmental temperature changes in the laboratory.

Cooling the spray chamber to as low as −20°C by either Peltier cooling or a recirculating system using ethylene glycol as the coolant is particularly useful when it comes to analyzing some volatile organic samples. It has the effect of reducing the amount of organic solvent entering the interface, and when combined with the addition of a small amount of oxygen into the nebulizer gas flow, it is beneficial in reducing the buildup of carbon deposits on the sampler cone orifice and also minimizing the problematic carbon-based spectral interferences.[43]

ULTRASONIC NEBULIZERS

Ultrasonic nebulization was first developed in the late 1980s for use with ICP optical emission.[44] Its major benefit was that it offered an approximately 10× improvement in detection limits, because of its more efficient aerosol generation. However, this was not such an obvious benefit for ICP-MS, because more matrix entered the system compared to a conventional nebulizer, increasing the potential for signal drift, matrix suppression, and spectral interferences. This was not such a major problem for simple aqueous-type samples, but was problematic for real-world matrices. The elements that showed the most improvement were the ones that benefited from lower solvent-based spectral interferences. Unfortunately, many of the other elements exhibited higher background levels, and as a result showed no significant improvement in detection limit. In addition, the increased amount of matrix entering the mass spectrometer usually necessitated the need for larger dilutions of the sample, which again negated the benefit of using a USN with ICP-MS for samples with a heavier matrix. This limitation led to the development of an ultrasonic nebulizer fitted with an additional membrane desolvator. This design virtually removed all the solvent from the sample, which dramatically improved detection limits for a large number of the problematic elements and also lowered metal oxide levels by at least an order of magnitude.[45]

The principle of aerosol generation using an ultrasonic nebulizer is based on a sample being pumped onto a quartz plate of a piezoelectric transducer. Electrical energy of 1–2 MHz frequency is coupled to the transducer, which causes it to vibrate at high frequency. These vibrations disperse the sample into a fine-droplet aerosol, which is carried in a stream of argon. With a conventional ultrasonic nebulizer, the aerosol is passed through a heating tube and a cooling chamber, where most of the sample solvent is removed as a condensate before it enters the plasma. If a membrane desolvation system is fitted to the ultrasonic nebulizer, it is positioned after the cooling unit. The sample aerosol enters the membrane desolvator, where the remaining solvent vapor passes through the walls of a tubular microporous membrane. A flow of argon gas removes the volatile vapor from the exterior of the membrane, while the analyte aerosol remains inside the tube and is carried into the plasma for

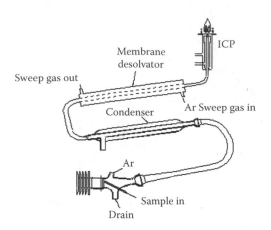

FIGURE 17.10   Schematic of an ultrasonic nebulizer fitted with a membrane desolvation system (courtesy of Cetac Technologies).

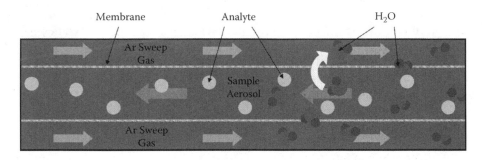

FIGURE 17.11   Principles of membrane desolvation showing the water molecules passing through a microporous membrane and being swept away by the argon gas, while the analyte is transported through the tube to the plasma (courtesy of Elemental Scientific Inc.).

ionization. Membrane desolvation systems also have the capability to add a secondary gas such as nitrogen, which has shown to be very beneficial in changing the ionization conditions to reduce levels of oxides in the plasma. The combination of membrane desolvation with an ultrasonic nebulizer can be seen more clearly in Figure 17.10, and Figure 17.11 shows the principles of membrane desolvation with water vapor as the solvent.

For ICP-MS, the system is best operated with both desolvation stages working, although for less demanding ICP-OES analysis, the membrane desolvation stage can be bypassed if required. The power of the system when coupled to an ICP mass spectrometer can be seen in Table 17.4, which compares the sensitivity (counts per second) and signal-to-background ratio of a membrane desolvation USN with a conventional cross-flow nebulizer for two classic solvent-based polyatomic interferences, $^{12}C^{16}O_2^+$ on $^{44}Ca^+$ and $^{40}Ar^{16}O^+$ on $^{56}Fe^+$, using a quadrupole ICP-MS system.

It can be seen that for the two analyte masses, the signal-to-background ratio is significantly better with the membrane-desolvated ultrasonic nebulizer than with

TABLE 17.4

Comparison of Sensitivity and Signal/Background Ratios for Two Analyte Masses—$^{44}Ca^+$, $^{56}Fe^+$—Using A Conventional Cross-Flow Nebulizer and A USN Fitted with A Membrane Desolvation System

Analytical Mass	Cross-Flow Nebulizer (cps)	Signal/BG	USN with Membrane Desolvation (cps)	Signal/BG
25 ppb $^{44}Ca^+$ (BG subtracted)	2,300	2,300/7,640	20,800	20,800/1730
$^{12}C^{16}O_2^+$ (BG)	7,640	= 0.30	1,730	= 12.0
10 ppb $^{56}Fe^+$ (BG subtracted)	95,400	95,400/868,00	262,000	262,000/8,200
$^{40}Ar^{16}O^+$ (BG)	868,000	= 0.11	8,200	= 32.0

Note: Signal/BG is calculated as the background subtracted signal divided by the background.
Source: Courtesy of Cetac Technologies.

the cross-flow design, which is a direct result of the reduction of the solvent-related spectral background levels. This approach is even more beneficial for the analysis of organic solvents because when they are analyzed by conventional nebulization, modifications have to be made to the sampling process, such as the addition of oxygen to the nebulizer gas flow, use of a low-flow nebulizer, and probably external cooling of the spray chamber. With a membrane desolvation USN system, volatile solvents such as isopropanol can be directly aspirated into the plasma with relative ease. However, it should be mentioned that depending on the sample type, this approach does not work for analytes that are bound to organic molecules. For example, the high volatility of certain mercury and boron organometallic species means that they could pass through the microporous PTFE membrane and never make it into the ICP-MS. In addition, samples with high dissolved solids, especially ones that are biological in nature, could possibly result in clogging the microporous membrane unless substantial dilutions are made. For these reasons, caution must be used when using a membrane desolvation system for the analysis of certain types of sample matrices.

SPECIALIZED MICROFLOW NEBULIZERS WITH DESOLVATION TECHNIQUES

Microflow or low-flow nebulizers, which were described in greater detail in Chapter 3, are being used more and more for routine applications. The most common ones used in ICP-MS are based on the microconcentric design, which operate at sample flows of 20–500 µL/min. Besides being ideal for small sample volumes, the major benefit of microconcentric nebulizers is that they are more efficient and produce smaller droplets than a conventional nebulizer. In addition, many microflow nebulizers use chemically inert plastic capillaries, which makes them well suited for the analysis of highly corrosive chemicals. This kind of flexibility has made low-flow nebulizers very popular, particularly in the semiconductor industry, where it is essential to analyze high-purity materials using a sample introduction system that is free of contamination.[46]

Such is the added capability and widespread use of these nebulizers across all application areas that manufacturers are developing application-specific integrated systems that include the spray chamber and a choice of different desolvation techniques to reduce the amount of solvent aerosol entering the plasma. Depending on the types of samples being analyzed, some of these systems include a low-flow nebulizer, Peltier-cooled spray chambers, heated spray chambers, Peltier-cooled condensers, and membrane desolvation technology. Some of the commercially available equipment include the following:

- **Microflow nebulizer coupled with a Peltier-cooled spray chamber:** An example of this is the PC3 from Elemental Scientific Inc. (ESI). This system is offered with or without the nebulizer and utilizes a Peltier-cooled cyclonic spray chamber made from either quartz, borosilicate glass, or a fluoropolymer. Options include the ability to reduce the temperature to −20°C for analyzing organics and a dual-spray chamber for improved stability.
- **Microflow nebulizer with heated spray chamber and Peltier-cooled condenser:** An example of this design is the Apex inlet system from ESI. This unit includes a microflow nebulizer, heated cyclonic spray chamber (up to 140°C), and a Peltier multipass condenser/cooler (down to −5°C). A number of different spray chamber and nebulizer options and materials are available, depending on the application requirements. Also, the system is available with Teflon or Nafion microporous membrane desolvation, depending on the types of samples being analyzed. Figure 17.12 shows a schematic of the Apex sample inlet system with the cross-flow nebulizer.
- **Microflow nebulizer coupled with membrane desolvation:** An example of this is the Aridus system from Cetac Technologies. The aerosol from the nebulizer is either self-aspirated or pumped into a heated PFA spray chamber (up to 110°C) to maintain the sample in a vapor phase. The sample vapor then enters a heated PTFE membrane desolvation unit, where a coun-

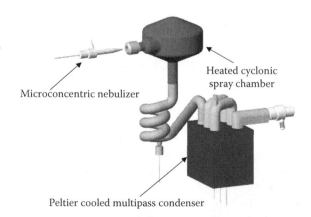

Microconcentric nebulizer

Heated cyclonic spray chamber

Peltier cooled multipass condenser

FIGURE 17.12   A schematic of the Apex sample inlet system (courtesy of Elemental Scientific Inc.).

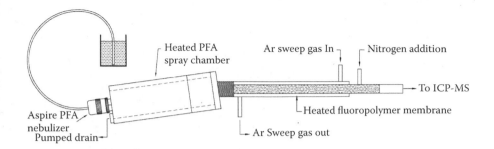

FIGURE 17.13   A schematic of the Aridus II microflow nebulizer with membrane desolvation (courtesy of Cetac Technologies).

terflow of argon sweep gas is added to remove solvent vapors that permeate the microporous walls of the membrane. Nonvolatile sample components do not pass through the membrane walls, but are transported to the ICP-MS for analysis. Figure 17.13 shows a schematic of the Aridus II microflow nebulizer with membrane desolvation.

There is an extremely large selection of these specialized sample introduction techniques, so it is critical that you talk to the vendors so they can suggest the best solution for your application problem. They may not be required for the majority of your application work, but there is no question they can be very beneficial for analyzing certain types of sample matrices and for elements that might be prone to solvent-based spectral overlaps.

DIRECT INJECTION NEBULIZERS

Direct injection nebulization is based on the principle of injecting a liquid sample under high pressure directly into the base of the plasma torch.[47] The benefit of this approach is that no spray chamber is required, which means that an extremely small volume of sample can be introduced directly into the ICP-MS with virtually no carryover or memory effects from the previous sample. Because they are capable of injecting less than 5 µL of liquid, they have found a use in applications where sample volume is limited or where the material is highly toxic or expensive.

They were initially developed over 15 years ago and found some success in certain niche applications that could not be adequately addressed by other nebulization systems, such as introducing samples from a chromatography separation device into an ICP-MS or the determination of mercury by ICP-MS, which is prone to severe memory effects. Unfortunately, they were not considered particularly user-friendly, and as a result became less popular when other sample introduction devices were developed to handle microliter sample volumes. More recently, a refinement of the direct injection nebulizer has been developed, called the direct inject high-efficiency nebulizer (DIHEN), which appears to have overcome many of the limitations of the original design.[48] The advantage of the DIHEN is its ability to introduce microliter volumes into the plasma at extremely low sample flow rates (1–100 µL/min), with an aerosol droplet size similar to a concentric nebulizer fitted with a spray chamber.

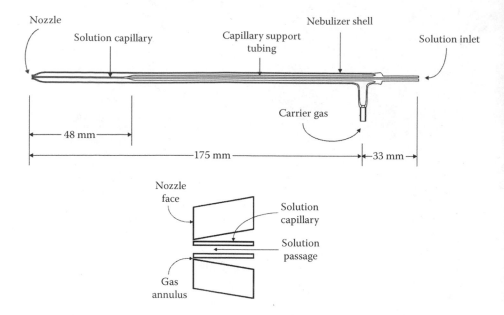

FIGURE 17.14 A schematic of a commercially available DIHEN system (courtesy of Meinhard Glass Products).

The added benefit is that it is almost 100% efficient and has extremely low memory characteristics. A schematic of a commercially available DIHEN system, is shown in Figure 17.14.

RAPID SAMPLING PROCEDURES

With the increasing demand to analyze more and more samples, manufacturers of autosamplers and sample introduction accessories are designing automated sampling systems to maximize sample throughput. This is being achieved in a variety of different ways by optimizing the sample delivery process to reduce the pre- and post-measurement time. Some typical optimization procedures include the following:

- **Autosampler response** is the time it takes for the instrument to send a signal to the autosampler to move the sample probe to the next sample. By moving the autosampler probe over to the next sample while the previous sample is being analyzed, a significant amount of time will be saved over the entire automated run.
- **Sample uptake** is the time taken for a sample to be drawn into the autosampler probe and pass through the capillary and pump tubing into the nebulizer. By using a small vacuum pump to rapidly fill the sample loop, which is positioned in close proximity of the sample loop to the nebulizer, sample uptake time is minimized.
- **Signal stabilization** is the time required to allow the plasma to stabilize after air has entered the line from the autosampler probe dipping in and

out of the sample tubes (this can also be exaggerated if the pump speed is increased to help in sample delivery). However, if the pump delivering the sample to the plasma remains at a constant flow rate, and the injection valve ensures no air is introduced into the sample line, very little stabilization time is required.

- **Rinse-out** is the time required to remove the previous sample from the sample tubing and sample introduction system. So, if the probe is being rinsed during the sample analysis, minimal rinse time is needed.
- **Overhead time** is the time spent by the ICP performing calculations and printing results, so if this time is used to ensure the previous sample has reached baseline, minimal rinse time is required for the next sample.

There is no doubt that by optimizing these steps, a significant improvement can be made to the overall analysis time, especially in high-workload routine environmental laboratories, where high sample throughput is an absolute requirement.[49,50]

Another slightly different approach is to use a rapid-rinse accessory in conjunction with an autosampler. With this system, the nebulizer and spray chamber are rinsed out as soon as the measurement cycle is completed and allowed to continue rinsing until the next sample is ready. By rinsing out the sample in the time that is usually spent waiting for the solution to flow from the autosampler to the nebulizer, a significant amount of time is being saved in a typical multielement analysis. In fact, vendors of both these types of sampling accessories are claiming a 30–40% savings in time over conventional autosamplers and sample introduction equipment.

REFERENCES

1. E. R. Denoyer, K. J. Fredeen, and J. W. Hager, *Analytical Chemistry*, **63**(8), 445–457A, 1991.
2. J. F. Ready, *Effects of High Power Laser Radiation*, Academic Press, NY, Chapters 3–4, 1972.
3. L. Moenke-Blankenburg, *Laser Microanalysis*, Wiley, NY, 1989.
4. E. R. Denoyer, R. Van Grieken, and F. Adams, D. F. S. Natusch, *Analytical Chemistry*, **54**, 26A, 1982.
5. J. W. Carr and G. Horlick, *Spectrochimica Acta*, **37B**, 1, 1982.
6. T. Kantor et al. *Talanta,* **23**, 585, 1979.
7. H. C. G. Human et al. *Analyst*, **106**, 265, 1976.
8. M. Thompson, J. E. Goulter, and F. Seiper, *Analyst*, **106**, 32, 1981.
9. A. L. Gray, *Analyst,* **110**, 551, 1985.
10. P. A Arrowsmith and S. K. Hughes, *Applied Spectroscopy*, **42**, 1231–1239, 1988.
11. T. Howe, J. Shkolnik, and R. Thomas, *Spectroscopy*, **16**, 2, 54–66 2001.
12. D. Günther and B. Hattendorf, *Mineralogical Association of Canada—Short Course Series*, **29**, 83–91, 2001.
13. T. E. Jeffries, S. E. Jackson, and H. P. Longerich, *Journal of Analytical Atomic Spectrometry*, **13**, 935–940, 1998.
14. R. E. Russo, X. L. Mao, O. V. Borisov, and L. Haichen, *Journal of Analytical Atomic Spectrometry*, **15**, 1115–1120, 2000.
15. H. Liu, O. V. Borisov, X. Mao, S. Shuttleworth, and R. Russo, *Applied Spectroscopy*, **54**(10), 1435, 2000.
16. SEM photo courtesy of Dr. Honglin Yuan, Northwest University, Xi'an, China.

17. S. E. Jackson, H. P. Longerich, G. R. Dunning, and B. J. Fryer, *Canadian Mineralogist,* **30**, 1049–1064, 1992.

18. D. Günther and C. A. Heinrich, *Journal of Analytical Atomic Spectrometry,* **14**, 1369, 1999.

19. D. Günther, I. Horn, and B. Hattendorf, *Fresenius Journal of Analytical Chemistry,* **368**, 4–14, 2000.

20. R. E. Wolf, C. Thomas, and A. Bohlke, *Applied Surface Science,* **127–129**, 299–303, 1998.

21. J. Gonzalez, X. L. Mao, J. Roy, S. S. Mao, and R. E. Russo, *Journal of Analytical Atomic Spectrometry* **17**, 1108–1113, 2002.

22. T. E. Jeffries, W. T. Perkins, and N. J. G. Pearce, *Analyst,* **120**, 1365–1371, 1995.

23. J. Roy and L. Neufeld, *Spectroscopy,* **19**(1), 16–28, 2004.

24. J. Ruzicka and E. H. Hansen, *Analytic Chimica Acta,* **78**, 145, 1975.

25. R. Thomas, *Spectroscopy,* **17**(5), 54–66, 2002.

26. A. Stroh, U. Voellkopf, and E. Denoyer, *Journal of Analytical Atomic Spectrometry,* **7**, 1201, 1992.

27. Y. Israel, A. Lasztity, and R. M. Barnes, *Analyst,* **114**, 1259, 1989.

28. Y. Israel and R. M. Barnes, *Analyst,* **114**, 843, 1989.

29. M. J. Powell, D. W. Boomer, and R. J. McVicars, *Analytical Chemistry,* **58**, 2864, 1986.

30. S. N. Willie, Y. Iida, and J. W. McLaren, *Atomic Spectroscopy,* **19**(3), 67, 1998.

31. R. Roehl and M. M. Alforque, *Atomic Spectroscopy,* **11**(6), 210, 1990.

32. J. W. McLaren, J. W. H. Lam, S. S. Berman, K. Akatsuka, and M. A. Azeredo, *Journal of Analytical Atomic Spectrometry,* **8**, 279–286, 1993.

33. L. Ebdon, A. Fisher, H. Handley, and P. Jones, *Journal of Analytical Atomic Spectrometry,* **8**, 979–981, 1993.

34. D. B. Taylor, H. M. Kingston, D. J. Nogay, D. Koller, and R. Hutton, *Journal of Analytical Atomic Spectrometry,* **11**, 187–191, 1996.

35. S. M. Nelms, G. M. Greenway, and D. Koller, *Journal of Analytical Atomic Spectrometry,* **11**, 907–912, 1996.

36. C. J. Park, J. C. Van Loon, P. Arrowsmith, and J. B. French, *Analytical Chemistry,* **59**, 2191–2196, 1987.

37. R. D. Ediger and S. A. Beres, *Spectrochimica Acta,* **47B**, 907, 1992.

38. C. J. Park and M. Hall, *Journal of Analytical Atomic Spectrometry,* **2**, 473–480, 1987.

39. C. J. Park and J. C. Van Loon, *Trace Elements in Medicine,* **7**, 103, 1990.

40. G. Chapple and J. P. Byrne, *Journal of Analytical Atomic Spectrometry,* **11**, 549–553, 1996.

41. U. Voellkopf, M. Paul, and E. R. Denoyer, *Fresenius Journal of Analytical Chemistry,* **342**, 917–923, 1992.

42. S. A. Beres, E. R. Denoyer, R. Thomas, and P. Bruckner, *Spectroscopy,* **9**(1), 20–26, 1994.

43. F. McElroy, A. Mennito, E. Debrah, and R. Thomas, *Spectroscopy,* **13**(2), 42–53, 1998.

44. K. W. Olson, W. J. Haas, Jr, and V. A. Fassel, *Analytical Chemistry,* **49**(4), 632–637, 1977.

45. J. Kunze, S. Koelling, M. Reich, and M. A. Wimmer, *Atomic Spectroscopy,* **19**, 5, 1998.

46. G. Settembre and E. Debrah, *Micro,* June 1998.

47. D. R. Wiederin and R. S. Houk, *Applied Spectroscopy,* **45**(9), 1408–1411, 1991.

48. J. A. McLean, H. Zhang, and A. Montaser, *Analytical Chemistry,* **70**, 1012–1020, 1998.

49. *Improving Throughput of Environmental Samples by ICP-MS Following EPA Method 200.8,* ESI Application Note, http://www.elementalscientific.com/products/SC-FAST_enviro.html.

50. M. P. Field, M. LaVigne, K. R. Murphy, G. M. Ruiz, and R. M. Sherrell, *Journal of Analytical Atomic Spectrom*etry, **22**, 1145, 2007.

18 Coupling ICP-MS with Chromatographic Techniques for Trace Element Speciation Studies

The specialized sample introduction techniques described in Chapter 17 were mainly developed as a result of a basic limitation of ICP-MS in carrying out elemental determinations on certain types of complex sample matrices. However, even though all of these sampling accessories significantly improved the flexibility of the technique, they were still being used to measure the total metal content of the samples being analyzed. If the requirement was to learn more about the valency state or speciated form of the element, the trace metal analytical community had to look elsewhere for answers. Then, in the early 1990s, researchers started investigating the use of ICP-MS as a detector for chromatography systems, which triggered an explosion of interest in this exciting new hyphenated technique, especially for environmental and biomedical applications. In this chapter we look at what drove this research and discuss, in particular, the use of high-performance liquid chromatography (HPLC) with ICP-MS in carrying out trace element speciation studies.

ICP-MS has gained popularity over the years, mainly because of its ability to rapidly quantitate ultratrace metal contamination levels. However, in its basic design, ICP-MS cannot reveal anything about the metal's oxidation state, alkylated form, how it is bound to a biomolecule, or how it interacts at the cellular level. The desire to understand in what form or species an element exists led researchers to investigate the combination of chromatographic separation devices with ICP-MS. The ICP mass spectrometer becomes a very sensitive detector for trace element speciation studies when coupled to a chromatographic separation device based on HPLC, ion chromatography (IC), gas chromatography (GC), size exclusion chromatography (SEC), capillary electrophoresis (CE), etc. In these hyphenated techniques, elemental species are separated based on their chromatographic retention, mobility, or molecular size, and then eluted/passed into the ICP mass spectrometer for detection.[1] The intensities of the eluted peaks are then displayed for each isotopic mass of interest, in the time domain, as shown in Figure 18.1. The figure shows a typical time-resolved chromatogram for a selected group of masses.

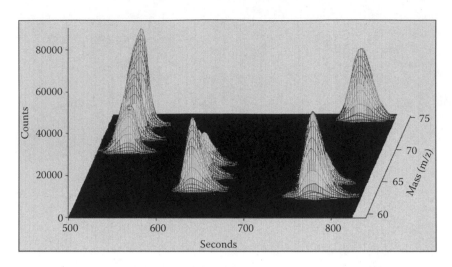

FIGURE 18.1   A typical time-resolved chromatogram generated using chromatography coupled with ICP-MS, showing a temporal display of intensity against mass (copyright © 2003–2007, all rights reserved, PerkinElmer Inc.).

There is no question that ICP-MS has allowed researchers in the environmental, biomedical, geochemical, and nutritional fields to gain a much better insight into the impact of different elemental species on humans and their environment. Even though elemental speciation studies were being carried out using other atomic spectrometry (AS) detection techniques, it was the commercialization of ICP-MS in the early 1980s, with its extremely low detection capability, that saw a dramatic increase in the number of trace element speciation studies being carried out. Today, the majority of these studies are being driven by environmental regulations. In fact, the U.S. EPA has published a number of speciation methods involving chromatographic separation with ICP-MS, including Method 321.8 for the speciation of bromine compounds in drinking water and wastewater, and Method 6800 for the measurement of various metal species in potable and wastewaters by isotope dilution mass spectrometry. However, other important areas of interest include nutritional and metabolic studies, toxicity testing, bioavailability measurements, and pharmaceutical research. Speciation studies cross over many different application areas, but the majority of determinations being carried out can be classified into three major categories:

- **Measurement of different oxidation states**—For example, hexavalent chromium, Cr (VI), is a powerful oxidant and is extremely toxic, but in soil and water systems it reacts with organic matter to form trivalent chromium, Cr (III), which is the more common form of the element and an essential micronutrient for plants and animals.[2]
- **Measurement of alkylated forms**—Very often the natural form of an element can be toxic, although its alkylated form is relatively harmless, or vice versa. A good example of this is the element arsenic. Inorganic forms of the element such as As (III) and As (V) are toxic, whereas many of its alkylated

---

TABLE 18.1

Some Typical Inorganic and Organic Species That Have Been Studied by Researchers Using Chromatographic Separation Techniques

Oxidation States	Alkylated Forms	Biomolecules
• Se^{+4}	• Methyl—Hg, Ge, Sn, Pb, As, Sb, Se, Te, Zn, Cd, Cr	• Organometallic complexes— As, Se, Cd
• Se^{+6}	• Ethyl—Pb, Hg	• Metalloporphyrines
• As^{+3}	• Butyl, Phenyl, Cyclohexyl—Sn	• Metalloproteins
• As^{+5}		• Metallodrugs
• Sn^{+2}		• Metalloenzymes
• Sn^{+4}		• Metals at the cellular level
• Cr^{+3}		
• Cr^{+6}		
• Fe^{+2}		
• Fe^{+3}		

---

forms such as monomethylarsonic acid (MMA) and dimethylarsonic acid (DMA) are relatively innocuous.[3]

• **Measurement of metallobiomolecules**—These molecules are formed by the interaction of trace metals with complex biological molecules. For example, in animal farming studies, the activity and mobility of an innocuous arsenic-based growth promoter are determined by studying its metabolic impact and excretion characteristics. So, measurement of the biochemical form of arsenic is crucial in order to know its growth potential.[4]

Table 18.1 represents a small cross section of both inorganic and organic species of interest classified under these three categories.

There are 400–500 speciation papers published every year, the majority of which are based on environmentally significant elements such as As, Cr, Hg, Se, and Sn.[5] The following is a small selection of some of the most recent research that can be found in the public domain:

• Determination of chromium (VI) in drinking water samples, using HPLC-ICP-MS[6]
• Determination of trivalent and hexavalent chromium in pharmaceutical and biological materials by IC-ICP-MS[7]
• The use of LC-ICP-MS in better understanding the role of inorganic and organic forms of selenium in biological processes[8]
• Identification of selenium compounds in contaminated estuarine waters using IC-ICP-MS[9]
• Determination of organoarsenic species in marine samples, using cation exchange HPLC-ICP-MS[10]
• Determination of biomolecular forms of arsenic in chicken manure by CE-ICP-MS[11]
• Analysis of tributyltin (TBT) in marine samples using HPLC-ICP-MS[12]

- Measurement of anticancer platinum compounds in human serum by HPLC-ICP-MS[13]
- Bioavailability of cadmium and lead in beverages and foodstuffs using SEC-ICP-MS[14]
- Investigation of sulfur speciation in petroleum products by capillary gas chromatography with ICP-MS detection[15]
- Analysis of methyl mercury in water and soils by HPLC-ICP-MS[16]
- Use of HPLC-ICP-MS for the analysis of phospholipids[17]

HPLC COUPLED WITH ICP-MS

It can be seen from this brief snapshot of speciation publications that by far the most common chromatographic separation techniques being used with ICP-MS are the many different types of liquid chromatography, such as adsorption, ion exchange, size exclusion, gel permeation, and normal- or reverse-phase chromatography. To get a better understanding of how the technique works, particularly when attempting to develop a routine method to simultaneously measure multiple species in the same analytical run, let us take a more detailed look at how the HPLC system is coupled to the ICP mass spectrometer. Figure 18.2 shows a typical setup of the hardware components.

The coupling of the ICP-MS system to the liquid chromatograph hardware components is relatively straightforward, connecting a capillary tube from the end of the HPLC column, through a switching valve, to the sample introduction system of the ICP mass spectrometer. However, matching the column flow with the uptake of the ICP-MS sample introduction system is not a trivial task. Therefore, to develop a successful trace element speciation method, it is important to optimize not only the chromatographic separation, but also selection of the nebulization process to match

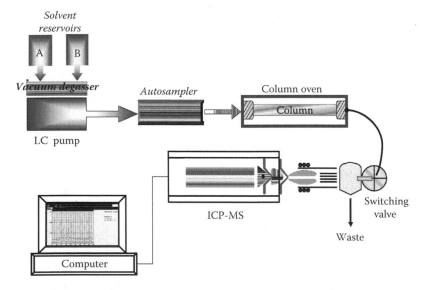

FIGURE 18.2 A typical configuration of an HPLC system interfaced with an ICP mass spectrometer (copyright © 2003–2007, all rights reserved, PerkinElmer Inc.).

the flow of the sample being eluted off the column, together with finding the best ICP-MS operating conditions for the analytes or species of interest. Let us take a closer look at this.

CHROMATOGRAPHIC SEPARATION REQUIREMENTS

Traditionally, the measurement of trace levels of elemental species by HPLC has been accomplished by separating the species using column separation technology and detecting them, one element at a time, as they elute. This approach works well for one element or species, but is extremely slow and time consuming for the determination of multiple species or elements, because the chromatographic separation process has to be optimized for each species being measured. Therefore, the limitation to multielement speciation is rarely the detection of the elements, but is usually the separation of the species. The inherent problem is that liquid chromatography works on the principle of equilibration between the species of interest, the mobile phase, and the column material. Because the chemistries of elements and their species differ, it is difficult to find common conditions capable of separating species of more than one element simultaneously. For example, toxic elements of environmental interest, such as arsenic (As), chromium (Cr), and selenium (Se), have different reaction chemistries, thus requiring different chromatographic conditions to separate them. So, to achieve the analytical goal of fast, simultaneous measurement of different species of these elements in a single sample injection, the chromatographic separation has to be optimized.

Let us first examine the chromatography, focusing on two common forms of separation (ion exchange and reversed-phase ion-pairing chromatography) together with two commonly employed HPLC elution techniques (isocratic and gradient). Each process will be described, and the advantages and disadvantages of each will be discussed in the context of choosing a separation scheme to achieve our analytical goals. One final note before we discuss the separation process: it is extremely important when preparing the sample that the speciated form of the element not be changed or altered in any way. This is not such a serious problem with a simple matrix such as drinking water, which typically involves straightforward acidification. However, when more complex samples such as soils or biological samples are being analyzed, it is quite common to use strong acids, oxidizing agents, or high temperatures to get the samples into solution. So, maintaining the integrity of the valency state or species should always be an extremely important decision when preparing a sample for speciation analysis.

Ion Exchange Chromatography (IEC)

In this technique, separation is based on the exchange of ions (anions or cations) between the mobile phase and the ionic sites on a stationary phase bound to a support material in the column. A charged species is covalently bound to the surface of the stationary phase. The mobile phase, typically a buffer solution, contains a large number of ions that have a charge opposite to that of the surface-bound ions. These mobile-phase ions, referred to as counterions, establish an equilibrium with the sta-

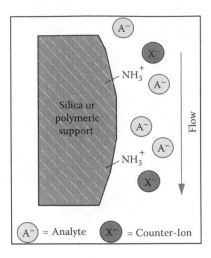

FIGURE 18.3 Principle of separation using anion exchange chromatography.

tionary phase. Sample ions passing through the column and having the same ionic charge as the counterion compete with the mobile-phase ions for sites on the column material, resulting in a disruption of the equilibrium and retention of that analyte. It is the differential competition of various analytes with the mobile-phase ions that ultimately produces species separation and chromatographic peaks, as detected by ICP-MS. Therefore, analyte retention is based on the affinity of different ions for the support material and on other solution parameters, including counterion type, ionic strength (buffer concentration), and pH. Varying these mobile-phase parameters changes the separation.

The principle of IEC is schematically represented in Figure 18.3 for an anion exchange separation. In this figure, the anions are denoted by A^- and represent the analyte species, whereas the counterions are symbolized by X^-. The main benefit of this approach is that it has a high tolerance to matrix components, and the main disadvantages are that these columns tend to be expensive and somewhat fragile.

Reversed-Phase Ion Pair Chromatography (RP-IPC)

Ion pair chromatography typically uses a reversed-phase column in conjunction with a special type of chemical in the mobile phase called an *ion-pairing reagent*. "Reversed phase" essentially means that the column's stationary phase is nonpolar (less polar, more organic) than the mobile-phase solvents. For RP-IPC, the stationary phase is a carbon chain, most often consisting of 8 or 18 carbon atoms (C8, C18) bonded to a silica support. The mobile-phase solvents usually consist of water mixed with a water-miscible organic solvent, such as acetonitrile or methanol.

The ion-pairing reagent is a compound that has both an organic and an ionic end. To promote ionic interaction, the ionic end should have a charge opposite to that of the analytes of interest. For anionic IPC, a commonly used ion-pairing reagent is tetrabutylammonium hydroxide (TBAOH); for cationic IPC, hexanesulfonic acid (HSA) is often chosen. In principle, the ionized/ionizable analytes interact with the

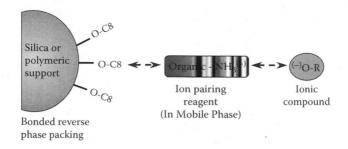

FIGURE 18.4   Principle of ion pair chromatography (anionic mode).

ionic end of the ion-pairing reagent, whereas the ion-pairing reagent's organic end interacts with the C18 stationary phase. Retention and separation selectivity are primarily affected by characteristics of the mobile phase, including pH, selection/concentration of ion-pairing reagent, and ionic strength, as well as the use of additional mobile-phase modifiers. With this scheme, a reversed-phase column acts like an ion exchange column.

The benefits of this type of separation are that the columns are generally less expensive and tend to be more rugged than ion exchange columns. The main disadvantages, compared to IEC, are that the separation may not be as good, and there may be less tolerance to high sample matrix concentrations. A simplified representation of this approach is shown in Figure 18.4.

COLUMN MATERIAL

As previous explained, several variables of the mobile phase can be modified to affect separations, both by IEC and RP-IPC. A very important consideration when choosing a column is the pH of the mobile phase used to achieve the separation. The pH is critical in the selection of the column for both IEC and RP-IPC because the support materials may be pH sensitive. One of the most common column materials is silica, because it is inexpensive, rugged, and has been used for many years. The main disadvantage of silica columns is that they are useful only over a pH range of 2–8. Outside this range, the silica dissolves. For applications requiring pHs outside of this range, polymer columns are commonly used. Polymer columns are useful over the entire pH range. The downside to these columns is that they are typically more expensive and fragile. Columns consisting of other support materials are also available.

ISOCRATIC OR GRADIENT ELUTION

Another consideration is how best to elute the species from the column. This can be accomplished using either an isocratic or gradient elution scheme. Isocratic refers to using the same solvent throughout the analysis. Because of this, samples can be injected immediately after the preceding one has finished eluting. This type of elution can be performed on all HPLC pumping systems. The advantages of isocratic elutions are simplicity and higher sample throughput.

Gradient elution involves variation of the mobile phase composition over time by a number of steps such as changing the organic content, altering the pH, changing concentration of the buffer, or using a completely different buffer. This is usually accomplished by having two or more bottles of mobile phases connected to the HPLC pump. The pump is then programmed to vary the amount of each mobile-phase component. The mobile-phase components are mixed on-line before reaching the column, and therefore, a more complex pumping system is required to handle this task.

The main reason for using gradient elution is that the mobile-phase composition can be varied to fine-tune the separation. As a result, component separations are better, and chromatograms are generally shorter than with isocratic elutions. However, overall sample throughput is much lower than with isocratic separations because of the variation in mobile-phase composition. So, after an elution is complete, the mobile phase must be changed back to its original composition as at the start of the chromatogram. This means that the equilibrium between the mobile phase and the column must be reestablished before the next sample can be injected. If the equilibrium is not established, the peaks will elute at different times compared to the previous sample. The equilibration time varies with column, but it is usually between 5 and 30 min.

Therefore, if a fast, automated, routine method for the measurement of multi-species/elements is the desired analytical goal, it is often best to attempt an isocratic separation method first, because of the complexity of method development and the low sample throughput of gradient elution methods. In fact, a simultaneous method for the separation of As, Cr, and Se species in drinking water samples was demonstrated by Neubauer and coworkers; they developed a method to determine inorganic forms of arsenic (As^{+3}, As^{+5}), chromium (Cr^{+3}, Cr^{+6}), and selenium (Se^{+4}, Se^{+6}, and $SeCN^-$) by reverse-phase ion-pairing chromatography with isocratic elution.[18] Details of the HPLC separation parameters/conditions they used are shown in Table 18.2.

TABLE 18.2

HPLC Separation Parameters/Conditions for Measuring Inorganic As, Cr, and Se Species in Potable Waters

HPLC configuration	Quaternary pump, column oven, and autosampler
Column	C8, reduced activity, 3.3 cm × 0.46 cm (3 μm packing)
Column temperature	35°C
Mobile phase	mM TBAOH + 0.15 mM NH_4CH_2COOH + 0.15 mM EDTA (K salt) + 5% MeOH
pH	7.5
pH adjustment	Dilute HNO_3, NH_4OH
Injection volume	50 mL
Sample flow rate	1.5 mL/min
Samples	Various potable waters
Sample preparation	Dilute with mobile phase (2–10×); heat at 50–55°C for 10 min

Source: K. R. Neubauer, P. A. Perrone, W. Reuter, R. Thomas, *Current Trends in Mass Spectroscopy*, May 2006.

It is important to emphasize that if these species were being measured on a single-element basis, the optimum chromatographic conditions would be different. However, the goal of this study was to determine all the species in a single multielement run so that the method could be applied to a routine, high-throughput environment. The 3.3-cm-long column was packed with a C8 hydrocarbon material (3 μm particle size). The mobile phase was a mixture of TBAOH, ammonium acetate (NH_4CH_2COOH), the potassium salt of EDTA, and methanol. The pH was adjusted to 7.5 (prior to the addition of methanol) using 10% nitric acid and 10% ammonium hydroxide. The sample preparation consisted of dilution with the mobile phase and heating at 50–55°C to speed the formation of the Cr III–EDTA complex. For each analysis, 50 μL of sample was injected into the column, which resulted in a sample flow of 1.5 mL/min being eluted off the column into the ICP mass spectrometer.

SAMPLE INTRODUCTION REQUIREMENTS

When coupling an HPLC system to an ICP mass spectrometer, it is very important to match the flow of sample being eluted off the column with the ICP-MS nebulization system. With today's choice of sample introduction components, there are specialized nebulizers and spray chambers on the market that can handle sample flows from 20 μL/min up to 3000 μL/min. The most common type of nebulizer used for chromatography applications is the concentric design, because it is self-aspirating and it generates an aerosol with extremely small droplets, which tends to produce better signal stability compared to a cross-flow design. The choice of which type of concentric nebulizer to use should therefore be based on the sample flow coming off the column. If the sample flow is on the order of 1 mL/min, a higher-flow concentric nebulizer should be used, and if the flow is much lower, such as in nano- or micro-flow LC work, a specialized low-flow nebulizer should be used.

It is also very important that the dead volume of the sample introduction process be kept to an absolute minimum to optimize peak integration of the separated species over the length of the transient signal. For this reason, the length of sample capillary from the end of the column to the nebulizer should be kept to a minimum; the internal volume of the nebulizer should be as small as possible; the connectors/fittings/valves should all have low dead volume; and a self-aspirating nebulizer should be used to avoid the need for peristaltic pump tubing. In addition, a spray chamber with a short aerosol path should be selected, which will not add additional dead volume to the method. However, depending on the total flow of the sample and the type of nebulizer, a spray chamber may not even be required. Cutting down on the sample introduction dead volume and minimizing peak broadening by careful selection of column technology will ultimately dictate the number of species that can be separated in a given time—an important consideration when developing a routine method for high sample throughput. Figure 18.5 shows a high-efficiency concentric nebulizer (HEN) designed for HPLC-ICP-MS work, and Figure 18.6 shows the difference between the capillary of this nebulizer (on the right) and a standard concentric type A nebulizer (on the left).

Another important reason to match the nebulizer with the flow coming off the column is that concentric nebulizers are mainly self-aspirating. For this reason,

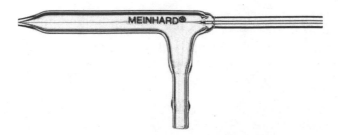

FIGURE 18.5 A high-efficiency concentric nebulizer (HEN) designed for HPLC-ICP-MS work (courtesy of Meinhard Glass Products).

FIGURE 18.6 The difference between the capillary of a high-efficiency concentric nebulizer for HPLC work (on the right) and a standard concentric type A nebulizer (on the left) (courtesy of Meinhard Glass Products).

the column flow must be high enough to ensure that the nebulizer can sustain a consistent and reproducible aerosol. On the other hand, if the column flow is too low, a makeup flow might need to be added to the column flow to meet the flow requirements of the nebulizer being used. This has the additional benefit of being able to add an internal standard after the column with another pump to correct for instrument drift or matrix effects in gradient elution work.

OPTIMIZATION OF ICP-MS PARAMETERS

In the early days of trace element speciation studies using chromatography coupled with ICP-MS, researchers had no choice but to interface their own LC pumps, columns, and autosamplers, etc., to the ICP mass spectrometer, because off-the-shelf systems were not commercially available. However, the analytical objectives of a research project are a little different from the requirements for routine analysis. With a research project, there are fewer time constraints to optimize the chromatography and detection parameters, whereas in a commercial environment, there are often financial penalties if the laboratory cannot be up, running real samples and generating revenue as quickly as possible. This demand, especially from commercial laboratories in the environmental and biomedical communities, for routine trace element speciation methods convinced the instrument manufacturers and vendors to develop fully integrated HPLC-ICP-MS systems.

The availability of these off-the-shelf systems rapidly drove the growth of this hyphenated technique, so much so that vendors who did not offer it with full application and hardware/software support were at a disadvantage. As this technique is maturing and is being used more as a routine analytical tool, it is becoming clear that the requirements of the ICP-MS system doing speciation analysis are different from those of an instrument carrying out trace element determinations using conventional nebulization. With that in mind, let us take a closer look at the typical requirements of an ICP mass spectrometer that is being utilized as a multielement detector for trace element speciation studies.

COMPATIBILITY WITH ORGANIC SOLVENTS

The requirements of the sample introduction system, and in particular the nebulization process, have been described earlier in this chapter. However, some reverse-phase HPLC separations use gradient elution with mixtures of organic solvents such as methanol or acetonitrile. If this is the case, consideration must be given to the fact that some volatile organic solvents will extinguish the plasma.[19] Therefore, modifications to the sample introduction might need to be made, such as adding small amounts of oxygen to the nebulizer gas flow, or perhaps using a cooled spray chamber or a desolvation device to stop the buildup of carbon deposits on the sampler cone. Other approaches such as direct injection nebulization have been used to introduce the sample eluent into the ICP-MS, but historically they have not gained widespread acceptance because of usability issues.

COLLISION/REACTION CELL OR INTERFACE CAPABILITY

Another requirement of the ICP-MS system for speciation work is the collision/reaction cell/interface capability. It is becoming clear that as more and more speciation methods are being developed, the ability to minimize polyatomic spectral interferences generated by the solvent, buffer, mobile phase, pH-adjusting acids/bases, and the plasma gas, etc., is of crucial importance. Take, for example, the elements discussed earlier. As, Cr, and Se are notoriously difficult elements for ICP-MS analysis because their major isotopes suffer from argon- and sample-based polyatomic interferences. Arsenic has only one isotope (m/z 75), which is difficult to quantify in chloride-containing samples because of the presence of $^{40}Ar^{35}Cl^+$. Low-level chromium analysis is difficult because of the presence of the $^{40}Ar^{12}C^+$ and $^{40}Ar^{13}C^+$ interferences, which overlap the two major isotopes of chromium at masses 52 and 53. These interferences are nearly always present, but are especially strong in samples with organic content. The argon dimers ($^{40}Ar^{40}Ar^+$, $^{40}Ar^{38}Ar^+$) at masses 80 and 78 interfere with the major isotopes of Se at mass 80 and 78, respectively, and bromine, which is usually present in natural waters, forms $^{79}BrH^+$ and $^{81}BrH^+$, which interferes with the Se masses at 80 and 82, respectively. The effects of these and other interferences have been reduced somewhat with conventional ICP-MS instrumentation, using alternate masses, interference correction equations, cool plasma technology, and desolvation techniques, but these approaches have not shown themselves to be particularly useful for these elements, especially at ultratrace levels.

For these reasons, it will be very beneficial if the ICP-MS instrumentation is fitted with a collision or reaction cell or interface and has the capability to minimize the formation of these undesired polyatomic interferences, using either collisional mechanisms with kinetic energy discrimination[20,21] or ion–molecule reaction kinetics with mass bandpass tuning,[22] or by introduction of a collision/reaction gas into the interface region.[23] The best approach will depend on the type of sample being analyzed, the number of species being determined, and the detection limit requirements, but there is no doubt that for multielement work, it is beneficial if one gas can be used for all the analyte species. This is exemplified by the research of Neubauer and coworkers described earlier, who used oxygen as the reaction gas to reduce the polyatomic spectral interferences described earlier to determine seven different species of As, Cr, and Se in potable water. In fact, even though the oxygen was used to remove the argon-carbide and argon dimer interferences to quantify the chromium and selenium species, respectively, it was used in a different way to quantify the arsenic species. It was used to react with the arsenic ion to form the arsenic-oxygen molecular ion ($^{75}As^{16}O$) at mass 91, and move it away from the argon-chloride interference at mass 75. This novel approach has been reported many times in the literature[22,24] and has in fact been approved by the EPA in the recent ILM05.4 analytical procedure update for their Superfund, Contact Laboratory Program (refer to Chapter 19 for more details).

The instrumental conditions for the speciation analysis are shown in Table 18.3, and a plot of signal intensity versus time of the simultaneous separation of a 1 µg/L standard of As, Cr, and Se is shown in Figure 18.7.

TABLE 18.3

ICP-MS Instrumental Conditions for the Speciation Analysis of As^{+3}, As^{+5}, Cr^{+3}, Cr^{+6}, Se^{+4}, Se^{+6}, and SeCN in Potable Water Samples

Nebulizer	Quartz concentric
Spray chamber	Quartz cyclonic
RF power	1500 W
Collision/reaction cell technology	Dynamic reaction cell
Reaction gas	$O_2 = 0.7$ mL/min
Analytical masses	AsO^+ (m/z 91), Se^+ (m/z 78), Cr^+ (m/z 52)
Analyte species	As^{+3}, As^{+5}, Cr^{+3}, Cr^{+6}, Se^{+4}, Se^{+6}, $SeCN^-$
Dwell time	330 ms (per analyte)
Analysis time	5.5 min

Source: K. R. Neubauer, P. A. Perrone, W. Reuter, R. Thomas, *Current Trends in Mass Spectroscopy*, May 2006.

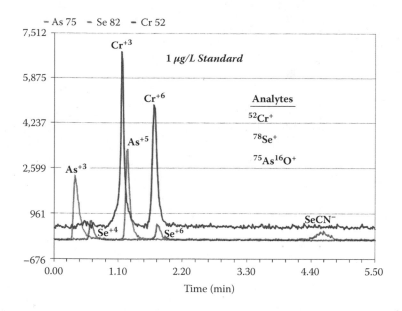

FIGURE 18.7 A plot of signal intensity versus time of the simultaneous separation of a 1 μg/L standard of As, Cr, and Se species. (From K. R. Neubauer, P. A. Perrone, W. Reuter, R. Thomas, *Current Trends in Mass Spectroscopy*, May 2006.)

It should be emphasized that even though oxygen was used as the reaction gas in this study, many other collision/reaction strategies using inert gases such as helium and low-reactivity gases such as hydrogen have been successfully used for the determination of As, Cr, and Se. The question is whether a single gas can be used to remove all the interferences to an acceptable level and allow quantitation at the trace level. If more than one gas is required, that is not such a major hardship, especially as all gas flows are under computer control and can be changed in a multielement run. However, if the method needs to be automated for a high-sample-throughput environment, it will be significantly faster if only one gas is used for interference removal, so that the collision/reaction cell conditions need not be changed for each element.

OPTIMIZATION OF PEAK MEASUREMENT PROTOCOL

It can be seen from the chromatogram that the total separation time for the seven inorganic species is on the order of 2 min for the common oxidation states of the elements, and about 5 min if there is any selenocyanide in the sample. This means that all the peaks have to be integrated and quantified in a transient signal lasting 2–5 min. So, even though this is not considered a short transient such as an ETV or small spot laser ablation work that typically lasts 5–10 s, it is not a continuous signal generated by a pneumatic nebulizer. For that reason, it is critical to optimize the measurement time in order to achieve the best multielement signal-to-noise ratio in the sampling time available. This is demonstrated in Figure 18.8, which shows the temporal separation of a group of elemental species in a chromatographic transient peak. The plot represents signal intensity against mass over the time period of the

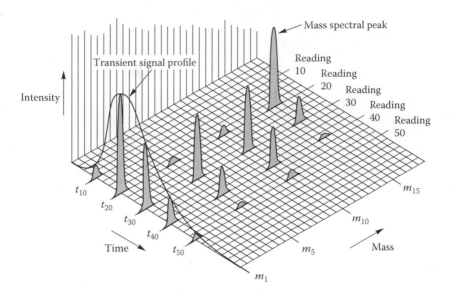

FIGURE 18.8 The temporal separation of a group of elemental species in a chromato-graphic transient peak.

chromatogram. To get the best detection limits for this group of elements or species, it is very important to spend all the available time quantifying the peaks of interest.

For this reason, the quadrupole scanning/settling time and the time spent measuring the analyte peaks must be optimized to achieve the highest signal quality. This is described in greater detail in Chapter 12 and basically involves optimizing the number of sweeps, selecting the best dwell time, and using short settling times to achieve the highest measurement duty cycle and maximize the peak signal-to-noise ratio over the duration of the transient event. If this is not done, there is a strong likelihood that the quality of the speciation data could be compromised. In addition, if the extended dynamic range is used to determine higher concentrations of elemental species, the scanning and settling time of the detector will also have an impact on the quality of the signal. For that reason, detectors that require two scans to characterize an unknown sample will use up valuable time in the quantitation process. This is somewhat of a disadvantage when doing multielement speciation on a chromatographic transient signal, especially if you have limited knowledge of the analyte concentration levels in your samples.

Full Software Control and Integration

In the early days of coupling HPLC components with ICP-MS, there were very few sophisticated communication protocols between the two devices. The sample was injected into the chromatographic separation system, and when the analyte species was close to being eluted off the column, the read cycle of the ICP mass spectrometer was initiated manually to capture the data using the instrument's time-resolved software. Processing and manipulation of the data was then carried out after the chromatogram had been captured, sometimes by a completely different software program. The HPLC and the ICP-MS were considered almost two distinctly different hardware and software devices, with very little communication between them. This was even more surprising considering that many of the ICP-MS vendors also offered LC equipment. As you can imagine, this was not the ideal scenario for a laboratory that wanted to carry out automated speciated analysis. As a result of this demand, manufacturers realized that unless they offered fully integrated hardware and software solutions for trace element speciation work, it was never going to be accepted as a routine analytical tool. Today, just about all the ICP mass spectrograph manufacturers offer fully integrated HPLC-ICP-MS systems. Some of the features available on today's instrumentation include the following:

- Full software and hardware control of both the chromatograph and the ICP-MS system from one computer
- Ability to check the status of the ICP-MS system when setting up the chromatography method or vice versa
- Computer control of the switching valve to allow the HPLC to be used in tandem with the ICP-MS normal sample introduction system
- Real-time display of spectral data, including peak identification and quantitation, while the chromatogram is being generated
- Full data-handling capability, including manipulation of spectral peaks and calibration curves
- Comprehensive reporting options or exportable data to third-party programs

It is also important to emphasize that most vendors also offer integrated systems with turnkey application methods that are almost ready to run samples as soon the instrument is installed. Although this is not available for all applications, it is becoming a standard offering for some of the more routine environmental applications. Manufacturers are also realizing that most analytical chemists who are experienced at trace element analysis have very little expertise in chromatography. For that reason, they are providing full backup and customer support with HPLC application specialists as well as with the traditional ICP-MS product specialists.

SUMMARY

Combining chromatography with ICP-MS has revolutionized trace metal speciation analysis. In particular, when HPLC is coupled with the selectivity and sensitivity of ICP-MS, many elemental species at sub-ppb levels can now be determined in a single sample injection. When interference reduction methods such as collision/reaction cell or interface technology are available, it is possible to separate and detect different inorganic species of environmental significance in one automated run. By using a fully integrated, computer-controlled HPLC/ICP-MS system and optimizing the chromatographic separation, sample introduction, and ICP-MS detection parameters, simultaneous quanitation can be carried out in a few minutes. There is no question that this kind of sample throughput will arouse the interest of the environmental, biomedical, nutritional, and other application communities interested in trace element speciation studies and help them realize that it is feasible to carry out this kind of analysis in a truly routine manner.

REFERENCES

1. R. Lobinski, I. R. Pereiro, H. Chassaigne, A. Wasik, and J. Szpunar, *Journal of Analytical Atomic Spectrometry*, **13**, 860–867, 1998.
2. A. G. Cox and C. W. McLeod, *Mikrochimica Acta*, **109**, 161–164, 1992.
3. S. Branch, L. Ebdon, and P. O'Neill, *Journal of Analytical Atomic Spectrometry*, **9**, 33–37, 1994.
4. J. R. Dean, L. Ebdon, M. E. Foulkes, H. M. Crews, and R. C. Massey, *Journal of Analytical Atomic Spectrometry*, **9**, 615–618, 1994.
5. Z. A. Grosser and K. Neubauer, *Today's Chemist at Work*, May 2004.
6. Y. L. Chang and S. J. Jiang, *Journal of Analytical Atomic Spectrometry*, **9**, 858, 2001.
7. K. E. Lokits, D. D. Richardson, and J. A. Caruso, *Handbook of Hyphenated ICP-MS Applications*, Agilent Technologies, August, 2007, http://www.chem.agilent.com/temp/rad5F0C7/00001681.PDF.
8. K. DeNicola, D. D. Richardson, and J. A. Caruso, *Spectroscopy*, **21**, 2, 18–24, 2006.
9. D. Wallschlager and N. Bloom, *Journal of Analytical Atomic Spectrometry*, **16**, 1322, 2001.
10. J. J. Sloth, E. H. Larsen, and K. Julshamn, *Journal of Analytical Atomic Spectrometry* **18**, 452–459, 2003.
11. C. Rosal, G. Momplaisir, and E. Heithmar, *Electrophoresis*, **26**, 1606–1614, 2005.
12. L. Yang, Z. Mester, and R. E. Sturgeon, *Analytical Chemistry*, **74**, 2968, 2002.
13. V. Vacchina, L. Torti, C. Allievi, and R. Lobinski, *Journal of Analytical Atomic Spectrometry*, **18**, 884–890, 2003.
14. S. Mounicou, J. Szpunar, R. Lobinski, D. Andrey, and C. J. Blake, *Journal of Analytical Atomic Spectrometry*, **17**, 880–886, 2002.
15. B. Bouyssiere, P. Leonhard, D. Pröfrock, F. Baco, C. Lopez Garcia, S. Wilbur, and A. Prange, *Journal of Analytical Atomic Spectrometry*, **5**, 700–702, 2004.
16. D. Chen, M. Jing, and S. Wang, *Handbook of Hyphenated ICP-MS Applications*, Agilent Technologies, August, 2007, http://www.chem.agilent.com/temp/rad5F0C7/00001681. PDF.
17. M. Kovacevic, R. Leber, S. D. Kohlwein, and W. Goessler, *Journal of Analytical Atomic Spectrometry*, **19**, 80–84, 2004.
18. K. R. Neubauer, P. A. Perrone, W. Reuter, and R. Thomas, *Current Trends in Mass Spectroscopy*, May 2006.
19. F. McElroy, A. Mennito, E. Debrah, and R. Thomas, *Spectroscopy*, **13**(2), 42–53, 1998.
20. M. Bueno, F. Pannier, M. Potin-Gautier, and J. Darrouzes, *Determination of Organic and Inorganic Selenium Using HPLC-ICP-MS*, Agilent Technologies Application Note—5989-7073EN, 2007, http://www.chem.agilent.com/temp/radBA32F/00001661.PDF.
21. S. McSheehy and M. Nash, *Determination of Selenomethionine in Nutritional Supplements Using HPLC Coupled to the XSeriesII ICP-MS with CCT*, Thermo Scientific Application Note—40745, 2005, http://www.thermo.com/eThermo/CMA/PDFs/Articles/articlesFile_26474.pdf.
22. J. Di Bussolo, W. Reuter, L. Davidowski, and K. Neubauer, *Speciation of Five Arsenic Compounds in Urine by HPLC-ICP-MS*, PerkinElmer LAS Application Note—D-6736, 2004, http://las.perkinelmer.com/content/applicationnotes/app_speciationfivearsenic-compounds.pdf.
23. M. Leist and A. Toms, *Low Level Speciation of Chromium in Drinking Waters Using LC-ICP-MS*, Varian Inc. Application Note—29, http://www.varianinc.com/image/vimage/docs/products/spectr/icpms/atworks/icpms29.pdf.
24. D. S. Bollinger and A. J. Schleisman, *Atomic Spectroscopy*, **20**(2), 60–63, 1999.

19 ICP-MS Applications

Today, there are approximately 8000 ICP-MS installations worldwide, performing a wide variety of applications, from routine, high-throughput multielement analysis to trace element speciation studies with high-performance liquid chromatography. Every year, as more and more laboratories invest in the technique, the list of applications is getting larger and more diverse. In this chapter, we take a look at the major market segments addressed by ICP-MS, and give detailed examples of the most common types of applications being carried out. I have focused mainly on quadrupole-based technology because it represents the vast majority of instruments doing routine analysis. However, even though double-focusing magnetic sector instrument installations might be more suited to research-type applications, their distribution across worldwide market segments is going to be similar to quadrupole technology. On the other hand, the smaller number of TOF-ICP-MS installations makes it difficult to make an accurate assessment of how it is being used, but it is clearly making an impact in transient peak applications that involve collecting large amounts of elemental data over a short period of time.

As a result of the widespread use and acceptability of ICP-MS, the cost of commercial instrumentation has dramatically fallen over the past 20 years. When the technique was first introduced, $250,000 was a fairly typical amount to spend, whereas today one can purchase a quadrupole or time-of-flight (TOF) system for about $150,000. Although it can cost a great deal more to invest in magnetic sector technology or a quadrupole instrument fitted with a collision/reaction cell or interface, most laboratories that are looking to invest in the technique should be able to justify the purchase of an instrument without price being a major concern. One of the benefits of this kind of price erosion is that, slowly but surely, the AA and ICP-OES user community are being attracted to ICP-MS, and as a result, the technique is being used in more and more diverse application areas. Figure 19.1 shows a percentage breakdown of the major market segments being addressed by ICP-MS on a worldwide basis. Three points should be emphasized here. First, these data can be significantly different on a geographical or regional basis, because of factors such as a country's commitment (or lack of it) to environmental concerns or the size of a region's electronics or nuclear industry, for example. Second, many laboratories carry out more than one type of application, and as a result can be represented in more than one market segment. Finally, the research market segment has been listed as a separate category to show the instruments that are being used in an academic environment or for nonroutine applications. However, many universities, federal organizations, or corporate R&D groups might be using their instrumentation for research purposes in a particular application segment. For these reasons, these data should only be considered an approximation for comparison purposes.

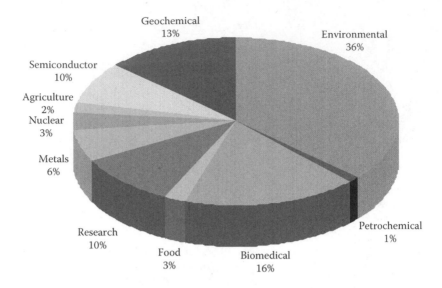

FIGURE 19.1   Global breakdown of major market segments addressed by ICP-MS.

Let us now take a look at each of these market segments in greater detail. The intent in this chapter is to present a broad cross section of what is being done on a routine basis in each market segment, together with some of the newer application work. They represent typical analytical problems, but in no way are meant to be a comprehensive list of all application work being addressed by ICP-MS. Wherever possible, I have suggested further reading, because new methodologies are being published on a regular basis, particularly applications using collision/reaction cell/ interface technology. For that reason, it is very important to keep up to date with application-based literature references in the public domain.

ENVIRONMENTAL

As can be seen by the pie chart, environmental applications represent the largest market segment for ICP-MS. In fact, about a third of all applications being carried out are environmental in nature. The most common type of environmental samples being analyzed today for trace element determinations include drinking water, groundwater, wastewater, river waters, estuarine water, seawater, solid waste, soils, sludges, sediments, and airborne particulates. There is no question that the enormous growth in environmental applications, especially in North America, is based on legislature driven by the U.S. Environmental Protection Agency (EPA) (www.epa. gov). Environmental users are generally not pushing the extreme detection capability of ICP-MS. This can be seen in Table 19.1, which compares the National Primary Drinking Water Regulations (NPDWR) maximum contaminant levels (MCLs) with typical ICP-MS detection limits for 12 primary contaminants in drinking water.

These levels are covered by EPA Method 200.8,[1] which is approved for all 12 primary contaminants in drinking water and most of the secondary ones, including

---

TABLE 19.1

Comparison of ICP-MS Detection Limits with NPDWR Maximum Contaminant Levels for the 12 Primary Trace Metal Contaminants in Drinking Water

Element	NPDWR MCL (µg/L)	Typical ICP-MS DLs (µg/L)
As	10	0.05
Ba	2000	0.01
Be	4	0.01
Cd	5	0.02
Cr	100	0.05
Hg	2	0.01
Pb	15	0.005
Ni	100	0.005
Cu	1300	0.005
Sb	6	0.002
Se	50	0.2
Tl	2	0.001

Source: EPA Method 200.8: December 5, 1994 (*Federal Register*—Vol. 59[232], p. 62,546).

Al, Mn, Ag, and Zn. It should be noted that in January 2001, the MCL goal for arsenic (As) in drinking water was set at zero.[2] This was a health-based initiative and was not actually enforceable. However, in February 2002, an enforceable MCL of 10 ppb was applied to community and noncommunity water systems, which are not presently subject to arsenic standards. In addition, the EPA Office of Water (www.epa.gov/ow) has stated that all water systems, nationwide, had to be fully compliant by January 2006. This extremely low level means that only ICP-MS or GFAA (under Method 200.9) methods can be used to determine arsenic, because the ICP-OES methodology (inc. Method 200.7) cannot meet the required limits of quantitation.

In addition to drinking water, Method 200.8 can also be used for trace elements in wastewater under the National Pollutant Discharge Elimination System (NPDES). It has had general approval since 1995, but full acceptance varies on a regional basis, which means that each laboratory must apply for an ATP (Alternate Test Procedure) to their local EPA Quality Assurance Officer. In addition, since January 2000, Method 200.8 can also be used under NPDES rules for the analysis of wastewater from industrial incinerators.[3] Other Office of Water ICP-MS–related methodology include the following:

- Method 1638, which is a variation of Method 200.8, for the determination of trace elements in ambient waters[4]
- Method 1640 for the determination of trace metals in ambient waters by on-line chelation and preconcentration[5]
- Method 1669 for the sampling of ambient water for the determination of trace metals at EPA water quality criteria levels[6]

These EPA-driven methods represent the bulk of the routine environmental analyses being carried out by ICP-MS today. However, there are many other types of samples being analyzed, which represent a much smaller but significant contribution to the environmental application segment. For example, to better understand industrial-based airborne pollution covered by the Clean Air Act, air quality is often monitored using air-filtering systems. These typically consist of small pumps (either static or personal) where the air is sucked through a special filter for extended periods. The filter paper is then removed, dissolved in a dilute acid, and analyzed by an appropriate technique. Because trace metal concentration levels are sometimes extremely low, ICP-MS has proved itself to be a very useful tool to analyze these airborne particulate samples and help pinpoint sources of industrial pollution.

Other important work involves the analysis and classification of certified reference materials on the environment produced by standards organizations such as NIST (National Institute of Standards and Technology) and NRC (National Research Council of Canada). Some of these standards include drinking water, river water, open-ocean seawater, coastal seawaters, estuarine waters/sediments, freeze-dried dogfish/muscle tissue, spinach/orchard leaves, and many, many more. These reference materials are often analyzed using isotope dilution methods (refer to Chapter 13), because traditional external calibration typically does not offer high enough accuracy.[18]

However, it should be emphasized that probably 90% of all routine environmental laboratories are using basic quadrupole ICP-MS instrumentation. That is not to say other types of mass analyzers are not suitable for environmental analysis, but when detection limit requirements, sample throughput demands, operator skill level, and financial considerations are taken into account, quadrupole technology is the logical choice. In fact, vendors are now beginning to offer turnkey systems containing all aspects of EPA methodology, including analyte masses, internal standards, integration times, and QC protocol, etc. These methods are designed specifically for environmental users, because the majority of instruments are being operated by technicians having limited experience in ICP-MS. For more information on the analysis of environmental samples by ICP-MS, the following references should also be helpful.[16–21]

BIOMEDICAL

The second largest market segment is biomedical. Compared to other markets such as environmental and geochemical, the biomedical community was relatively late in realizing the benefits of ICP-MS as a routine tool. Although early biomedical researchers showed the capabilities of ICP-MS,[22,23] it was not until the early 1990s that it was first used as a technique for routine nutritional and toxicity studies.[24] Since then, it has probably become the fastest-growing market segment for ICP-MS, because it provides a fast, cost-effective way to carry out trace element studies in critically important areas of biomedical research such as toxicology, pathology, nutrition, forensic science, occupational hygiene, and environmental contamination. And when coupled with chromatographic separation techniques, ICP-MS becomes an ultrasensitive detector for carrying out trace element speciation studies (refer to

Chapter 18 for more details about this technique). Examples of biomedical analyses being carried out by ICP-MS include the following:

- Determination of toxic elements, in whole blood, as an indication of whether a person could be exposed to some kind of contamination in their home or from industrial-based pollution of the environment.[25]
- The analysis of urine can give a better understanding of the toxicity or bio-availability characteristics of an element or its speciated form.[26]
- It is important to know the levels of nutritional elements in human serum to understand how they are absorbed into the bloodstream.[27]
- Analysis of organ tissue samples to confirm the diagnosis of certain diseases.[28]
- Monitoring aluminum in patients who are undergoing kidney dialysis treatment.[29]
- The determination of trace elements in bones and teeth as an indicator of heavy metal exposure.[30]
- The multielement analysis of hair samples can help determine whether a person is lacking in essential vitamins and nutrients.[31]
- Studies of the effects of different inorganic and organic species on human health using chromatographic separation coupled with ICP-MS.[32]

More recently there has been an explosion of interest in using collision/reaction cell/interface technology for the analysis of biomedical samples, because of the benefits it brings to the determination of many of the toxologically and nutritionally significant elements such as As, Se, Cr, Fe, and Cu. Traditionally, these elements have been very difficult to analyze by ICP-MS because of the spectral interferences derived from a combination of the matrix, solvent/acid, and plasma gas ions. This approach is allowing significant improvements in detection capability for both the total and speciated forms of these elements in biomedical-related samples, such as blood serum and tissue samples.

As you would expect, the analysis of clinical-type samples is not that straightforward, because of the complex nature of blood, urine, serum, and body tissue samples, etc. Unlike environmental samples, which often require just simple acidification or maybe an acid digestion, the matrix components of biomedical samples can pose some unique problems for ICP-MS in the areas of sample preparation, interference correction, calibration, and long-term stability.[33] Let us take a look at these in greater detail.

SAMPLE PREPARATION

Ideally, the sample preparation methods must be simple, straightforward, and capable of being carried out in a routine manner. The more complex the sample preparation, the greater the chance of contamination, which ultimately affects accuracy and spike recoveries. The preferred method of sample preparation is by simple dilution with a suitable diluent such as dilute nitric acid for urine or 5–10% tetra methyl ammonium hydroxide (TMAH) for blood. However, this is not always possible with all types of biological materials. In these cases, a digestion with concentrated HNO_3

acid, followed by filtration or centrifuging, may be required to leach all elements into solution. For this type of sample preparation, microwave digestion apparatus has simplified the digestion of difficult samples and is usually the preferred approach over conventional hot plate acid digestion.

INTERFERENCE CORRECTIONS

During method development, special attention must be given to correct for matrix and spectral interferences. Matrix suppression and sample transport interferences are compensated very well by the selection of suitable internal standards, which are matched to the ionization properties of the analyte elements. This is a routine and well-understood method for compensating for matrix-related interferences. However, a more serious problem in the analysis of clinical samples is that analytes of interest can be affected by isobaric, polyatomic, and molecular spectral interferences resulting from plasma and matrix species. Table 19.3 shows some common interferences seen in clinical samples.

To get around this problem using a basic quadrupole system, either another isotope of the element of interest has to be monitored or an elemental correction equation needs to be applied. This is common methodology used to analyze clinical samples. However, if the trace metal levels in the sample are extremely low or sample preparation necessitates the use of an acid/solvent that contains one of the interfering ions (e.g., Cl^+ or N^+), this approach struggles. For that reason, ultratrace levels in some clinical samples require the use of either a high-resolution magnetic sector instrument to resolve the interference away or collision/reaction cell/interface technology to stop the formation of the interference, using ion molecule reaction and collision mechanisms. As mentioned previously, collision/reaction cells and interfaces are expanding the role of ICP-MS for the routine analysis of biomedical samples, because it is proving a cost-effective way of reducing many of the problematic spectral interferences that have traditionally limited the detection capability of some of the clinically relevant elements.[34,35]

TABLE 19.3

Some Common Spectral Interferences Seen in Clinical Matrices

Element	Interference
$^{24}Mg^+$	$^{12}C^{12}C^+$
$^{27}Al^+$	$^{13}C^{14}N^+$
$^{51}V^+$	$^{16}O^{35}Cl^+$
^{52}Cr	$^{40}Ar^{12}C^+, ^{16}O^{35}ClH^+$
$^{56}Fe^+$	$^{40}Ar^{16}O^+$
$^{58}Ni^+$	$^{58}Fe^+, ^{42}Ca^{16}O^+$
$^{63}Cu^+$	$^{40}Ar^{23}Na^+$
$^{75}As^+$	$^{40}Ar^{35}Cl^+$
$^{80}Se^+$	$^{40}Ar^{2+}$

CALIBRATION

Because of the differences in the matrix components of samples such as urine, blood, or serum, simple external calibration can often produce erroneous results. For that reason, it is common to use other calibration methods such as standard additions or addition calibration to achieve accurate data. These methods have been described in detail in Chapter 13, but they are required because of the matrix suppression effects caused by large variations in patients' biological fluid samples. The sample preparation method used will often dictate the type of calibration curve to use, but all three methods are absolutely necessary to achieve good, accurate data when analyzing clinical samples by ICP-MS.

STABILITY

Today, ICP-MS has proved rugged enough to be used routinely in high-throughput clinical laboratories. However, complex blood, urine, serum, and digested body tissue matrices can affect signal stability, resulting in the need for frequent recalibration. One of the major problems is that matrix components (salts, carbon, proteins, etc.) can deposit either on the tip of the plasma torch sample injector or on the orifice of the sampler, or the skimmer cone, which over time can eventually lead to blockage and signal instability. Another negative impact of clinical matrices on an ICP mass spectrometer is that material can deposit on the ion optics system, leading to instability and the likelihood of reoptimizing the lens voltages. Even though some instrument designs will be affected less than others, it is well accepted that routine maintenance (including regular cleaning and replacing parts) is absolutely essential to keep up with the harsh demands of running clinical samples, especially if the instrument is being used on a routine basis.

Although levels of interest are generally lower than those required by the environmental ICP-MS community, the biomedical market segment is interested in a similar suite of elements and also has similar sample throughput and productivity demands. This has been driven by a growing demand to bring down the cost of analysis to lessen the financial burden on hospitals and health authorities. All these factors have contributed to the overwhelming acceptance of ICP-MS for the trace element analysis of biomedical samples, in preference to slower, less productive techniques such as GFAA.

It is also worth emphasizing again that understanding the effects of different elemental forms and species on human health and its impact on the environment has sparked an enormous growth in speciation studies using ICP-MS and chromatography separation devices, such as HPLC,[36–38] SEC,[39] supercritical fluid extraction (SFEC),[40] and capillary zone electrophoresis (CZE).[41] This hyphenated approach has been described in greater detail in Chapter 18, where other references on the toxicological impact of elemental species on the environment can be found.

GEOCHEMICAL

Geochemists were some of the first researchers to realize the enormous benefits of ICP-MS for the determination of trace elements in digested rock samples.[42] Until then, they had been using a number of different techniques, including neutron activation analysis (NAA), thermal ionization mass spectrometry (TIMS), ICP-OES, x-ray techniques, and GFAA. Unfortunately, they all had certain limitations, which meant that no one technique was suitable for all types of geochemical samples. For example, NAA was very sensitive, but when combined with radiochemical separation techniques for the determination of rare earth elements, it was extremely slow and expensive to run.[43] TIMS was the technique of choice for carrying out isotope ratio studies because it offered excellent precision, but unfortunately was painfully slow.[44] Plasma emission was very fast and excellent for multielement analysis, but it was not very sensitive. In addition, because the technique suffered from spectral interferences, ion exchange techniques often had to be used to separate the analyte elements from the rest of the matrix components.[45] X-ray techniques such as XRF (x-ray fluorescence) and XRD (x-ray diffraction) were rapid, but generally not suited for ultralow levels and also struggled with some of the lighter mass elements.[46] Although GFAA had good sensitivity, it was predominantly a single-element technique and was therefore very slow.[47] It was also not suitable for low levels of refractory or rare earth elements, because the low atomization temperature of the electrothermal heating device (<3000°C) did not produce sufficiently high numbers of ground-state atoms. Even though all these techniques are still used to some degree, all these factors led to the very rapid acceptance of ICP-MS by the geochemical user community.

Geochemists represent some of the most demanding users of ICP-MS. Invariably, they are looking for ultratrace levels in the presence of large concentrations of major elemental components in digested rock samples, such as Ca, Mg, Si, Al, and Fe. This alone presents difficulties for the sample introduction and interface region because of the potential for signal drift caused by the geological material depositing itself on the cones and ion lens system. In addition, if there are large concentrations of high-mass elements such as Tl, Pb, or U present in the sample, they can cause severe space charge, matrix suppression on the analyte masses. Another potential problem is that major and trace components in the sample can combine with argon-, solvent-, and acid-based species to produce quite severe polyatomic, isobaric, doubly charged, and oxide-based spectral interferences. When this is combined with the extremely demanding sample preparation methods using highly corrosive chemicals such as concentrated aqua regia (HCl/HNO_3), hydrofluoric acid (HF), and fusion mixtures to dissolve the samples, geological samples become some of the most difficult to analyze by ICP-MS. And because of the complexity of some of these matrices and sample digestion procedures, collision/reaction cell/interface technology is usually not the best approach to use, because ion–molecule collision and reaction mechanisms are not ideally suited for the reduction of sample and matrix-induced polyatomic spectral interferences. To illustrate these difficulties, let us take a look at some typical geochemical applications being carried out by ICP-MS.

DETERMINATION OF RARE EARTH ELEMENTS

The determination of rare earth elements was one of the very first applications that attracted geochemists to ICP-MS, mainly because of the lengthy sample preparation and analysis times involved with previously used techniques such as ICP-OES and NAA. However, even though ICP-MS offered significant benefits over these techniques, it was not without its problems, because of the potential of spectral interferences from other rare earth elements in rocks or natural water samples. For that reason, instrument parameters have to be optimized, depending on the rare earth elements being determined and the kinds of interferents present in the sample. For example, plasma power and nebulizer gas flows must be adjusted to minimize the formation of oxide species. This is necessary because an oxide or hydroxide species of one rare earth element can spectrally interfere with another rare earth element at

TABLE 19.4

Examples of Rare Earth Elements That Readily Form Oxide and Hydroxide Species in ICP-MS

Rare Earth Oxide/Hydroxide	Interferes With
$^{135}Ba^{16}O^+$	$^{151}Eu^+$
$^{136}Ba^{16}O^+$, $^{136}Ce^{16}O^+$	$^{152}Sm^+$
$^{141}Pr^{16}O^+$, $^{140}Ce^{16}OH^+$	$^{157}Gd^+$
$^{143}Nd^{16}O^+$, $^{142}Ce^{16}OH^+$	$^{159}Tb^+$
$^{146}Nd^{16}OH^+$, $^{147}Sm^{16}O^+$	$^{163}Dy^+$
$^{149}Sm^{16}O^+$	$^{165}Ho^+$
$^{152}Sm^{16}O^+$	$^{168}Er^+$
$^{153}Eu^{16}O^+$, $^{152}Sm^{16}OH^+$	$^{169}Tm^+$
$^{158}Gd^{16}O^+$	$^{174}Yb^+$
$^{158}Gd^{16}OH^+$, $^{159}Tb^{16}O^+$	$^{175}Lu^+$

Source: D. J. Douglas, *Canadian Journal of Spectroscopy,* **34**(2), 1989.

16 or 17 amu higher. The problem can be alleviated by using a sample desolvation device such as a chilled spray chamber to reduce oxide formation, but unfortunately cannot be completely eliminated. For that reason, to get the best detection capability for rare earth elements in geological matrices, instrument sensitivity must often be sacrificed for low oxide performance, and even then, mathematical correction equations need to be applied. Examples of rare earth elements that readily form oxides/hydroxides, and the elements they interfere with, are shown in Table 19.4.[48]

It is also worth pointing out that in addition to the formation of oxide species, some rare earth elements can generate high levels of doubly charged ions (ions with two positive charges as opposed to one). This is not so much of a problem with the determination of other rare earth elements, but their spectral impact on other lower-mass analytes. Examples of rare earth elements that easily form doubly charged species include barium, cerium, samarium, and europium, as shown in Table 19.5. If these elements are present in high enough concentrations, certain isotopes can interfere with analytes at one-half of their mass. Parameter optimization can help, but even more important is to minimize the effects of high plasma potential (a secondary discharge at the interface is known to increase doubly charged species) with a well-grounded RF coil.[49] However, with certain geological matrices, no matter what precautions are taken, doubly charged species are unavoidable, depending on the analytes of interest.

TABLE 19.5

Examples of Rare Earth Elements That Readily Form Doubly Charged Species and the Analyte Masses They Interfere With

Doubly Charged Species	Interferes With
$^{138}Ba^{2+}$	$^{69}Ga^+$
$^{140}Ce^{2+}$	$^{70}Ge^+$, $^{70}Zn^+$
$^{151}Eu^{2+}$	$^{75}As^+$
$^{152}Sm^{2+}$	$^{76}Ge^+$, $^{76}Se^+$

ANALYSIS OF DIGESTED ROCK SAMPLES USING FLOW INJECTION

The benefits of FI techniques for ICP-MS have been described in detail in Chapter 17. The main advantage of FI for the analysis of geological samples is the ability to aspirate high concentrations of dissolved solids into the mass spectrometer. With continuous nebulization, it is generally accepted that to maintain good stability, the total dissolved solids (TDS) in the sample should not exceed 0.2% w/v, which can be a severe limitation if analyte concentrations are extremely low. However, using the microsampling capability of FI, where small volumes (typically <500 µL) of the sample are transported into the ICP-MS in a continuous flow of carrier liquid, much larger levels of dissolved solids can be tolerated. In fact, it is fairly common to put in excess of 1% w/v dissolved into the ICP-MS system using this approach and still maintain good accuracy and precision for geological matrices. This is exemplified in Table 19.6, which shows the determination of a group of elements in a USGS (United States Geological Survey) standard reference rock (andesite)—AGV-1, using USGS SRM BEN (basalt) for calibration. Both sample and calibration standard were dissolved using a lithium tetraborate ($Li_2B_4O_7$) fusion mixture, which, including weight

TABLE 19.6
Determination of A Group of Elements in a USGS Standard Reference Rock (Andesite)—AGV-1 Using Flow Injection Microsampling (TDS in Solution Was 1.2% w/v)

Element	USGS AGV-1 Reference Value (mg/kg)	Measured Value by FI-ICP-MS (mg/kg)	Precision (% RSD)
Ba	1226.0	1204.0	1.0
Be	2.1	2.1	2.1
Ce	67.0	70.5	0.6
Co	15.3	14.4	1.2
Cs	1.3	1.7	1.8
Cu	60.0	52.3	0.9
Eu	1.6	1.5	4.6
Ga	20.0	20.2	0.4
La	38.0	33.7	1.4
Lu	0.3	0.2	4.1
Mo	2.7	2.0	2.3
Sr	662.0	628.3	0.6
Yb	1.7	1.4	0.6
Zn	88.0	111.3	0.8
W	0.6	0.8	2.5
V	121.0	121	0.0
U	1.9	1.8	6.6

Source: A. Stroh, U. Voellkopf, and E. R. Denoyer, *Journal of Analytical Atomic Spectrometry*, **7**, 1201–1205, 1992.

of sample, represented 1.2% w/v total dissolved solids—six times more material than is typically aspirated into an ICP mass spectrometer.[50]

It should also be emphasized that when these levels of dissolved solids are being introduced into the ICP-MS on a routine basis, it is very important that the nebulization system be able to aspirate the sample without clogging the nebulizer orifice/tips. For that reason, a nebulizer that is designed for high solids should be used. It can be a high-solids concentric nebulizer, a cross-flow nebulizer, or a Babbington design nebulizer, but it should not be a low-flow or microflow concentric nebulizer, which would not be able to handle the flow injection introduction of samples with such high levels of dissolved solids.

GEOCHEMICAL PROSPECTING

Exploring for deposits of the platinum group elements (PGEs), commonly known as precious metals, is typically carried out by sampling large areas to establish a concentration contour map. The occurrence, distribution, and concentration of these precious metal deposits are then used to ascertain whether mining is economically feasible in that area. Analytical methodology developed for the determination of precious metals in geological samples must therefore have sufficient sensitivity to quantify individual PGEs at the ng/g (ppb) level, but also be fast enough to cost-effectively handle such a large number of samples.[51]

Determination of precious metals in geological samples is generally a three-step process. The first step involves preparing a representative sample and then isolating the analyte from the ore matrix using established methods such as the fire assay technique. This typically involves fusion with a flux, producing a lead or nickel sulfide button that is then ground into a powder. The second step separates the precious metals from the rest of the matrix by a process called *cupellation*. This process involves heating the powdered sample in a cupel made of bone ash (phosphate of lime), where the matrix components are oxidized into the porous cupel, leaving the precious metals separated out from the rest of the sample.[52] The final step involves dissolution of the precious metals with a suitable acid and measurement of the analyte concentrations by some kind of instrumental technique. There are slight variations to the fire assay procedure (based on how the PGE is extracted from the lead/nickel sulfide button), which are often dictated by the type of sample being collected and elemental requirements of the analysis. The result is that a number of different trace element techniques have been used for this type of analysis, including GFAA, ICP-OES, and NAA. All three approaches have been used to quantify PGEs with good accuracy in fire assay samples, but as mentioned earlier, ICP-OES will struggle with low concentrations, and in the case of GFAA and NAA, sample throughput is severely restricted because of its slow speed of analysis. All these factors have contributed to the rapid acceptance of ICP-MS for the determination of PGEs by fire assay and other sample preparation methods.[53,54] This is emphasized in Figure 19.2, which shows the superior detection capability of ICP-MS over both GFAA and ICP-OES for the PGEs.

The ultralow detection capability of ICP-MS, combined with its rapid speed of analysis, high sample throughput, and excellent accuracy and precision make it ideally

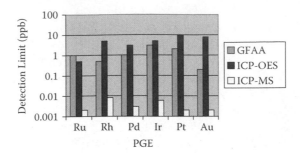

FIGURE 19.2   Detection capability improvement of ICP-MS over GFAA and ICP-OES for the platinum group elements (PGEs).

suited for this type of work. In fact, in countries such as Australia and Canada that have large mineral deposits, large commercial laboratories have sprung up that use ICP-MS on a 24-hour basis, to support their country's extensive mining exploration business.[51]

ISOTOPE RATIO STUDIES

The study of isotope ratios is extremely important to geochemists and environmentalists, both as a means of approximating the age of rock formations (geochronology)[55] and tracing the source of metallic pollutants on the environment and ecosystems.[56] However, one of the main requirements for this kind of analysis is the ability of the method to generate data with extremely high precision. For this reason, the most widely used instrumental approach has involved the use of TIMS.[57] Unfortunately, even though TIMS is capable of producing isotope ratio precision down to 0.005% RSD, the analytes have to be isolated from the matrix, making the sample pretreatment cumbersome and time consuming. In addition, the sample solutions obtained have to be preconcentrated and loaded onto a filament, which are then mounted onto a sample turret and subsequently inserted into the vacuum-pumped chamber of the TIMS instrument.

These sample throughput limitations led geochemical researchers to investigate ICP-MS as a possible solution to their problem. Unfortunately, it soon became clear that although quadrupole ICP-MS demonstrated excellent throughput, the best isotope precision it could offer on a routine basis was 0.2–0.5% RSD. It was not until the commercialization of double-focusing magnetic sector ICP-MS technology in the early 1990s that geochemists realized they had an analytical tool that could perhaps compete with TIMS for carrying out isotope ratio studies. The extremely high sensitivity, low background, fast scanning, and flat-topped peaks of this technique consistently demonstrated precision data on the order of 0.05–0.10% RSD, as can be seen in Table 19.7, which shows ^{206}Pb$^+$/^{207}Pb$^+$ isotope ratio precision data, taken from a paper by Vanhaecke and coworkers.[58] It should be noted that the lead concentration data were varied slightly in each measurement set to produce a peak height of ~200,000 cps, and it can be clearly seen that the experimental data for each set is approaching its statistical counting limits.

TABLE 19.7

Typical $^{206}Pb^+/^{207}Pb^+$ Isotope Precision Data for a Single-Collector, Double-Focusing Magnetic Sector ICP-MS, Compared to Its Statistical Counting Limits

Measurement Set	Experimental RSD for $^{206}Pb^+/^{207}Pb^+$ ($n = 10$)	Theoretical RSD (based on counting statistics)
1	0.11	0.062
2	0.044	0.062
3	0.12	0.065
4	0.063	0.053

Source: F. Vanhaecke, L. Moens, R. dams and P. Taylor, *Analytical Chemistry*, **68**(3), 567–569, 1996.

Even though this kind of data was much better than quadrupole ICP-MS, it still was not as good as TIMS. For this reason, the geological community wanted even better isotope ratios by ICP-MS, and as a result, instrument manufacturers eventually answered their demands with the development and commercialization of multi-collector, magnetic sector ICP-MS systems. This design, which utilized multiple detectors instead of just one, allowed for the simultaneous measurement of each mass, offering the capability to generate isotope ratios equivalent to TIMS.[59]

Such are the extreme demands of the geochemical application sector that researchers are looking for techniques and sampling accessories that offer a high level of performance and flexibility. For that reason, the high resolution, high sensitivity, and excellent precision capability of magnetic sector systems makes them ideally suited to this kind of work. In addition, collision/reaction cell/interface technology, although not ideal for many geological matrices, is becoming more popular with geochemists because of its ability to reduce many of the traditional geochemical-based spectral interferences using collision and reaction chemistries.[60] Vanhaecke and coworkers have shown that $^{87}Rb^+$ can be "chemically resolved" from $^{87}Sr^+$ using a mixture of methyl fluoride (CH_3F) and neon (Ne) gas in a dynamic reaction cell, so that $^{87}Sr^+/^{86}Sr^+$ isotope ratios can be measured with good enough precision for perhaps screening samples for geochronological dating studies.[61] The attraction of this technology over a high-resolution approach is that it would take a resolving power of 290,000 to separate $^{87}Rb^+$ from $^{87}Sr^+$ using magnetic sector technology. In addition, if there is a significant amount of krypton in the argon supply (which is fairly common), it would require a resolving power of 66,000 to separate $^{86}Kr^+$ from $^{86}Sr^+$. Unfortunately, this is way beyond the capabilities of commercial magnetic sector instruments, which typically offer resolving power up to 10,000.

In addition, Beres and coworkers showed the benefit of using a collision/reaction cell to determine a group of PGEs in copper-bearing ores and minerals rich in refractory elements. In the case of copper ores, which are digested in concentrated hydrochloric acid, high levels of chloride, copper, and nickel ions together with argon ions from the plasma form polyatomic complexes, which interfere with the determination of rhodium, ruthenium, and palladium. On the other hand, in the analysis of refractory element minerals, which are digested in aqua regia, the refractory metals hafnium, zirconium, and tantalum readily form oxides and interfere with the isotopes of palladium, gold, and platinum. They showed that by using ammonia as a reaction gas, the major polyatomic spectral interferences generated by the copper/nickel chloride matrix and the argon gas could be significantly reduced. Methyl chloride gas was successfully used to minimize refractory oxides and allow the determination of ppb levels of palladium, gold, and platinum.[62]

LASER ABLATION

Laser ablation as a sampling tool and its applicability for use with ICP-MS has been described in great detail in Chapter 17. There is no question that after many years of being considered a "novel and interesting" technique, it has now been refined to become an extremely useful sampling technique for many types of materials.[63] However, it was primarily geochemists and mineralogists who drove the development of laser ablation for ICP-MS because of their desire for ultratrace analysis of optically challenging materials, such as calcite, quartz, glass, and fluorite, combined with the capability to characterize small spots and microinclusions on the surface of the sample. For this reason, most of the fundamental studies on the ablation process have been based on the analytical demands of the geochemical community.[64]

However, in the early days there was much debate as to the optimum design to use for the many diverse types of geochemical samples.[65] Based on literature in the public domain, the general consensus today is that 266 nm Nd:YAG technology is extremely good for bulk analysis, and for some of the less challenging micro-inclusion work. The shorter-wavelength 213 nm Nd:YAG technology couples much better with UV-transparent material such as silicates, fluorites, and calcites, and is also better suited for the study of minute fluid inclusions, because of its more controlled ablation process. On the other hand, 193 nm ArF excimer laser offers the most precise ablation characteristics of all three designs and excels with the most optically challenging materials.[66] It is also the best tool to use for precise and accurate depth measurement studies. The main disadvantage of the ArF excimer design is that, because of its optical complexity and the requirement to use a toxic gas, a more skilled person is required to operate and maintain it. It is also the most expensive of all three designs to purchase.

So, although the development of laser ablation for ICP-MS has gone in many different directions, it is now generally accepted that shorter wavelengths are considered more suited for geological matrices, especially for the analysis of small spots and inclusions in UV-transparent materials. However, it is still a very active area of research, which sees new developments and refinements on a regular basis. Although arguments can be made for the benefits of one specific design over another, it is not that straightforward, especially when the capabilities of the ICP mass spectrometer are taken into consideration. For example, when it is being applied to the analysis of microinclusions, it is absolutely critical that the ICP-MS system be capable of very high sensitivity, because you may only be able to fire one laser pulse to ablate the area of interest. In addition, when analyzing a fast transient (~10 s) of fine particles generated by a single laser pulse, it is very important that the scanning and settling times of the mass analyzer be kept to an absolute minimum. For these reasons, it is fairly common to see double-focusing magnetic sector ICP-MS technology used with laser ablation systems because of its extremely high sensitivity.[67] Also, TOF is beginning to show its benefits for laser ablation work, because of its ability to simultaneously sample the ion beam and capture the maximum amount of data in the limited duration of the short transient peak.[68]

So, the optimum combination of laser ablation system and ICP mass spectrometer can often be sample and application specific. There is no question that in the hands of a good operator, most laser ablation systems should work well with any commercial ICP-MS systems and be capable of generating data of the highest quality on complex rock samples, small inclusions in rocks, or carrying out elemental surface-mapping studies. It is not the intent of this book to show bias toward any design, but just to emphasize that if there is a need for this kind of solid-sampling capability, each integrated system should be evaluated on a sample-by-sample basis. In addition, good literature references should be read to get a better understanding of what instrument features (bother ICP-MS and laser ablation systems) are important to carry out high-quality, interference-free geochemical analysis.[69–71]

SEMICONDUCTOR

The semiconductor industry is probably the most demanding user of ICP-MS, with regard to its detection capability requirements. Consumer demand for smaller electronic devices and more compact integrated circuits has resulted in the need for ultratrace metal contamination levels on the surface of silicon wafers and also in the high-purity chemicals and gases used in various stages of the semiconductor manufacturing process. In order to reduce costs and increase yield, chip manufacturers are making larger-diameter wafers with even narrower line widths. This trend, which is being driven by initiatives such as the ITRS (International Technology Roadmap for Semiconductors),[72] is setting the course for the next generation of semiconductor devices and has resulted in progressively lower trace element contamination levels in all semiconductor-related materials. Whereas 10 years ago, the SEMI (Semiconductor Equipment and Materials International) organization deemed that 1 ppb purity levels were adequate for many of the process chemicals, today 100 ppt is typical—and for some of the more critical materials, 10 ppt guideline levels are currently being proposed.[73]

The SEMI BOSS (Book of Semiconductor Standards) has approved ICP-MS for the determination of trace metals in a number of chemicals at the 10 ppb (Grade 3) and 100 ppt (Grade 4) levels and is looking into the feasibility of approving the technique for some chemicals at the 10 ppt (Grade 5) level. Table 19.8 shows typical specification levels for all the semiconductor-significant elements. Some element specifications are different for different chemicals, but this table represents a good approximation of the trend for comparison purposes.[73]

However, the BOSS states that GFAA can also be used if ICP-MS does not have the required detection limit. The inherent problem lies in the fact that many of these corrosive chemicals have to be diluted 10 times, or even 100 times, to aspirate them

TABLE 19.9

Typical Detection Limits and BEC Values Achievable with a Dynamic Reaction Cell, for All the 21 Semiconductor-Significant Elements Defined in SEMI Grade 5 (10 ppt) Standard

Element	Detection Limit (ppt)	BEC (ppt)
Aluminum[a]	0.23	0.42
Antimony	0.08	0.08
Arsenic	0.48	1.60
Barium	0.06	0.04
Boron	3.60	7.10
Cadmium	0.08	0.11
Calcium[a]	0.27	0.63
Chromium[a]	0.14	0.29
Copper[a]	0.06	0.68
Iron[a]	0.49	2.60
Lead	0.07	0.09
Lithium	0.26	0.22
Magnesium	0.23	0.18
Manganese	0.17	0.54
Nickel	0.43	0.66
Potassium[a]	0.27	2.60
Sodium	0.20	0.22
Tin	0.12	0.88
Titanium[a]	0.92	1.70
Vanadium[a]	0.12	0.04
Zinc[a]	0.63	1.20

[a] These elements were obtained using NH_3 as the reaction gas, whereas the other elements were determined in standard mode with no reaction gas.

Source: J. M. Collard, K. Kawabata, Y. Kishi, and R. Thomas, *Micro*, January 2002.

stimulate ion–molecule collisions and reactions. Depending on the design of the cell, the by-products of the reactions and collisions are either removed by reaction mechanisms and bandpass filtering, or collision mechanisms and KED. The capabilities of a dynamic reaction cell have been demonstrated by Collard and coworkers, who showed that method detection limits and spike recovery data for all 21 elements in Grade 5 (10 ppt) hydrogen peroxide could be achieved using strict SEMI methodology.[75] Table 19.9, which was taken from that study, shows typical detection limits and BEC (background equivalent concentration) values achievable with a dynamic reaction cell for all 21 elements defined in a SEMI Grade 5. BEC values are included, because many analysts in the semiconductor community believe that it gives a better indication as to how efficient the background reduction technique is. BEC is defined as the intensity of the spectral background at a particular mass, expressed as a concentration value. The lower the BEC, the easier it is to distinguish an analyte signal from its background.

One of the application areas in which collision/reaction cell/interface technology is showing enormous potential is in the reduction of polyatomic interferences that are not addressed by cool plasma or high-resolution technology. For example, the determination of arsenic and chromium in high-purity hydrochloric acid matrix is a very difficult analysis, because of the $^{40}Ar^{35}Cl^+$ and $^{35}Cl^{16}OH^+$ polyatomic spectral interferences on $^{75}As^+$ and $^{52}Cr^+$, respectively. High-resolution systems do not offer good detection capability, because it requires high resolving power to separate $^{75}As^+$ from $^{40}Ar^{35}Cl^+$ and $^{52}Cr^+$ from $^{35}Cl^{16}OH^+$, which results in a significant loss of sensitivity. Cool plasma technology has shown limited success for chromium in a high chloride matrix because of matrix suppression effects, and is not really suited for arsenic because of its high ionization potential, thus making it very difficult to ionize in a low-temperature plasma. Schleisman and coworkers have demonstrated a detection limit of <2 ppt for arsenic and 7 ppt for chromium in a 10% hydrochloric acid using a dynamic reaction cell.[76] Some of the other successful interference reduction studies reported in the literature using collision/reaction cell technology include $^{40}Ar^{12}C^+$ on the determination of $^{52}Cr^+$ in an organic matrix[77] and $^{31}P^{16}O^{16}O^+$ on the determination of $^{63}Cu^+$ in phosphoric acid.[78]

There appears to be a definite trend toward the use of collision/reaction cell/ interface technology for the analysis of other high-purity process chemicals used in the semiconductor industry, including ultrapure water,[79] hydrochloric acid,[80] sulfuric acid,[78] isopropyl alcohol,[81] and photoresist.[82] The technology is also showing a great deal of promise for implementing remote, automated on-line monitoring of semiconductor process baths using customized sampling streams and components.[83]

It is important to emphasize that the semiconductor industry is unique in its demands on instrument manufacturers, because unlike any other application area, it is constantly chasing "zero." Although this is an unrealistic demand, zero means as little contamination as possible during the manufacturing process, which translates into fewer defects, and therefore a higher yield of semiconductor devices. This is what drives the industry, and the choice of analytical techniques used reflects this. For this reason, any trace element equipment that is applied to contamination control—whether it is the analysis of ultra-high-purity water, the determination of trace metals in chemicals and gases, or carrying out vapor phase decomposition (VPD) studies on the surface of silicon wafers[84]—must be designed specifically for these demands. For example, the surface of the instrument should be as smooth as possible, so that it does not attract particles of dust. Instrumental components such as spray chambers, nebulizers, and pump tubing must be clean and free of contamination. Roughing pumps should be capable of remote operation from the instrument to minimize the effects of vibration. In addition to instrument cleanliness issues, the instrument and sample preparation areas should be housed in at least a Class 1000 clean room, and for some applications, a Class 100, Class 10, or even a Class 1 might be required. All the volumetric flasks, beakers, and storage bottles, etc., need to be of the highest quality with regard to trace metal content. Finally, the calibration standards and acids used to prepare the samples must be of the highest purity available. The bottom line is that no matter what type of ICP-MS is used for trace element determinations, even the most sophisticated and high-performance instrument will generate bad data, unless all the sample preparation stages and cleanliness issues are taken into consideration.

NUCLEAR

The types of samples generated by the nuclear industry, including bulk nuclear materials; high- or low-level radioactive waste; water-, soil-, and biota-based remediation samples; and environmental impact studies and human health monitoring, put unique demands on any analytical technique used for isotopic quantitation. Even though traditional ionizing radiation counting techniques have worked exceptionally well over the years, they are painstakingly slow. The inherent problem lies in the fact that to ensure that radiometric-derived interferences from other sample components are kept to a minimum, time-consuming chemical separations have to be carried out. In addition, the half-life of the analyte isotope has a significant impact on the method detection limit, which means that to get meaningful data in a realistic amount of time, they are better suited for the determination of short-lived radioisotopes. They have been successfully applied to the quantitation of long-lived radionuclides, but unfortunately require a combination of extremely long counting times and large amounts of sample in order to achieve low levels of quantitation.[85]

Limitations in the traditional α (alpha) spectrometry, γ (gamma) spectroscopy, and scintillation and proportional counting technology, especially at extremely low levels, led to the use of atom-counting techniques for radiochemical analysis, such as TIMS (thermal ionization mass spectrometry), SIMS (secondary ion mass spectrometry), AMS (accelerator mass spectrometry), and FTA (fission track analysis). In addition, other techniques were being developed, such as FT-ICR (Fourier transform ion cyclotron resonance), RIMS (resonance ionization mass spectrometry), and TOF-SIMS (time-of-flight SIMS), which were primarily being driven by the specific application demands of nuclear-based industry. However, even though all these approaches worked very well, depending on the application, they were primarily being used for specific tasks and were not considered truly routine analytical tools. In addition, many of these techniques utilized very complex components, such as dedicated nuclear reactors and linear accelerators, which meant they were extremely expensive to manufacture.

The drive in the nuclear industry for a more routine approach that was faster, had less interferences, required easier sample preparation, generated less waste, had good calibration standards, and importantly, offered lower cost per sample analysis, led researchers to investigate the use of ICP-MS. It was ironic that even though one of the first ICP-MS systems was built at a U.S. Department of Energy (DOE) site in 1980,[86] the nuclear community was relatively slow in accepting its routine use for radionuclide analysis. However, it soon became clear that the technique was going to be very complementary to traditional radiation counting technology used by the nuclear industry.[85] This can be seen in Table 19.10, which compares sensitivity, maturity status, typical use, and the advantages/disadvantages of ICP-MS with some

TABLE 19.10

Comparison of Atom-Counting Techniques for the Radionuclides

Technique	Sensitivity	Maturity of Technique for Radionuclides	Typical Use in Nuclear Industry	Advantages	Disadvantages
TIMS	10^6 atoms	Routine	Isotope ratios for many elements	Quantitative, high precision	Ultraclean sample preparation, slow, expensive, interferences from hydrocarbons
FTA	10^6 atoms	Routine	^{239}Pu	Quantitative	Need a nuclear reactor, interference from ^{235}U, expensive
SIMS	10^9 atoms	Routine	Isotope ratios for depth and surface profiling	High spatial resolution and ion imaging	Interferences from hydrocarbons, semiquantitative
AMS	10^5 atoms	Routine	^{10}Be, ^{14}C, ^{26}Al, ^{129}I	10^{15} abundance sensitivity	Complex technology, expensive
ICP-MS	10^6 atoms	Developing/ routine	Isotope ratios for many elements	Rapid, low cost, simple sample preparation,	Isobaric and polyatomic spectral and matrix interferences
LAMMA	10^9 atoms	Developing/ routine	Isotope ratios	High spatial resolution	Semiquantitative
FT-ICR	10^9 atoms	Research/ developing	Isotope ratios	High resolution, several ion sources	Isobaric and polyatomic spectral interferences
RIMS	10^9 atoms	Research/ developing	Isotope ratios	High selectivity, GFAA, and GD (glow discharge) ion sources	Nonquantitative

Source: Counsel of Ionizing Radiation Measurements Workshop on Standards, Intercomparisons, and Performance Evaluations for Low-Level and Environmental Radionuclide Mass Spectrometry— Meeting Proceedings, NIST, Gaithersburg, MD, April 1999.

of the more established atom-counting techniques and also some of the ones that are still considered to be in the research stage of development.[87]

There is no question that one of the major reasons for the success of ICP-MS in the nuclear industry is that the DOE is changing the mission of many of its facilities from defense-related nuclear materials production to site remediation and monitoring. This change has resulted in the need to fully characterize hazardous wastes and environmental samples, combined with the necessity to routinely monitor workers' exposure to harmful radiation. For this reason, nuclear facilities in the United States and elsewhere are strongly emphasizing these determinations, and are demanding better and faster analytical techniques to ensure the quality of the materials that they supply for the production of nuclear energy and other nuclear-related technologies. These factors, which have primarily been driven by DOE initiatives for cost-effective radiochemical analyses, have significantly increased the number of samples from nuclear waste management and nuclear facility cleanups since the mid-1990s, and as a result, the use and applications of ICP-MS have seen a dramatic increase in this field. Such is the interest in exploring its full potential that the ASTM Committee on Nuclear Fuel Cycle (C-26) has put together a subcommittee (C26-05) to primarily focus on the testing of samples generated by the nuclear industry. One of their main tasks is to not only develop methods using traditional particle-counting techniques, but also to explore the use of plasma spectrochemistry instrumentation such as ICP-MS to handle the diverse needs of the industry's application demands. Many of the ICP-MS applications developed over the past 20 years have come out of this subcommittee, including *Standard Test Method for Analysis of Total and Isotopic Uranium and Total Thorium in Soils by Inductively Coupled Plasma-Mass Spectrometry* (C-1345-96) and *Standard Test Method for Determining Radionuclides in Soils by Inductively Coupled Plasma-Mass Spectrometry Using Flow Injection Preconcentration* (C-1310-01). For a full list of all the methods and standards approved by this committee, visit the Nuclear Fuel Cycle home page on the ASTM Web site.[88]

To get a better understanding of the benefits that ICP-MS brings to the analysis of nuclear-fuel-related samples, let us take a closer look at the application demands of the industry.

APPLICATIONS RELATED TO THE PRODUCTION OF NUCLEAR MATERIALS

Typical analyses carried out in this category include the determination of various radionuclides and the measurement of isotope ratios in enriched uranium compounds such as uranium dioxide (UO_2) powder, hydrolyzed uranium hexafluoride (UF_6), and uranyl nitrate liquor (UNL). Depending on the isotopes of interest and the type of mass analyzer used, the problems associated with the analysis of uranium compounds include spectral interferences from actinides and other trace elements in the sample. For example, the determination of $^{99}Tc^+$ using quadrupole technology can be problematic because of the presence of an isobaric interference from $^{99}Ru^+$ and a molecular interference from $^{98}MoH^+$.[89] If these spectral interferences are not that severe, they can be corrected by mathematical equations; otherwise, some kind of high-resolution mass analyzer must be used to resolve the interfering species away from the analyte isotope. In addition, the high uranium matrix has the potential to cause severe space-charge-induced matrix suppression, especially if low-mass elements are also being determined. To a certain degree, this kind of interference is unavoidable, but can be minimized by careful optimization of ion lens voltages to reject the maximum number of uranium ions.[90] The use of collision/reaction cell/interface technology may have limited applicability to nuclear-fuel-type samples, because the spectral problems appear to be more related to isobaric overlaps produced by other radionuclide matrix component ions, and not so much by the acid or solvent combining with argon ions. There are very few publications on the use of collision/reaction cells/interfaces applied to these types of samples, so it is very difficult to know if the technology is being used to address the industry's trace element application problems.

On the other hand, if the requirement is for good isotope ratios, double-focusing magnetic sector technology offers the best solution. For example, if the isotope ratio of $^{235}U/^{238}U$ is being monitored in UF_6, it has shown that multicollector (MC) magnetic sector ICP-MS technology will give the best precision data. In fact, this is the preferred methodology over the more traditional TIMS approach, because unlike TIMS, the fluoride matrix does not have to be removed. For this reason, the complete analysis using MC-ICP-MS is completed in about one fifth of the time of TIMS, and achieves very similar isotopic ratio precision data.[91]

APPLICATIONS IN THE CHARACTERIZATION OF HIGH-LEVEL NUCLEAR WASTE

Some of the many applications in this category include the use of ICP-MS to support the processing, stabilization, and long-term storage of high-level waste (HLW). Common matrices encountered in this kind of work include sludges, slurries, and in particular, the glass waste forms that will be used for the isolation of nuclear waste in underground geological repositories. Analyses usually require the detection of low levels and isotopic content of uranium, in addition to small amounts of actinides and fission products, including ^{237}Np (neptunium), $^{239,\ 240}Pu$ (plutonium), ^{241}Am (americium), and ^{244}Cm (curium). The isotopic data for uranium generally does not need to be of the highest accuracy and precision, but to know primarily if the uranium is depleted, natural, or enriched, and if so, an estimate of its enrichment level. These types of samples are further complicated by the fact that they are typically contained in high-salt matrices, so they generally have to be diluted to aspirate into the ICP mass spectrometer.[92] Other uses for ICP-MS in the application area involve uranium and plutonium solubility studies in groundwater and related samples and also to help determine the efficiency of the separation process when carrying out traditional radiochemical counting. It is also important to point out that because of the dangers associated with characterizing high-level nuclear waste by ICP-MS, most of the work carried out is done with instrumentation that is either completely enclosed in a radiologically controlled glove box, or at least with the torch box, sample introduction system, and interface cones positioned inside a radiologically controlled hood.[93]

APPLICATIONS INVOLVING THE MONITORING OF THE NUCLEAR INDUSTRY'S IMPACT ON THE ENVIRONMENT

It is crucial that the nuclear industry not only be able to safely dispose of its low-level waste and monitor its impact on the environment, but also be responsible for cleaning up old sites related to nuclear power and the production of nuclear weapons. These types of environmental remediation and monitoring activities can generate an

alloys used by the aerospace industry or low-carbon steel strip made specifically for the auto industry, is that they are solid-sampling techniques. In other words, a multi-element analysis can be carried out with very little or no sample preparation.

For this reason, there has generally been very little demand for the multielement analysis of solutions in the metallurgical industry. Usually, it was only required if there was some elemental heterogeneity or segregation problems with the sample itself, or if there was a need to confirm an abnormal result generated by one of the solid-sampling techniques. In these situations, the sample had to be dissolved in some acid medium and analyzed by either flame AA, if only a few elements were required, or ICP-OES, if many elements were needed. Only in extreme cases, when the analytes in solution were below AA/ICP-OES detection limits, would GFAA be required. For all these reasons, there was no real demand for ICP-MS in the metallurgical industry, not because it was not a suitable technique, but because most of the trace element determinations in the industry were being adequately addressed by the other, well-established approaches.

However, over the past 5 years, we are beginning to see a growing trend in the use of ICP-MS in this application segment.[97] This is partly driven by the fact that high-purity metals and complex alloys are often very challenging to analyze by emission or absorption-based techniques, such as AA, ICP-OES, or GD-OES, because of spectral and matrix interferences generated by the high levels of major elements in the sample.[98] This has a major impact on detection capability, especially in the aircraft and aerospace industries, which use very-high-purity metals and high-temperature alloys. However, I believe the major reason for the recent growth in ICP-MS in the metallurgical industry is the exciting potential of coupling laser ablation with ICP-MS. It is clear that modern 266 nm Nd:YAG laser ablation systems, especially the ones optimized for bulk analysis, are now capable of ablating just about any metal and producing a continuous stream of fine particles suitable for an ICP-MS system. As a result, LA-ICP-MS is not only offering metallurgical chemists the ability to directly analyze solid samples with good stability, detection limits, and precision, but when switched back to solution nebulization, is also capable of producing superior detection capability compared to FAA, GFAA, or ICP-OES.

The very low detection capability of LA-ICP-MS therefore makes it an ideal measurement device for ultratrace levels in high-purity metals and high-temperature alloys, which are very difficult to get into solution. Traditionally, this analysis has been done by GD-MS, which, unfortunately, is a slow technique and also requires a complex standardization procedure requiring many calibration standards of a similar matrix to the samples. This is demonstrated in Table 19.12, which shows the determination of a group of elements in a Ni/Mo/W high-temperature alloy by both techniques.[99] The benefit of LA-ICP-MS is that it offers the potential for similar detection capability as GD-MS, but significantly faster throughput. It should be emphasized that no certified reference material was available for this matrix, so the aim was not to evaluate the accuracy of the two techniques, but just to show that they are capable of producing similar results.

The added benefit of LA-ICP-MS is that with optimum selection of the laser wavelength, the sampling area can be as low as 10 μm.Therefore, by rastering across

TABLE 19.12

The Determination of a Group of Elements in an Al/Mo/W High-Temperature Alloy by GD-MS and LA-ICP-MS (Courtesy of Cetac Technologies)

Element	GD-MS (ppb)	LA-ICP-MS (ppb)
Na	0.14	0.08
Mg	78	79
Si	323	255
Zr	354	314
Nb	218	170
Sn	1.4	1.1
Hf	125	110

Source: T. Howe, J. Shkolnik, and R. Thomas, *Spectroscopy*, **16**(2), 54–66, 2001.

the surface, it can also detect any heterogeneity or segregation on the surface of the sample. This kind of sampling precision is beyond the capability of GD-MS, because it is used predominantly as a bulk sampling technique. In fact, elemental segregation on the surface of the sample could be the reason why the data in this table does not agree for all elements in the alloy.

LA-ICP-MS is also ideal as a depth-profiling tool for the characterization of metal coatings. By optimization of the ablation process for high spatial resolution, the thickness of a metal coating can be determined by slowly ablating a few microns below the surface and monitoring the signal of the metal coating and the bulk material over time. This is another reason why laser ablation ICP-MS is such a flexible tool for metallurgical applications.[99]

PETROCHEMICAL AND ORGANIC-BASED SAMPLES

In the production of petrochemicals and related products, it is critical for refineries and chemical plants to closely monitor trace element contamination levels at various stages of the manufacturing process. For example, in the refining of crude oil, some elements such as Ni and V, even at ppb levels, can act as catalyst poisons and cause enormous problems owing to the volumes of hydrocarbons that are processed.[100] In addition, if the final product is intended for use by the food industry or the manufacture of electronic devices, the specifications for trace element contamination are even more stringent.

The problem is that the analysis of petrochemical samples can be extremely difficult because of the complex nature of crude oils, distillates, residues, fuel oils, petroleum products, organic solvents, and all the various by-products from refining crude oil. These complex oil-based samples pose major problems for any analytical technique, owing to the difficulty of introducing them directly into the instrument. So, the analytical challenge for any trace element technique being used in the petrochemical industry is to be able to carry out fast, reliable determinations of total and also speciated forms of critical metals, in a wide variety of complex samples, with the minimum of sample preparation.

Unfortunately, some of the traditional ways of getting petrochemical samples into solution are extremely slow and labor intensive. Common sample preparation methods include digestion with strong acids/oxidizing agents, and ashing the sample in a muffle furnace and redissolving the residue in a suitable solvent. The acid digestion procedure alone tends to lead to an incomplete dissolution because of the high level of carbonaceous material, so the ashing procedure or a combination of oxidation and ashing is preferred. This also allows for preconcentration of the sample to provide adequate amounts of the test analytes to be analyzed, if they are present at ultratrace levels. The choice of which of these traditional sample preparation approaches to use is often determined by the final instrumental technique. However, they all have a number of characteristics in common—apart from taking up a considerable amount of time in the total analytical process, they can also lead to loss of sample, loss of volatile analytes, and major contamination problems.

Avoiding these kinds of problems was among the reasons why the petrochemical community became interested in ICP-MS. Previously, ICP-OES was one of the preferred techniques for the multielement analysis of oil-based samples. However, because of the achievable detection limits of ICP-OES, a sample preparation technique known as the sulfated ash method (SASH) usually had to be used.[101] This approach, which involved oxidation of the oil sample with concentrated sulfuric acid and high-temperature ashing, took approximately 3 days to ensure all the analytes were in solution. When the use of ICP-MS was investigated, they found that because of its extremely high sensitivity, a simple dilution of the sample with a solvent such as toluene could be used. In other words, the lengthy sulfated ash method used to get the analytes into solution could be avoided, which represented an enormous time savings.

Unfortunately, there was a slight downside to the ICP-MS methodology. To directly aspirate oil that has been diluted in a volatile organic solvent, a special chilled spray chamber has to be used to desolvate the sample. This reduces the solvent loading and allows organic samples to be aspirated without adversely affecting the stability of the plasma. In addition, high RF power is necessary, together with a small amount of oxygen in the nebulizer gas flow to "burn off" any remaining solvent. This has the effect of stopping carbon deposits from building up on the interface cones and also minimizing the formation of carbon-based spectral interferences. If the appropriate modifications are made to the sample introduction system, the analysis of most organic-based samples is relatively straightforward. Besides the enormous time savings, contamination problems are dramatically reduced and the loss of volatile elements is avoided, compared to the complex SASH sample preparation procedure. Table 19.13 compares ICP-OES using SASH and ICP-MS using a simple 1:1000 dilution in toluene, for the determination of Ni and V in NIST 1618–certified reference fuel oil.[102] It should be noted that large dilutions are typical for the analysis of oil-based samples by ICP-MS or ICP-OES to minimize sample transport and viscosity effects. It can be seen that the accuracy and precision of both

TABLE 19.13

Comparison Data for the Determination of Ni and V in NIST 1618–Certified Reference Fuel Oil by ICP-OES, Using the Sulfated Ash Method and ICP-MS, Using a Simple 1:1000 Dilution in Toluene

NIST 1618 CRM	Total Sample Preparation/ Analysis Time	Sample Weight (g)	Ni (ppm)	RSD (%)	V (ppm)	RSD (%)
ICP-OES/SASH	72 h	5	76.2	1.7	426	1.3
ICP-MS/Dilution	45–60 min	3	75.9	1.5	424	1.9
Certificate Value	—	—	75.2 ± 0.4	—	423.1 ± 3.4	—

Source: F. McElroy, A. Mennito, E. Debrah, and R. Thomas, *Spectroscopy,* **13**(2), 42–53, 1998.

methods is similar and in good agreement with the certificate, but the ICP-MS determination is almost 100 times faster.

However, even if all these precautions are taken, some analytes are still problematic because the carbon-based polyatomic spectral interferences can never be totally eliminated. In addition, the small amount of oxygen in the nebulizer gas flow will add to the spectral complexity of the background by generating oxide-based interferences. Some of the elements that suffer from these types of spectral interferences in an organic matrix are shown in Table 19.14.

For this reason, collision/reaction cell/interface technology is almost a necessity when analyzing petrochemical- or oil-based samples, especially if the analytes are at extremely low levels. To emphasize this, Abu-Shakra showed the benefits of using collision/reaction cell technology together with an optimized sample introduction system for the analysis of oil samples diluted in kerosene. By using ammonia gas in a dynamic reaction cell ICP mass spectrometer, he showed that sub-ppb BEC values in the oil were achievable for a group of 20 elements. The interference reduction capability of the DRC was aided by the use of a Peltier-cooled cyclonic spray chamber with a low-flow PFA concentric nebulizer and the addition of 20 mL/min

TABLE 19.14

Some of the Problematic Elements and Their Potential Spectral Interferences in an Organic Matrix

Analyte	Interference
$^{24}Mg^+$	$^{12}C^{12}C^+$
$^{27}Al^+$	$^{12}C^{14}NH^+$
$^{44}Ca^+$	$^{12}C^{16}O^{16}O^+$
$^{51}V^+$	$^{38}Ar^{13}C^+$
$^{52}Cr^+$	$^{40}Ar^{12}C^+$
$^{56}Fe^+$	$^{40}Ar^{16}O^+$

of oxygen to minimize carbon buildup on the cones and also to reduce the effect of carbon-based spectral interferences in the determination of the notoriously difficult elements, such as Mg, Al, Ca, Cr, V, and Fe.[103]

FOOD AND AGRICULTURE

The trace element analysis of foodstuff and plant material has always been important, because the nature and concentration of many elements are related to the biological role they play in the physiology (biological study of the functions) of the living organism. Factors that influence trace element levels in food, crops, and plant materials include natural processes, inadvertent contamination during growth, and manufacturing and preparation processes. Some elements such as As, Cd, Hg, and Pb are considered toxic, whereas others such as Se, Cr, Zn, Mn, and Ni have a dual personality, because in some forms they are essential and in other forms they are toxic. Therefore, there is a need to classify two groups of trace elements in foodstuffs—toxic elements, which are typically present at trace levels, and nutritional elements, which are mostly, but not exclusively, present at higher levels. Therefore, the challenge of any technique used in the food industry is not only to be able to determine ultratrace levels (sub-ppb), but also to be able to determine higher concentration levels (typically, ppm). This has traditionally been done by a combination of FAA, GFAA, and ICP-OES, but clearly, if many elements need to be classified, it can be very time consuming, especially if conventional acid digestion methods are used to get the sample into solution.

For these reasons, ICP-MS has proved to be a very attractive option for the analysis of foodstuffs, especially as modern instruments fitted with collision/reaction cell or interface technology have the capability to extend the dynamic range to determine higher levels, as well as the ability to determine low levels of the traditionally difficult elements. There have been a number of publications on the use of ICP-MS for the analysis of foodstuffs,[104,105] but they mainly focused on elements at the trace level, because earlier technology was not able to handle such a wide spread in analyte concentrations with one sample preparation method. However, we are now beginning to see more applications in the open literature on the multielement analysis of food, at both high and low levels, using a single sample preparation. For example, Zhou and coworkers showed that 17 elements (V, Cr, Mn, Co, Ni, Zn, As, Se, Mo, Pd, Cd, Sn, Hg, Tl, Pb, Rh, and Re) from low-ppt to high-ppb levels, could be determined in 16 varieties of foodstuff, with good accuracy and precision using a simple external calibration.[106] The benefit of this methodology is that all elements can be measured at the same time in one solution, prepared by digesting the sample with concentrated nitric acid in a microwave oven. This is demonstrated in Table 19.15, which shows the determination of a group of selected elements in various food-based Chinese CRMs (National Research Center for Certified Reference Materials, Beijing, China). All results are expressed in ng/g, in the food.

More recently, Yamanaka and Fryer showed the use of collision/reaction cell technology ICP-MS to determine elements at the trace and macro levels in various plant materials. They demonstrated that by using both helium and hydrogen gases in an octapole-based collision cell, they could determine ppm levels of Cr, Cu, and Fe

TABLE 19.15

Determination of a Group of Selected Elements (in ng/g in Food) in Various Food-Based Chinese Certified Reference Materials[106]

Element	Rice (ng/g)		Pork Liver (ng/g)		Mussels (ng/g)	
	Found	Cert.	Found	Cert.	Found	Cert.
Mn	9.5	9.8	9.37	8.32	10.7	10.2
Co	—	—	—	—	1.18	0.94
Ni	—	—	—	—	0.90	1.03
Zn	14.8	14.1	180	172	136	138
As	0.051	0.051	0.066	0.044	5.5	6.1
Se	0.050	0.045	—	—	—	—
Mo	—	—	—	—	0.62	0.6
Cd	0.018	0.020	0.077	0.067	4.0	4.5
Hg	—	—	—	—	0.073	0.067
Pb	—	—	0.59	0.54	—	—

and ppb levels of As and Se, together with the measurement of percentage levels of Ca and K in spinach and tomato leaves in the same method. This shows the capability of modern instrumentation to reduce problematic spectral interferences to allow determinations at the trace level, while at the same time extending its dynamic range and also measuring higher concentrations in plant materials digested in a concentrated nitric–perchloric acid mix. Some of the polyatomic interferences that were being suppressed in this analysis included ArC^+ on chromium, $ArNa^+$ on copper, ArO^+ on iron, $ArCl^+$ on arsenic, and Ar_2^+ on selenium.[107]

In addition to estimating the total metal content of food-related samples, ICP-MS, coupled with various chromatography separation devices, is proving an invaluable detection technique to characterize extremely low levels of various elemental species in foodstuffs. An example of using ICP-MS in this way is provided by a study on the role of selenium as an anticarcinogen.[108] Se is both an essential and a toxic element. On the one hand, it is thought to have anticancer properties because it protects cell membranes from damage due to oxidation, and on the other hand, selenium deficiency causes skeletal and cardiac muscle dysfunction; indeed, at high levels, some forms of selenium are considered extremely toxic. It is therefore very important to know the biodegradation process of different selenium compounds in plants such as garlic, onions, and broccoli to get a better understanding of their anticancer properties. The research groups in this study used ICP-MS in conjunction with HPLC to separate various organoselenium compounds in plant material. They showed that trace levels of selenoamino acids, including selenocysteine, selenomethionine, methylselenocysteine, and propylselenocysteine, could be determined, even in the presence of large amounts of sulfur. This is particularly significant, because selenium predominantly follows the chemistry of sulfur, which can present considerable separation

35. Y. Abdelnour and J. Murphy, *The Analysis of Whole Blood Samples by Collision Inter-face ICP-MS*, Varian Inc. Application Note—28, 2007, http://www.varianinc.com/image/vimage/docs/products/spectr/icpms/atworks/icpms28.pdf.
36. Y. Shibata and M. Morita, *Analytical Science*, **5**, 107, 1989.
37. D. Beauchemin, K. W. M. Siu, J. W. McLaren, and S. Berman, *Journal of Analytical Atomic Spectrometry*, **4**, 285, 1989.
38. H. M. Crews, J. R. Dean, L. Ebdon, and R. C. Massey, *The Analyst*, **114**, 895, 1989.
39. B. Gercken and R. M. Barnes, *Analytical Chemistry*, **63**, 283, 1991.
40. N. P. Vela and J. A. Caruso, *Journal of Analytical Atomic Spectrometry*, **11**, 1129, 1996.
41. V. Majidi and N. J. Miller-Ihli, *The Analyst,* **123**, 803, 1998.
42. A. R. Date and A. L. Gray, *The Analyst,* **108**, 159, 1983.
43. L. A. Haskin, T. R. Wilderman, and M. A. Haskin, *Radioanalytical Chemistry*, **1**, 337–348, 1968.
44. A. J. Walder and P. A. Freeman, *Journal of Analytical Atomic Spectrometry*, **7**, 571–575, 1992.
45. K. E. Jarvis and I. Jarvis, *Geostandards Newsletter*, **12**(1), 1988.
46. A. M. G. Figueiredo and L. S. Marques, *Geochimica Brasiliensis*, **3**(1), 1989.
47. E. Pruszkowski and P. Barrett, *Spectrochimica Acta,* **39B**, 485, 1994.
48. F. E. Lichte, A. L. Meier, and J. G. Crock, *Analytical Chemistry*, **59**(8), 1150–1157, 1987.
49. D. J. Douglas, *Canadian Journal of Spectroscopy,* **34**(2), 1989.
50. A. Stroh, U, Voellkopf, and E. R. Denoyer, *Journal of Analytical Atomic Spectrometry*, **7**, 1201–1205, 1992.
51. E. R. Denoyer, R. Ediger, and J. Hager, *Atomic Spectroscopy*, **10**(4), 97–102, 1989.
52. F. E. Beamish and J. C. Van Loon, *Analysis of Noble Metals,* Academic Press, NY, 178, 1977.
53. C. Riddle, A. Vander Voet, and W. Doherty, *Geostandards Newsletter*, **12**(1), 203, 1988.
54. H. P. Longerich, G. A. Jenner, and S. E. Jackson, *Chemical Geology*, **83**(105), 1990.
55. A. P. Dickin, *Radiogenic Isotope Geology,* Cambridge University Press, Cambridge, UK, 1995.
56. L. Halicz, Y. Erel, and A. Veron, *Atomic Spectroscopy*, **17**(5), 186–189, 1996.
57. K. J. R. Rosman, W. Chisholm, C. F. Boutron, J. P. Candelone, and U. Gorlach, *Nature,* **362**, 333, 1993.
58. F. Vanhaecke, L. Moens, R. dams, and P. Taylor, *Analytical Chemistry*, **68**(3), 567–569, 1996.
59. A. N. Halliday, D. D. Lee, J. N. Christensen, M. Rehkamper, W. Yi, X. Luo, C. M. Hall, C. J. Ballentine, T. Pettke, and C. Stirling, *Geochimica and Cosmochimica Acta,* **62**, 919–940, 1998.
60. The Benefits of the X Series 2 ICP-MS for the Analysis of Geological Samples Pre-pared Using the Lithium Metaborate Fusion Method, Thermo Scientific Application Note—40790, 2007, http://www.thermo.com/eThermo/CMA/PDFs/Articles/articles-File_2375.pdf.
61. F. Vanhaecke, L. Moens, S. D. Tanner, V. I. Baranov, and D. R. Bandura, *PerkinElmer Sciex Application Note,* Chemical Resolution of ^{87}Rb/^{87}Sr Isobaric Overlap: Fast Rb/Sr Geochronology by Means of DRC ICP-MS, D-6538, 2001.
62. S. Beres, L. Dione, K. Neubauer, and R. Thomas, *Current Trends in Mass Spectrom-etry*, 44–49, May 2005.
63. T. Howe, J. Shkolnik, and R. Thomas, *Spectroscopy,* **16**(2), 54–66, 2001.
64. S. E. Jackson, H. P. Longerich, G. R. Dunning, and B. J. Fryer, *Canadian Mineralogist*, **30**, 1049–1064, 1992.
65. D. Günther and C. A. Heinrich, *Journal of Analytical Atomic Spectrometry*, **14**, 1369, 1999.

66. J. Gonzalez, X. L. Mao, J. Roy, S. S. Mao, and R. E. Russo, *Journal of Analytical Atomic Spectrometry*, **17**, 1108–1113, 2002.
67. S. Shuttleworth and D. Kremser, *Journal of Analytical Atomic Spectrometry*, **13**, 697–699, 1999.
68. P. P. Mahoney, G. Li, and G. M. Hieftje, *Journal of Analytical Atomic Spectrometry*, **11**, 401–405, 1996.
69. D. Günther, I. Horn, and B. Hattendorf, *Fresenius Journal of Analytical Chemistry*, **368**, 4–14, 2000.
70. L. Neufeld and J. Roy, *Spectroscopy*, **19**(1), 16–28, 2004.
71. B. Hattendorf and D. Günther, *Journal of Analytical Atomic Spectrometry*, **19**(5), 600–606, 2004.
72. International Technology Roadmap for Semiconductors (ITRS), www.public.itrs.net, 2001.
73. *Book of SEMI Standards (BOSS)*, Semiconductor Equipment and Materials International, San Jose, CA.
74. S. J. Chang and S. L. Chen, *Instruments Today*, **20**, 51–58, 1998.
75. J. M. Collard, K. Kawabata, Y. Kishi, and R. Thomas, *Micro*, January 2002.
76. D. S. Bollinger and A. J. Schleisman, *Atomic Spectroscopy*, **20**, 2, 60–63, 1999.
77. K. Neubauer and U. Voellkopf, *Atomic Spectroscopy*, **20**(2), 64–68, 1999.
78. Y. Kishi, K. Kawabata, and R. Thomas, *Spectroscopy*, **18**, 1, 2003.
79. B. McKelvey, S. McIvor, and W. Wiltse, *Polymer Comparisons for the Storage of Trace Metal Analysis of Ultrapure Water with the Agilent 7500cs ICP-MS*, Agilent Technologies Application Note—5989-5782EN, 2006, http://www.chem.agilent.com/temp/rad1C16C/00001511.PDF.
80. K. Takahashi, *The Determination of Impurities in Semiconductor Grade Hydrochloric Acid Using the Agilent 7500cs ICP-MS*, Agilent Technologies Application Note 5989-4348EN, 2006, http://www.chem.agilent.com/temp/rad961D9/00001358.PDF.
81. C. M. Ping, Y. Kishi, K. Kawabata, and R. Thomas, *Micro*, **21**(3), 37–42, 2003.
82. K. Kawabata, Y. Kishi, and H. Shi, *Spectroscopy*, **19**(9), 14–21, 2004.
83. K. Kawabata, Y. Kishi, D. Palsulich, D. Wiederin, and D. Armstrong, *Semiconductor Manufacturing*, July 2005.
84. M. Radle, H. Lian, B. Nicoley, and A. J. Howard, *Semiconductor International*, July 2001.
85. *Applications of Inductively Coupled Plasma to Radionuclide Determinations*, R. W. Morrow and J. S. Crain, Eds., ASTM, West Conshohocken, PA, 1995.
86. R. S. Houk, V. A. Fassel, G. D. Flesch, H. J. Svec, A. L. Gray, and C. E. Taylor, *Analytical Chemistry*, **52**, 2283, 1980.
87. Counsel of Ionizing Radiation Measurements Workshop on Standards, Intercomparisons, and Performance Evaluations for Low-Level and Environmental Radionuclide Mass Spectrometry—Meeting Proceedings, NIST, Gaithersburg, MD, April 1999.
88. ASTM Committee for Nuclear Fuel Cycle (C26): Subcommittee on Test Methods and Standards (C26-05), http://www.astm.org/cgibin/SoftCart.exe/COMMIT/SUBCOMMIT/C2605.htm?L+mystore+hics2935+119325022.
89. P. R. Makinson, *Applications of ICP-MS to Radionuclide Determinations*, R. W. Morrow and J. S. Crain, Eds., ASTM, West Conshohocken, PA, pp. 7–19, 1995.
90. E. R. Denoyer, D. Jacques, E. Debrah, and S. D. Tanner, *Atomic Spectroscopy*, **16**(1), 1, 1995.
91. A. J. Walder and T. Hodgson, *Applications of ICP-MS to Radionuclide Determinations*, R. W. Morrow and J. S. Crain, Eds., ASTM, West Conshohocken, PA, pp. 20–25, 1995.

92. W. F. Kinard, N. E. Bibler, C. J. Coleman, R.A. Dewberry, W. T. Boyce, and S. B. Wyrick, *Applications of* ICP-MS *to Radionuclide Determinations*, Ed. R. W. Morrow and J. S. Crain, ASTM, West Conshohocken, PA, pp. 48–58, 1995.

93. J. M. Barrero Moreno, M. Betti, and J. I. Garcia Alonso, *Journal of Analytical Atomic Spectrometry*, **12**, 355–361, 1997.

94. M. Hollenbach, J. Grohs, S. Mamic, and M. Koft, *Applications of ICP-MS to Radionuclide Determinations*, R. W. Morrow and J. S. Crain, Eds., ASTM, West Conshohocken, PA, pp. 99–115, 1995.

95. G. Price-Russ III and J. M. Bazan, *Applications of* ICP-MS *to Radionuclide Determinations*, R. W. Morrow and J. S. Crain, Eds., ASTM, West Conshohocken, PA, pp. 131–140, 1995.

96. O. A. Vita and K. C. Mayfield, *Applications of ICP-MS to Radionuclide Determinations*, R. W. Morrow and J. S. Crain, Eds., ASTM, West Conshohocken, PA, pp. 141–147, 1995.

97. H. M. Kuss, D. Bossmann, and M. Muller, *Proceedings of the 3rd International Conference on Progress of Analytical Chemistry in the Iron and Steel Industry*, R. Nauche, Ed., EUR14113 EN, pp. 302–307, 1992.

98. H. M. Kuss, D. Bossmann, and M. Muller, *Atomic Spectroscopy,* **15**(6), 148–150, 1994.

99. T. Howe, J. Shkolnik, and R. Thomas, *Spectroscopy*, **16**(2), 54–66, 2001.

100. R. I. Botto and J. J. Zhu, *Journal of Analytical Atomic Spectrometry*, **9**, 905, 1994.

101. Sulfated Ash Sample Preparation Method—ASTM Method D-874.

102. F. McElroy, A. Mennito, E. Debrah, and R. Thomas, *Spectroscopy,* **13**(2), 42–53, 1998.

103. F. Abou-Shakra, *Analysis of Petroleum Samples by DRC-ICP-MS,* PerkinElmer LAS Application Note—007848-01, 2007, http://las.perkinelmer.com/content/application-notes/far_analysisofpetroleumsamplesdrcicpms.pdf.

104. H. M. Crews, *International Laboratory,* **23**, 38, 1993.

105. B. S. Sheppard, *The Analyst,* **119**, 1683, 1994.

106. H. Zhou and J. Liu, *Atomic Spectroscopy*, **18**(4), 115–118, 1997.

107. K. Yamanaka and F. Fryer, *Measurement of Macro and Trace Elements in Plant Digests Using the 7500c ICP-MS System,* Agilent Technologies Application Note-5988-4450EN, 2001, http://www.chem.agilent.com/temp/rad5F8DE/00000506.PDF.

108. P. Uden, J. Tyson, M. Kotrebai, and E. Block, Paper No. 870, *FACSS Conference,* Vancouver, BC, 1999.

109. K. Neubaur and R. E. Wolf, *Low Level Selenium Determination,* PerkinElmer LAS Application Note, D-6358, 2000.

110. R. D. Koons, *Journal of Forensic Sciences*, **43**(4), 748–754, 1998.

111. R. E. Wolf, C. Thomas, and A. Bohlke, *Applied Surface Science,* 127–129, 299–303, 1998.

112. R. E. Wolf, *Atomic Spectroscopy*, **18**(6), 169–174, 1997.

113. M. Bettinelli, U. Baroni, F. Bilei, and G. Bizzarri, *Atomic Spectroscopy*, **18**(3), 77–79, 1997.

114. M. Chaudhary-Webb, D. C. Pascal. W. C. Elliott, H. P. Hopkins, A. M. Ghazi, W. C. Ting, and I. Romieu, *Atomic Spectroscopy*, **19**(5), 156–163, 1998.

20 Comparing ICP-MS with Other Atomic Spectroscopic Techniques

Now that I have presented the basic principles of ICP-MS and its major application strengths, let us turn our attention to comparing it with other AS techniques used for trace element analysis. ICP-MS is a very powerful technique, but is it the right one for your laboratory? Do you need its multielement capability? Are the detection limits of your current techniques good enough? Will your operators be able to handle the more complicated method development of ICP-MS? Are you prepared for its increased running costs? In other words, have you considered the implications of owning an ICP mass spectrometer? To help you answer these questions, Chapter 20 takes a look at the strengths and weaknesses of ICP-MS and compare them with those of other trace element techniques like flame atomic absorption (FAA), electrothermal atomization (ETA), and inductively coupled optical emission spectrometry (ICP-OES), in order to help you decide if ICP-MS is really the right fit for your laboratory.

Since the introduction of the first commercially available atomic absorption spectrophotometer (AAS) in the early 1960s, there has been an increasing demand for better, faster, easier-to-use, and more flexible trace element instrumentation. A conservative estimate shows that today's market for atomic spectroscopy (AS)-based instruments, such as atomic absorption (AA), inductively coupled plasma optical emission (ICP-OES), and inductively coupled plasma mass spectrometry (ICP-MS), represents over $700 million in annual revenue. After market sales and service costs are added to this number, it is probably close to $1 billion.[1] As a result of this growth, we have seen a rapid emergence of more sophisticated equipment and easier-to-use software. Moreover, with an increase in the number of manufacturers of both instrumentation and sampling accessories, the choice of which technique to use is often unclear.

In order to select the best technique for a particular analytical problem, it is important to understand exactly what the problem is and how it is going to be solved. For example if the requirement is to monitor copper at percentage levels in a copper plating bath and it is only going to be done once per shift, it would be inappropriate to choose a rapid ultratrace multielement technique such as ICP-MS. A single-element technique such as flame AA would probably be adequate for this application. Although this might be an exaggerated example, it emphasizes that there is an optimum atomic spectroscopic technique for every application problem. When choosing a technique, it is important to understand not only the application problem, but also the strengths and weaknesses of the technology being applied to solve the problem. However, there are many overlapping areas between the major atomic spectroscopy

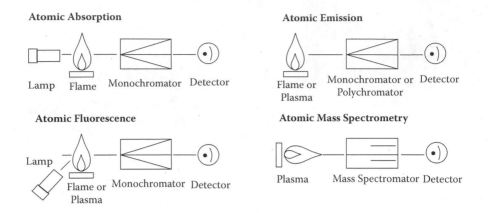

FIGURE 20.1 Simple schematic diagrams of the principles of atomic absorption, emission, fluorescence, and mass spectrometry (courtesy of PerkinElmer Life and Analytical Sciences).

techniques, so it is highly likely that for some applications, more than one technique would be suitable. For that reason it is important to go through a carefully thought-out evaluation process when selecting a piece of equipment.

First of all, let us take a brief look at the most commonly used atomic spectroscopy techniques—atomic absorption, ICP optical emission, and ICP mass spectrometry. There are different variations of each technique, but basically, atomic absorption uses the principle of generating free atoms (of the element of interest) in a flame or electrothermal atomizer (sometimes referred to as graphite furnace or GFAA), and measuring the amount of light absorbed from a wavelength-specific light source. ICP emission uses the principle of exciting the atoms in a plasma and measuring the amount of light emitted when they fall back to a ground (stable) state. And, as discussed in the previous chapters, ICP mass spectrometry uses the plasma to generate ions and measures the number of ions produced at a particular mass-to-charge ratio. Simple schematics of atomic absorption, emission, fluorescence, and mass spectrometry are shown in Figure 20.1.

Even though atomic fluorescence is considered an atomic spectroscopic technique, it is not covered in this chapter. Let us take a look at the other AS techniques in greater detail.

FLAME ATOMIC ABSORPTION

This is predominantly a single-element technique that uses a flame to generate ground-state atoms. The sample is aspirated into the flame via a nebulizer and a spray chamber. The ground-state atoms of the sample absorb light of a particular wavelength, either from an element-specific, hollow cathode lamp or a continuum source lamp. The amount of light absorbed is measured by a monochromator (optical system) and detected by a photomultiplier or solid-state detector, which converts the photons into an electrical pulse. This absorbance signal is used to determine the concentration of that element in the sample. Flame AA typically uses about 2–5 mL/min of liquid sample and is capable of parts per million detection limits.

ELECTROTHERMAL ATOMIZATION

This is also mainly a single-element technique, although multielement instrumentation is now available. It works on the same principle as flame AA, except that the flame is replaced by a small heated tungsten filament or graphite tube. The other major difference is that in ETA, a very small sample (typically, 50 μL) is injected onto the filament or into the tube, and not aspirated via a nebulizer and a spray chamber. Because the ground-state atoms are concentrated in a smaller area than a flame, more absorption takes place. The result is that ETA offers about 100 times lower detection limits than FAA.

RADIAL-VIEW ICP OPTICAL EMISSION

ICP-OES is a multielement technique that uses a traditional radial (side-view) inductively coupled plasma to excite ground-state atoms to the point where they emit wavelength-specific photons of light that are characteristic of a particular element. The number of photons produced at an element-specific wavelength is measured by high-resolving-power optics and a photon-sensitive device such as a photomultiplier or a solid state detector. This emission signal is directly related to the concentration of that element in the sample. The analytical temperature of an ICP is about 6000–7000°K, compared to that of a flame, which is typically 2500–4000°K. A radial ICP can achieve similar detection limits to flame AA for the majority of elements, but has the advantage of offering much better performance for the refractory and rare earth elements. The sample requirement for ICP-OES is approximately 1 mL/min.

AXIAL-VIEW ICP OPTICAL EMISSION

The principle is exactly the same as radial ICP-OES, except that in axial-view ICP-OES, the plasma is viewed horizontally (end-on). The benefit is that more photons are seen by the detector and as a result, detection limits can be as much as 2–10 times lower, depending on the design of the instrument. The disadvantage is that more severe matrix interferences are observed with an axial ICP. Sample requirements are the same as for radial ICP-OES.

INDUCTIVELY COUPLED PLASMA MASS SPECTROMETRY

This has been described in great detail in the previous chapters. The fundamental difference between ICP-OES and ICP-MS is that in ICP-MS, the plasma is not used to generate photons, but to generate positively charged ions. The ions produced are transported and separated by their atomic mass-to-charge ratio using a mass-filtering device such as a quadrupole. The generation of such large numbers of positively charged ions allows ICP-MS to achieve detection limits at the part per trillion level compared to that of ICP-OES, which is typically in the part per billion range.

   This is not meant to be a detailed description of the fundamental principles of each technique, but a basic understanding as to how they differ from each other. In the process of deciding whether ICP-MS is the best technique for your needs, there are basically four steps to consider[2]:

- Define the analytical objective.
- Establish selection criteria.
- Define the application tasks.
- Compare the techniques.

Each step in the process should serve to focus your attention on the techniques that best suit the requirements of the analytical task. Let us take a closer look at these steps.

DEFINE THE OBJECTIVE

In this step, the analytical objective should be broadly defined. For example, what is the concentration of iron in high-purity hydrochloric acid, or how much arsenic is in contaminated soil? However, it is important not to lose sight of what you are actually trying to accomplish with this analysis. In other words, for the previous example, one should not forget what decisions will be made based on knowing the trace element composition of the sample. Before proceeding to specifics, you should have a basic understanding of your objective when you finish the evaluation of several different analytical techniques. Once that has been done, one can proceed to focus on the techniques that could possibly accomplish this task.

ESTABLISH PERFORMANCE CRITERIA

Compiling performance criteria helps make clear the right techniques for the task. The field should now be narrowed down to establish a set of practical criteria, which might eliminate some of the less suitable techniques for a particular application. Some of these criteria will include (but not be limited to) detection limits, precision requirements, quality of data, sample throughput capability, ease of use, instrument reliability, operator training needs, or availability of application material.

DEFINE THE APPLICATION TASK

By rigorously defining the task, it will become relatively clear what techniques to evaluate. By comparing and contrasting the attributes of each of the techniques, one begins to appreciate the value of each and starts determining how the instrumentation will be used in the laboratory. The factors/issues that influence this decision will vary depending on the individual situation. They may not all be valid, but some will be of more importance than others. However, before an informed decision can be reached, each one should be considered to some degree. These issues can be broken down into four major categories—application, installation, user, and financial considerations. Let us take a closer look at them.

Application

This will include information about the elemental requirements, such as what detection limits and concentration ranges are expected, and how much accuracy and precision are required. It will also include sample information, such as how many samples are expected and at what frequency, how much time can be spent on sample preparation and how quickly they must be analyzed. Sometimes the amount of available sample will dictate the selection, or whether interferences from the matrix components have a major impact on analysis.

Installation

Installation factors might include the size of the instrument, how much laboratory space is required, what services are necessary, or how clean the laboratory and the sample preparation environment should be. As mentioned in Chapter 15, this is a major consideration if ICP-MS is the technique of choice.

User

This will tell you the required skill level of the operator, how easy the instrument is to use, or what training is required. The expertise of the operator should not be underestimated if ICP-MS is being seriously considered, because it generally requires an analyst with a higher skill level to develop good methodology.

Financial

Financial factors must be considered because the funds available might have to cover the cost of instrumentation, a specialized laboratory, or the salary of a dedicated expert to run the instrument. Sometimes financial issues can be the dominant reason why a technique is purchased, and it certainly has been a big factor in the relatively slow acceptance of ICP-MS.

COMPARISON OF TECHNIQUES

Going through these basic steps could possibly have narrowed the field to one technique or another. At this point, it may become clear that ICP-MS is the right technique. However, if this is not the case and there is still more than one candidate technique, a detailed comparison should be made to make the final selection. The following criteria should be used as a guideline to help in this final selection process:

- Detection limits
- Analytical working range
- Sample throughput
- Interferences
- Usability issues
- Cost of ownership

Detection Limits

The detection limits achievable for individual elements represent a significant criterion of the selection of an analytical technique for a given application problem. Without adequate detection limit capabilities, lengthy analyte concentration procedures

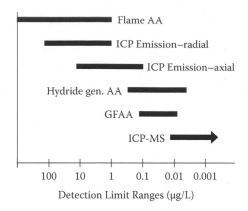

FIGURE 20.2 Typical detection limit ranges for major atomic spectroscopy techniques.

may be required prior to analysis. Typical detection limit ranges for major atomic spectroscopy techniques are shown in Figure 20.2.

There is no question that the best detection limits are obtained using ICP-MS followed by graphite furnace AA (ETA). Axial ICP-OES offers very good detection limits for most elements, but generally not as good as ETA. Radial ICP-OES and flame AA show approximately the same detection limits performance, except for the refractory and the rare earth elements, for which performances are much better by ICP-OES as it is very difficult to produce enough ground-state atoms by flame AA. For mercury and those elements that form volatile hydrides, such as As, Bi, Sb, Se, and Te, the cold vapor or hydride generation techniques offer exceptional detection limits. It is also worth mentioning that the detection capability of quadrupole-based ICP-MS is continuously being improved. Used in conjunction with collision/reaction cell/interface, it is now capable of low parts per quadrillion (ppq) detection limits for many elements, as demonstrated in Table 20.1, shows a recent comparison of detection limits in µg/L (ppb) between an ICP-MS fitted with a collision/reaction cell, and the other modern, state-of-the-art AS instrumentation. It is worth pointing out that ICP-MS detection limits for many of the elements such as Fe, K, Ca, Se, Cr, Mg, and Mn, which suffer from plasma- and solvent-based polyatomic interferences, are 1–2 orders of magnitude worse for conventional ICP-MS instrumentation compared to an instrument that has collision/reaction cell/interface capability.

Analytical Working Range

The analytical working range can be considered as the concentration range over which quantitative results can be obtained without having to recalibrate the instrument. Selecting a technique with an analytical working range (and detection limits) based on the expected analyte concentrations minimizes analysis times by allowing samples with varying analyte concentrations to be analyzed together. For example, ICP-MS, once considered exclusively an ultratrace element technique, can now handle concentration ranges from low ppt to high ppm. A wide analytical working range can also reduce sample-handling problems and minimize potential errors. It should also be emphasized that although the dynamic range of radial and axial ICP-OES

TABLE 20.1

A Comparison of Detection Limits between ICP-MS and the Other AS Instrumentation in µg/L ppb

Element	Flame AA	Hg/Hydride	GFAA	ICP-OES	ICP-MS	Element	Flame AA	Hg/Hydride	GFAA	ICP-OES	ICP-MS
Ag	1.5		0.005	0.6	0.002	Mo	45		0.03	0.5	0.001
Al	45		0.1	1	0.005	Na	0.3		0.005	0.5	0.0003
As	150	0.03	0.05	2	0.006	Nb	1500			1	0.0006
Au	9		0.15	1	0.0009	Nd	1500			2	0.0004
B	1000		20	1	0.003	Ni	6		0.07	0.5	0.0004
Ba	15		0.35	0.03	0.00002	Os				6	
Be	1.5		0.008	0.09	0.003	P	75,000		130	4	0.1
Bi	30	0.03	0.05	1	0.0006	Pb	15		0.05	1	0.00004
Br					0.2	Pd	30		0.09	2	0.0005
C					0.8	Pr	7500			2	0.00009
Ca	1.5		0.01	0.02	0.05	Pt	60		2.0	1	0.002
Cd	0.8		0.008	1	0.003	Rb	3		0.03	5	0.0004
Ce				1.5	0.0002	Re	750			0.5	0.0003
Cl					12	Rh	6			5	0.0002
Co	9		0.15	0.2	0.0002	Ru	100		1.0	1	0.0002
Cr	3		0.004	2	0.02	S				10	28
Cs	15				0.0003	Sb	45	0.15	0.05	2	0.0009
Cu	1.5		0.014	0.4	0.0002	Sc	30			0.1	0.004
Dy	50			0.5	0.0001	Se	100	0.03	0.05	4	0.0007
Er	60			0.5	0.0001	Si	90		1.0	10	0.03
Eu	30			0.2	0.00009	Sm	3000			2	0.0002
F					372	Sn	150		0.1	2	0.0005
Fe	5		0.06	0.1	0.0003	Sr	3		0.025	0.05	0.00002

(continued on next page)

TABLE 20.1 (continued)

A Comparison of Detection Limits between ICP-MS and the Other As Instrumentation in μg/L ppb

Element	Flame AA	Hg/Hydride	GFAA	ICP-OES	ICP-MS	Element	Flame AA	Hg/Hydride	GFAA	ICP-OES	ICP-MS
Ga	75			0.5	0.0002	Ta	1500			1	0.0005
Gd	1800			0.9	0.0008	Tb	900			2	0.00004
Ge	300			1	0.001	Te	30	0.03	0.1	2	0.0008
Hf	300	0.009		0.5	0.0008	Th				2	0.0004
Hg	300		0.6	1	0.016	Ti	75		0.35	0.4	0.003
Ho	60			0.4	0.00006	Tl	15		0.1	2	0.0002
I					0.002	Tm	15			0.6	0.00006
In	30			1	0.0007	U	15,000			10	0.0001
Ir	900		3.0	1	0.001	V	60			0.5	0.0005
K	3		0.005	1	0.0002	W	1500			1	0.005
La	3000			0.4	0.001	Y	75			0.2	0.0002
Li	0.8		0.06	0.3	0.0001	Yb	8			0.1	0.0002
Lu	1000			0.1	0.00005	Zn	1.5		0.02	0.2	0.0003
Mg	0.15		0.004	0.04	0.0003	Zr	450			0.5	0.0003
Mn	1.5		0.005	0.1	0.00007						

Note:　If no DL data is shown, it means the AS technique is not ideally suited to determine that element.

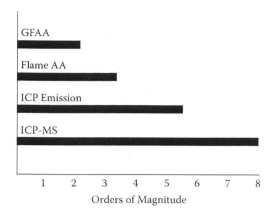

FIGURE 20.3   Analytical working ranges for major atomic spectroscopy techniques.

are the same, the working range of an axial ICP is shifted down approximately by one order of magnitude because the detection limits are 2–10 times lower. However, there are dual-view systems on the market that offer the benefits of both radial and axial viewing. Figure 20.3 shows typical analytical working ranges.

Sample Throughput

Sample throughput is typically the number of samples that can be analyzed in a given amount of time. For most techniques, analyses performed at the limit of detection or analyses for which the best precision is required will be more time consuming than the less demanding ones. In cases where this is not the limiting factor, the number of elements to be determined per sample and the analytical technique will determine the sample throughput. Let us take a brief look at the sample throughput capability of each technique.

Flame AA

Flame AA provides exceptional sample throughput when analyzing a large number of samples for just a few elements. A typical determination of a single element requires only 5–10 s. However, flame AA requires specific light sources (hollow cathode or continuum source lamps) and optical parameters for each element to be determined, and may require different flame gases for different elements. In automated multielement flame AA systems, all samples are usually analyzed for one element, following which the system automatically changes conditions for the next element, and so on until all the elements have been determined. As a result, even though it is used for multielement analysis, flame AA is generally considered to be a single-element technique.

It should be pointed out that there are now flame AA instruments on the market that are achieving higher sample throughput by carrying out element sequential analysis. Most traditional AA instrumentation is operated in sample sequential mode where every sample in an autosampler run is analyzed for one element at a time until all the elements in the multielement run are determined. However, by using the instrument in element sequential mode, all the elements are determined one

Usability

It is often said that the strength of any technique is the time it takes to set up methods and run routine samples. The three criteria that determine a technique's ability to be considered truly routine are ease of use, the skill level of the operator, and whether the application methodology is readily available. Here is a brief comparison of the four techniques with regard to usability.

Flame AA

FAA is very easy to use. It is now considered truly routine and requires minimal operator skill level. Extensive applications information is available. Excellent precision makes it a preferred technique for the determination of major constituents and higher concentration analytes.

Electrothermal Atomization

Graphite furnace applications are well-documented, though not as complete as flame AA. It has exceptional detection limit capabilities but with a limited analytical working range. Sample throughput is less than that of other atomic spectroscopy techniques. Operator skill requirements are much more extensive than for flame AA.

ICP-OES

This is the most widely used multielement atomic spectroscopy technique, with excellent sample throughput and very wide analytical range. Operator skill requirements are somewhere between those of flame AA and ETA. ICP-OES is now a mature technique, which means that sufficient applications literature is available.

ICP-MS

ICP-MS is a relatively new technique compared to the others. It has exceptional multielement capabilities at trace and ultratrace levels and also has the unique ability to perform isotopic analyses. Application information is not as readily available as the other techniques, but is growing rapidly. However, ICP-MS probably requires operators with a higher skill level to achieve good-quality data.

Cost of Ownership

The initial purchasing cost is obviously a big factor in regard to the cost of ownership; moreover, the running costs, and the cost of consumables and chemicals will also have a major impact, particularly over the 10-year lifetime of owning the instrument. Let us first take a look at the typical purchase price of each technique. There is no question that single-element techniques (flame AA and ETA) are generally less expensive than the multielement ones (ICP emission and ICP-MS). There can also be a considerable variation in cost among instrumentation of the same technique. Instruments offering only basic features are generally less expensive than more versatile systems, which frequently also offer a greater degree of automation. Figure 20.4 provides a comparison of typical cost ranges for the major atomic spectroscopy techniques. As a rough guideline, the scale starts at about $10,000–30,000 for FAA, $25,000–50,000 for ETA, $60,000–100,000 for ICP-OES, $140,000–200,000 for quadrupole (collision/reaction cell instruments will be at the higher end of this

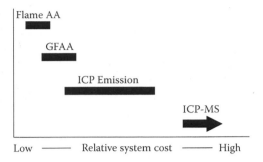

FIGURE 20.4 Relative purchasing costs of different AS equipment.

range) or TOF ICP-MS, and about $250,000 and above for top-of-the-range magnetic sector systems (prices will also vary based on different geographical regions of the world).

Let us now take a look at the cost of running each of the techniques. The initial purchase price is important, but the operating costs and the price of consumables and chemicals/standards should have a much bigger impact on the decision of which technique to invest in—because most laboratories typically keep an instrument for 8–10 years before they replace it. So when calculating the overall cost of owning an instrument, it is absolutely essential that this is factored into the decision. Therefore, to help you decide whether you can actually afford to run and operate an ICP-MS, here is a basic comparison between the running costs of the major AS techniques.[3] The assumption is that the instrument will run in its standard configuration with an autosampler, but with no other sampling accessories such as laser ablation, flow injection, or chromatography system attached.

For the purpose of this study, let us make the assumption that the major operating costs associated with running AS instrumentation are the gases, electricity, and consumable supplies. Although the salary of the operator, laboratory space, and sample preparation can legitimately be called operating expenses, they will not be used for this exercise. For comparison purposes, the evaluation will be based on a typical laboratory running their instrument 2½ days (20 h) per week and 50 weeks a year (1000 h per year). (Note: These financial data are based on the cost of gases, electricity, and instrument consumables in the United States as of October 2007. They have been obtained from a number of commercial sources, including two suppliers of industrial and high-purity gases,[4,5] a local utilities company,[6] ICP-MS instrument vendors,[7–10] and a variety of sample introduction and consumables suppliers.[11–13])

Gases
Flame AA   Most flame AA systems use acetylene (C_2H_2) as the combustion gas, and air or nitrous oxide (N_2O) as the oxidant. Air is usually generated by an air compressor, but the C_2H_2 and N_2O come in high-pressure cylinders. Normal atomic absorption grade C_2H_2 cylinders contain 380 ft^3 (10,760 L) of gas. N_2O is purchased by weight and comes in cylinders containing 56 lb of gas, which is equivalent to 490 ft^3 (13,830 L). A cylinder of C_2H_2 costs $200, whereas a cylinder of N_2O costs

about $70. Normal C_2H_2 gas flows in FAA are typically 2 L/min when air is the oxidant and 5 L/min when N_2O is the oxidant. N_2O gas flows are on the order of 10 L/min.

Air–C_2H_2 mixtures are used for the majority of elements, whereas an N_2O–C_2H_2 mixture has traditionally been used for the more refractory elements. So, for this costing exercise, we will assume that half the work is done using air–C_2H_2, and for the other half N_2O–C_2H_2 is being used. Therefore, a typical laboratory running the instrument for 1000 h per year will consume 16 cylinders of C_2H_2, which is equivalent to $3200 per year, and 22 cylinders of N_2O costing $1500, making a total of $4700.

Electrothermal Atomization The only gas that the electrothermal atomization process uses on a routine basis is high-purity argon, which costs $115 for a 340 ft³ (9630 L) cylinder. Typically, argon gas flows of up to 300 mL/min are required to keep an inert atmosphere in the graphite tube. At these flow rates, 540 h of use can be expected from one cylinder. Therefore, a typical laboratory running their instrument for 1000 h per year would consume almost two cylinders for $230.

ICP-OES and ICP-MS The consumption of gases in ICP-OES and ICP-MS is very similar. They both use approximately 15–20 L/min of gaseous argon, which means a cylinder of argon would last only about 10–12 h. For this reason, most users install a Dewar vessel containing a liquid supply of argon. Liquid argon tanks come in a variety of different sizes, but a typical Dewar system used for ICP-OES/ICP-MS holds about 230 L of liquid gas, which is equivalent to 6100 ft³ (174,000 L) of gaseous argon. (Note: The Dewar vessel can be bought outright, but are normally rented.) It costs about $400 to fill a 230 L Dewar vessel with liquid argon. At a typical argon flow rate 17 L/min total gas flow, a full vessel would last for almost 170 h. Again, assuming a typical laboratory runs their instrument for 1000 h per year, this translates to 6 fills at approximately $400 each, which is equivalent to about $2400 per year. If cylinders were used, over 100 would be required, which would elevate the cost to about $12,000 per year. (Note: When liquid argon is stored in a Dewar vessel, there is a natural bleed-off to the atmosphere when the gas reaches a certain pressure. For this reason, a bank of argon cylinders is probably the best option for laboratories that do not use their instruments on a regular basis.)

Another added expense with ICP-MS is that if it is fitted with collision/reaction cell technology, the cost of the collision or reaction gas will have to be added to the running costs of the instrument. Fortunately, for most applications, the gas flow is usually less than 5 mL/min, but for the collision/reaction interface approach, typical gas flows are 100–150 mL/min. The most common collision/reaction gases used are hydrogen, helium, and ammonia. The cost of high-purity helium is on the order of $400 for a 300 ft³ (8500 L) cylinder, whereas that of a cylinder of hydrogen/ammonia is approximately $250. One cylinder of either gas should be enough to last 1000 h at these kinds of flow rates. So, for this costing exercise, we will assume that the laboratory is running a collision/reaction cell/interface instrument.

It should also be pointed out that some collision/reaction cells require high-purity gases with extremely low impurity levels, because of the potential of the contaminants in the gas to create additional by-product ions (refer to Chapter 10).

This can be achieved either by purchasing laboratory-grade gases and cleaning them up with a gas purification system (getter), or by purchasing ultra-high-purity gases directly from the gas supplier. If the latter option is chosen, you should be aware that ultra-high-purity helium (99.9999%) is approximately twice the price of laboratory-grade helium (99.99%), whereas, ultra-high-purity hydrogen is approximately four times the cost of laboratory-grade hydrogen.

Electricity
Calculations for power consumption are based on the cost of electricity, which is about 15 cents per kilowatt per hour (kW/h) in the United States. This will vary depending on the location and demand, but it represents a good approximation for this costing exercise.

Flame AA   The power in a flame AA system is basically used for the hollow cathode lamps and the onboard microprocessor that controls functions like burner head position, lamp selection, photo multiplier tube voltage, grating position, etc. A typical instrument requires less than 1000 watts of power. If it is used for 1000 h per year, it will be drawing less than 1000 kW total power, which is about $150 per year.

Electrothermal Atomization   A graphite furnace system uses considerably more power than a flame AA system because a separate power unit is used to heat the graphite tube. In routine operation, there is a slow ramp heating of the tube for ~3 min until it reaches an atomization temperature of 2700°C. At this temperature, a maximum power of ~3.5 kW is required for 10–20 s. This heating cycle combined with the power requirements for the rest of the instrument costs ~$300, for a system that is run 1000 h per year.

ICP-OES and ICP-MS   Both these techniques can be considered the same with regard to power requirements as the RF generators are of very similar design. Based on the voltage, magnitude of the electric current, and the number of lines used, the majority of modern instruments draw about 5 kW total power. This works out to be about $750 for an instrument that is run 1000 h per year.

Consumables
Because of the fundamental differences between the four AS techniques, it is important to understand that there are considerable differences in the cost of consumables. In addition, the cost of the same component used in different techniques can vary significantly between different vendors and suppliers. So, where appropriate, I have extracted the data from a number of different sources and averaged the cost, based on a lower and upper range.

Flame AA   The major consumable supplies used in flame AA are the hollow cathode lamps. Depending on usage, you should plan to replace three of them every year, at a cost of $400–600 for a good quality, single-element lamp. However, if a continuum source AA system is being used, there will not be a requirement to replace lamps on a regular basis. Other minor costs are nebulizer tubing and autosampler tubes. These are relatively inexpensive, but should be planned for. The total cost

TABLE 20.5

**Operating Costs for a Sample
Requiring 10 Analytes, Based on the
Instrument Being Used 1000 h per Year**

Technique	Operating Cost for 10 Analytes per Sample ($)
FAA	0.31
ETA	6.00
ICP-OES	0.31
ICP-MS	0.71

year. Based on an annual operating cost of $5960, this equates to $6 per sample. (Note: If a multielement GFAA is being used, these costs will be reduced, but the actual cost will depend on how many elements are being determined simultaneously.)

ICP-OES   A duplicate ICP-OES analysis for as many analytes as you require takes about 3 min. So for 10 analytes, this is equivalent to 20 samples per hour or 20,000 samples per year. Based on an annual operating cost of $6250, this equates to $0.31 per sample.

ICP-MS   ICP-MS also takes about 3 min to carry out a duplicate analysis for 10 analytes, which is equivalent to 20,000 samples per year. Based on an annual operating cost of $14,150, this equates to $0.71 per sample.

Operating costs for all four AS techniques for the determination of 10 analytes/ sample are summarized in Table 20.5.

For laboratories with extremely high sample workloads requiring more than 20 analytes/sample, a single-element technique such as ETA, even though cost-prohibitive compared to ICP-MS, becomes less of a practical option. Although the running costs of FAA are very competitive with the multielement techniques, it is impractical in a high workload environment. On the other hand, when the elemental requirements are less demanding, FAA and ETA will look much more attractive if the running costs are based on cost per analyte. For example, for a laboratory that is running a set of samples that require just one analyte, the cost/sample for FAA and ETA will be $0.03 and $0.60, respectively, whereas the costs for ICP-OES and ICP-MS will basically remain the same as that of determining 10 analytes/sample. This cost difference is shown in Table 20.6.

It must also be emphasized that this comparison does not take into account the detection limit requirements, but is based on instrument-operating costs alone. These figures have been generated for a typical workload using what would be considered the average cost of gases, power, and consumables in the United States. Every laboratory's situation is unique, especially outside the United States, so this costing exercise should be treated with caution and only be used as a guideline for comparison purposes. However, it is a good exercise to show that there are differences among the running costs of the major AS techniques. If required, it can be taken a step further by also including the purchase price of the instrument, the cost of installing a clean

TABLE 20.6
Running Costs for a Sample Requiring
Just 1 Analyte, Based on the Instrument
Being Used 1000 h per Year

Technique	Operating Cost for 1 Analyte per Sample ($)
FAA	0.03
ETA	0.60
ICP-OES	0.31
ICP-MS	0.71

room, the cost of sample preparation, and the salary of the operator. This would be a very useful exercise as it would give a good approximation of the overall cost of analysis, and therefore it could be used as a guideline for calculating what a laboratory might charge for running samples on a commercial basis.

CONCLUSION

It is important to remember that there are many criteria to consider when selecting a trace element technique. You have to decide which are most suited to your application and your laboratory. This chapter is not meant to be an exhaustive comparison of all elemental techniques. It should be used as a guideline to evaluate the most commonly used trace and ultratrace atomic spectroscopy–based techniques. The comparison has been done in a very simplified way, and there has been no attempt to compare the many variations, features, and sampling accessories offered by the different manufacturers. However, it is clear that there is no single technique suitable for all applications. They all have their own strengths and weaknesses. It is therefore important when making the comparison that all these avenues are explored. Maybe ICP-MS is a technique that you would really like to have in your laboratory. True, it is a very powerful piece of equipment, but at the end of the day, it must be considered whether the purchase can be really justified. In most cases I believe it can be, but it is definitely worth investing the time and effort to collect the evidence in order to support that justification. Hopefully, this chapter has given you some insight into this process.

REFERENCES

1. Strategic Directions International, Inc., Los Angeles, CA, Analytical Instruments Global Assessment Report, 9th edition, July 2006.
2. R. J. Thomas, *Today's Chemist at Work*, **8**(10), 42–48, 1999.
3. R. J. Thomas, *Today's Chemist at Work*, **9**(9), 19–25, 2000.
4. Scott Specialty Gases, Plumsteadville, PA, http://www.scottgas.com.
5. Praxair, Inc., Danbury, CT, http://www.praxair.com.
6. Baltimore Gas and Electricity, Baltimore, MD, http://www.bge.com.
7. Agilent Technologies, Wilmington, DE, http://www.chem.agilent.com.

8. PerkinElmer Life and Analytical Sciences, Shelton, CT, http://las.perkinelmer.com.

9. Thermo Fisher Scientific, Inc., Waltham, MA, http://www.thermo.com/com.

10. Varian, Inc., Palo Alto, CA, http://www.varianinc.com.

11. Glass Expansion, Pocasset, MA, http://www.geicp.com.

12. Meinhard Glass Products, Golden, CO, http://www.meinhard.com.

13. Spectron, Inc., Ventura, CA, http://www.spectronus.com.

21 How to Select an ICP Mass Spectrometer

Some Important Analytical Considerations

Understanding the basic principles of ICP mass spectrometry is important but not absolutely essential to operate and use an instrument on a routine basis. However, understanding how these basic principles affect the performance of an instrument is a real benefit when evaluating the analytical capabilities of the technique. There is no question that the better informed you are going into an evaluation of commercial instrumentation, the better chance you have of selecting the right one for your application. Having been involved in demonstrating ICP-MS equipment and running customer samples for over 10 years, I know the mistakes that people make when they get into the evaluation process. So, in Chapter 21, I have presented a set of evaluation guidelines that hopefully will help you make the right decision.

OK, you have convinced your boss that ICP-MS is perfect for your laboratory. Hopefully, the chapters on the fundamental principles have given you the basic knowledge and a good platform on which to go out and evaluate the marketplace. However, they do not really give you an insight into how to compare instrument designs, hardware components, and software features, which are of critical importance when you have to make a decision regarding which instrument to purchase. There are a number of commercial systems available in the marketplace, which look very similar and have very similar specifications, but how do you know which is the best one that fits your needs? This chapter, in conjunction with the other chapters in the book, presents a set of evaluation guidelines to help you decide the most important analytical figures of merit for your application.

EVALUATION OBJECTIVES

It is very important before you begin the selection process to decide what your analytical objectives are. This is particularly important if you are part of an evaluation committee. It is fine to have more than one objective, but it is essential that all the members of the group begin the evaluation process with the objectives clearly defined. For example, is detection limit (DL) performance an important objective for your application, or is it more important to have an instrument that is easy to use? If the instrument is being used on a routine basis, maybe good reliability is also very critical. On the other hand, if the instrument is being used to generate revenue, perhaps sample throughput and cost of analysis are of greater importance. Every laboratory's scenario is unique, so it is important to prioritize before you begin the evaluation process. So, as well as looking at instrument features and components, the comparison should also be made with your analytical objectives in mind. Let us take a look at the most common ones that are used in the selection process. They typically include the following:

- Analytical performance
- Usability aspects
- Reliability issues
- Financial considerations

Let us examine these in greater detail.

ANALYTICAL PERFORMANCE

Analytical performance can mean different things to different people. The major reason that the trace element community was attracted to ICP-MS almost 20 years ago was its extremely low multielement DLs. Other multielement techniques, such as ICP-OES, offered very high throughput but just could not get down to ultratrace levels. Even though ETA offered much better detection capability than ICP-OES, it did not offer the sample throughput capability that many applications demanded. In addition, ETA was predominantly a single-element technique and so was impractical for carrying out rapid multielement analysis. These limitations quickly led to the commercialization and acceptance of ICP-MS as a tool for rapid ultratrace element analysis. However, there are certain areas where ICP-MS is known to have weaknesses. For example, dissolved solids for most sample matrices must be kept below 0.2%, otherwise it can lead to serious drift problems and poor precision.

Polyatomic and isobaric interferences, even in simple acid matrices, can produce unexpected spectral overlaps, which will have a negative impact on your data. High-resolution instrumentation and collision/reaction cell/interface technology are helping to alleviate these spectral problems, but they also have their limitations. Depending on the types of samples being analyzed, matrix components can dramatically suppress analyte sensitivity and affect accuracy. These potential problems can all be reduced to a certain extent, but different instruments approach and compensate for these problem areas in different ways. With a novice, it is often ignorance or a basic lack of understanding of how a particular instrument works that makes the selection process more complicated than it really should be. So, any information that can help you prepare for the evaluation will put you in a much stronger position.

It should be emphasized that these evaluation guidelines are based on my personal experience and should be used in conjunction with other material in the open literature that has presented broad guidelines to compare figures of merit for commercial instrumentation.[1–3] In addition, you should talk with colleagues in the same industry or application segment as yourself. If they have gone through a lengthy evaluation process, they can give you valuable pointers or even suggest an instrument that is better suited to your needs. Finally, before we begin, it is strongly suggested that you narrow the actual evaluation to two, or maybe three, commercial products. By carrying out some preevaluation research, you will have a better understanding of what ICP-MS technology or instrument to focus on. For example, if funds are limited and you are purchasing ICP-MS for the very first time to carry out high-throughput environmental testing, it is probably more cost-effective to focus on quadrupole technology. On the other hand, if you are investing in a second system to enhance the capabilities of your quadrupole instrument, it might be worth taking a look at collision/reaction cell or magnetic sector technology. Or, if fast multielement transient peak analysis is your major reason for investing in ICP-MS, TOF technology should be given serious consideration. One final note I would like to add, although it is not strictly a technical issue. If you are prepared to forego an instrument demonstration or do not need any samples run, you will be in a much stronger position to negotiate price with the instrument vendor. You should keep that in mind before you decide to get involved in a lengthy selection process.

So, let us begin by looking at the most important aspects of instrument performance. Depending on the application, the major performance issues that need to be addressed include the following:

- Detection capability
- Precision/signal stability
- Accuracy
- Dynamic range
- Interference reduction
- Sample throughput
- Transient signal capability

Detection Capability

Detection capability is a term used to assess the overall detection performance of an ICP mass spectrometer. There are a number of different ways of looking at detection capability, including instrument detection limit (IDL), elemental sensitivity, background signal, and background equivalent concentration (BEC). Of these four criteria, the IDL is generally thought to be the most accurate way of assessing instrument detection capability. It is often referred to as signal-to-background noise, and for a 99% confidence level is typically defined as 3× standard deviation (SD) of n replicates ($n = \sim 10$) of the sample blank and is calculated in the following manner:

$$IDL = \frac{3 \times \text{Standard deviation of background signal}}{\text{Analyte intensity} - \text{background signal}} \times \text{analyte concentration}$$

However, there are slight variations of both the definition and calculation of IDLs, so it is important to understand how different manufacturers quote their DLs if a comparison is to be made. They are usually run in single-element mode, using extremely long integration times (5–10 s) to achieve the highest-quality data. So, when comparing DLs of different instruments, it is important to know the measurement protocol used.

A more realistic way of calculating analyte DL performance in your sample matrices is to use method detection limit (MDL). The MDL is broadly defined as the minimum concentration of analyte that can be determined from zero with 99% confidence. MDLs are calculated in a similar manner to IDLs, except that the test solution is taken through the entire sample preparation procedure before the analyte concentration is measured multiple times. This difference between MDL and IDL is exemplified in EPA Method 200.8, where a sample solution at 2–5 times the estimated IDL is taken through all the preparation steps and analyzed. The MDL is then calculated in the following manner:

$$MDL = (t) \times (S)$$

Where t = Student's "t" value for a 95% confidence level and specifies a standard deviation estimate with $n - 1$ degrees of freedom ($t = 3.14$ for 7 replicates) and S = the SD of the replicate analyses.

Both IDL and MDL are very useful to understand the capability of ICP-MS. However, whatever method is used to compare DLs of different manufacturers' instrumentation, it is essential to carry out the test using realistic measurement times that reflect your analytical situation. For example, if you are determining a group of elements across the mass range in a digested rock sample, it is important to know how much the sample matrix suppresses the analyte sensitivity, because the DL of each analyte will be impacted by the amount of suppression across the mass range. On the other hand, if you are carrying out high-throughput multielement analysis of drinking or wastewater samples, you probably need to be using relatively short integration times (1–2 s per analyte) to achieve the desired sample throughput. Or if you are dealing with a laser ablation or flow injection transient peak that lasts 10–20 s, it is absolutely critical that you understand the impact that the time has on DLs compared to a continuous signal generated with a conventional nebulizer. (In fact, analysis time and DLs are very closely related to each other and will be discussed later on in this chapter.) In other words, when comparing IDLs, it is absolutely critical that the tests represent your real-world analytical situation.

Elemental sensitivity is also a useful assessment of instrument performance, but it should be viewed with caution. It is usually a measurement of background-corrected intensity at a defined mass and is typically specified as counts per second (cps) per concentration (ppb or ppm) of a midmass element such as $^{103}Rh^+$ or $^{115}In^+$. However, unlike DL, raw intensity usually does not tell you anything about the intensity of the background or the level of the background noise. It should be emphasized that instrument sensitivity can be enhanced by optimization of operating parameters such as RF power, nebulizer gas flows, torch-sampling position, interface pressure, and sampler/skimmer cone geometry, but usually comes at the expense of other performance criteria, including oxide levels, matrix tolerance, or background intensity. So, be very cautious when you see an extremely high sensitivity specification, because there is a strong probability that the oxide or background specs might also be high. For this reason, it is unlikely there will be an improvement in DL unless the increase in sensitivity comes with no compromise in the background level. It is also important to understand the difference between background and background noise when comparing specifications (the background noise is a measure of the stability of the background and is defined as the square root of the background signal). Most modern quadrupole instruments today specify 20–50 million cps per ppm of rhodium ($^{103}Rh^+$) or indium ($^{115}In^+$) and <10 cps of background (usually at 220 amu), whereas magnetic sector instrument sensitivity specifications are typically 10–20 times the higher with 10 times the lower background.

Another figure of merit that is being used more routinely nowadays is BEC. BEC is defined as the intensity of the background at the analyte mass, expressed as an apparent concentration and is typically calculated in the following manner:

$$BEC = \frac{\text{Intensity of background signal}}{\text{Analyte intensity} - \text{background intensity}} \times \text{analyte concentration}$$

It is considered more of a realistic assessment of instrument performance in real-world sample matrices (especially if the analyte mass sits on a high background), because it gives an indication of the level of the background—defined as a concentration value. DLs alone can sometimes be misleading because they are influenced by the number of readings taken, integration time, cleanliness of the blank, and at what mass the background is measured—and are rarely achievable in a real-world situation. Figure 21.1 emphasizes the difference between DL and BEC. In this example, 1 ppb of an analyte produces a signal of 10,000 cps and a background of 1000 cps. Based on the calculations defined earlier, the BEC is equal to 0.11 ppb because it is expressing the background intensity as a concentration value. On the other hand, the DL is 10 times lower because it is using the SD of the background (i.e., the noise) in the calculation. For this reason, BECs are particularly useful when it comes to comparing the detection capabilities of techniques such as cool plasma and collision/reaction cell technology, because it gives you a very good indication of how efficient the background reduction process is.

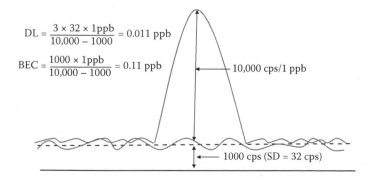

$$DL = \frac{3 \times 32 \times 1ppb}{10,000 - 1000} = 0.011 \text{ ppb}$$

$$BEC = \frac{1000 \times 1ppb}{10,000 - 1000} = 0.11 \text{ ppb}$$

10,000 cps/1 ppb

1000 cps (SD = 32 cps)

FIGURE 21.1 DL is calculated using the noise of the background, whereas BEC is calculated using the intensity of the background.

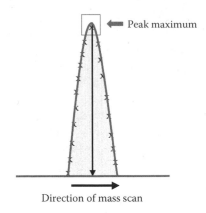

Direction of mass scan

FIGURE 21.2   There are typically two approaches to peak quantitation—peak hopping (usually at peak maximum) and multichannel scanning (across the full width of the peak).

It is also important to remember that peak measurement protocol will also have an impact on detection capability. As mentioned in Chapter 12, there are basically two approaches to measuring an isotopic signal in ICP-MS. There is the multichannel scanning approach, which uses a continuous smooth ramp of 1–20 channels per mass across the peak profile, and peak-hopping approach, where the mass analyzer power supply is driven to a discrete position on the peak, allowed to settle, and a measurement taken for a fixed period of time. This is usually at the peak maximum but can be as many points as the operator selects. This process is simplistically shown in Figure 21.2.

The scanning approach is best for accumulating spectral and peak shape information when doing mass calibration and resolution scans. It is traditionally used as a classical method development tool to find out what elements are present in the sample and to assess spectral interferences on the masses of interest. However, when the best possible DLs are required, it is clear that the peak-hopping approach is best. It is important to understand that to get the full benefit of peak hopping, the best DLs are achieved when single-point peak hopping at the peak maximum is chosen. It is well accepted that measuring the signal at the peak maximum will always give the best signal-to-background noise for a given integration time, and there is no benefit in spreading your available integration time over more than one measurement point per mass.[4] Instruments that use more than one point per peak for quantitation are sacrificing measurement time on the sides of the peak, where the signal-to-noise ratio is worse. However, the ability of the mass analyzer to repeatedly scan to the same mass position every time during a multielement run is of paramount importance for peak hopping. If multiple points per peak are recommended, it is a strong indication that the spectrometer has poor mass calibration stability, because it cannot guarantee that it will always find the peak maximum with just one point. Mass calibration specification, which is normally defined as a shift in peak position (in amu) over an 8 h period, is a good indication of mass stability. However, it is not always the best way to compare systems, because peak algorithms using multiple points are often used to calculate the peak position. A more accurate way is to assess the short-term

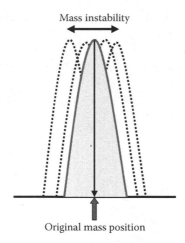

Mass instability

Original mass position

FIGURE 21.3   Good mass stability is critical for single-point, peak-hopping quantitation.

and long-term mass stability by looking at relative peak positions over time. The short-term stability can be determined by aspirating a multielement solution containing four elements (across the mass range) and recording the spectral profiles using multichannel ramp scanning of 20 points per peak. Now repeat the multielement scan 10 times and record the peak position of every individual scan. Calculate the average and relative standard deviation (RSD) of the scan positions. The long-term mass stability can then be determined by repeating the test 8 h later to see how far the peaks have moved. It is important, of course, that the mass calibration procedure not be carried out during this time. Figure 21.3 shows what might happen to the peak position over time, if the analyzer's mass stability is poor.

Precision

Short- and long-term precision specifications are usually a good indication of how stable an instrument is (refer to Chapter 12). Short-term precision is typically specified as %RSD of 10 replicates of 1–10 ppb of three elements across the mass range using 2–3 s integration times, whereas long-term precision is a similar test, but normally carried out every 5–10 min over a 4–8 h period. Typical short-term precision, assuming an instrument warm-up time of 30–40 min, should be approximately 1–3%, whereas long-term precision should be on the order of 3–5%—both determined without using internal standards. However, it should be emphasized that under these measurement protocols, it is unlikely you will see a big difference in the performance between different instruments in simple aqueous standards. A more accurate reflection of the stability of an instrument is to carry out the tests using a typical matrix that would be run in your laboratory at the concentrations you expect. It is also important that stability should be measured without the use of an internal standard. This will enable you to evaluate the instrument drift characteristics, without any type of signal correction method being applied.

It is recognized that the major source of drift and imprecision in ICP-MS, particularly with real-world samples, is associated with either the sample introduction area, design of the interface, or the ion optics system. Some of the common problems encountered are as follows:

- Pulsations and fluctuations in the peristaltic pump, leading to increased signal noise
- Blockage of the nebulizer over time, resulting in signal drift—especially if the nebulizer does not have a tolerance for high dissolved solids
- Poor drainage, producing pressure changes in the spray chamber and resulting in spikes in the signal
- Buildup of solids in the sample injector, producing signal drift
- Changes in the electrical characteristics of the plasma, generating a secondary discharge and increasing ion energies
- Blockage of the sampler and skimmer cone orifice with sample material, causing instability
- Erosion of the sampler and skimmer cone orifice with high-concentration acids
- Coating of the ion optics with matrix components, resulting in slight changes in the electrical characteristics the of ion lens system

These are all relative problems depending on the types of samples being analyzed. However, the most common and potentially serious problem with real-world matrices is the deposition of sample material on the interface cones and the ion optics over time. It does not impact short-term precision that much, because careful selection of internal standards matched to the analyte masses can compensate for slight instability problems. However, sample material, particularly matrix components found in environmental and geochemical samples, can have a dramatic effect on long-term stability. The problem is exaggerated even more in a high-throughput laboratory, because poor stability will necessitate more regular recalibration and might even require some samples to be rerun if QC standards fall outside certain limits. There is no question that if an instrument has poor drift characteristics, it will take much longer to run an autosampler tray full of samples, and in the long term, this will result in much higher argon consumption.

For these reasons, it is critical that when short- and long-term precision are evaluated, you know all the potential sources of imprecision and drift. It is therefore important that you choose either a matrix that is representative of your samples or one that will genuinely test the instrument out. Typical sample matrices include the following:

- **Drinking waters** containing calcium and magnesium salts at a few hundred ppm
- **Concentrated acids** that are representative of samples being run in the semiconductor industry
- **Rock digests** containing calcium, magnesium, iron, and aluminum at a few hundred ppm, with maybe some alkaline peroxide/borate fusion mixtures
- **Biological fluid** samples such as blood or urine, containing carbonaceous, organic, and saline components
- **Saline samples** containing sodium, magnesium, and calcium chlorides

- **Metallurgical alloys** containing concentrations of various metals dissolved in 1–5% mineral acids
- **Organic samples** such as diluted oils, alcohols, ketones, or aromatic solvents

Whatever matrices are chosen, it must be emphasized that for the stability test to be meaningful, no internal standards should be used, the sample should contain less than 0.2% total dissolved solids, and the representative elements should be at a reasonably high concentration (1–10 ppb) and be spread across the mass range. In addition, no recalibration should be carried out for the length of the test, which should reflect your real-world situation.[5] For example, if you plan to run your instrument in a high-throughput environment, you might want to carry out an 8 h or even an overnight (12–16 h) stability test. If you are not interested in such long runs, a 2–4 h stability test will probably suffice. However, just remember, plan the test beforehand and make sure you know how to evaluate the vast amount of data that will be generated. It will be hard work, but I guarantee it will be worth it in order to fully understand the short- and long-term drift characteristics of the instruments you are evaluating.

Isotope Ratio Precision

An important aspect of ICP-MS is its ability to carry out fast isotope ratio precision data. With this technique, two different isotopes of the same element are continuously measured over a fixed period of time. The ratio of the signal of one isotope to the other isotope is taken, and the precision of the ratios is then calculated. Analysts interested in isotope ratios are usually looking for the ultimate in precision. The optimum way to achieve this to get the best counting statistics would be to carry out the measurement simultaneously with a multicollector magnetic sector instrument or a TOF ICP-MS system. However, a quadrupole mass spectrometer is a rapid sequential system, so the two isotopes are never measured at exactly the same moment. This means that the measurement protocol must be optimized to get the best precision. As discussed earlier, the best and most efficient use of measurement time is to carry out single-point peak hopping between the two isotopes. In addition, it is also beneficial to be able to vary the total measurement time of each isotope, depending on their relative abundance. The ability to optimize the dwell time and the number of sweeps of the mass analyzer ensures that the maximum amount of time is being spent on the top of each individual peak where the signal-to-noise ratio is at its highest.[6]

It is also critical to optimize the efficiency cycle of the measurement. With every sequential mass analyzer, there is an overhead time called a *settling time* to allow the power supply to settle before taking a measurement. This time is often called nonanalytical time, because it does not contribute to the quality of the analytical signal. The only time that contributes to the analytical signal is the *dwell time*, or the time that is actually spent measuring the peak. The measurement efficiency cycle (MEC) is a ratio of the dwell time to the total analytical time (which includes settling time) and is expressed as follows:

$$\text{MEC (\%)} = \frac{\text{No. sweeps} \times \text{dwell time} \times 100}{\left\{ \text{No. sweeps} \times \left(\text{dwell time} + \text{settling time} \right) \right\}}$$

It is therefore obvious that to get the best precision over a fixed period of time, the settling time must be kept to an absolute minimum. The dwell time and the number of sweeps are operator selectable, but the settling time is usually fixed because it is a function of the quadrupole electronics. For this reason, it is important to know what the settling time of the mass spectrometer is when carrying out peak hopping. Remember, a shorter settling time is more desirable because it will increase the measurement efficiency cycle and improve the quality of the analytical signal.[7]

In addition, if isotope ratios are being determined on vastly different concentrations of major and minor isotopes using the extended dynamic range of the system, it is important to know the settling time of the detector electronics. This settling time will affect the detector's ability to detect the analog and pulse signals (or in dynamic attenuation mode with a pulse-only EDR system) when switching between measurement of the major and minor isotopes, which could have a serious impact on the accuracy and precision of the isotope ratio. So, for that reason, no matter how the higher concentrations are handled, shorter settling times are more desirable, so that the switching or attenuation can be carried out as quickly as possible.

This is shown in Figure 21.4, which shows a spectral scan of $^{63}Cu^+$ and $^{65}Cu^+$ using an automated pulse/analog EDR detection system. The natural abundance of these two isotopes is 69.17% and 30.83%, respectively. However, the ratio of these isotopes has been artificially altered to be 0.39% for $^{63}Cu^+$ and 99.61% for $^{65}Cu^+$. The intensity of ^{63}Cu is about 70,000 cps, which requires pulse counting, whereas the intensity of the $^{65}Cu^+$ is about 10 million cps, which necessitates analog counting. There is no question that the counting circuitry would miss many of the ions and generate erroneous concentration data if the switching between pulse and analog modes was not fast enough.

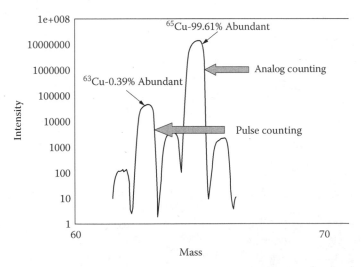

FIGURE 21.4 The detector electronics must be able to switch fast enough to detect isotope ratios that require both pulse and analog counting modes (data copyright © 2003–2007, all rights reserved, PerkinElmer Inc.).

So, when evaluating isotopic ratio precision, it is important that the measurement protocol and peak quantitation procedure are optimized. Isotope precision specifications are a good indication regarding what the instrument is capable of, but once again, these will be defined in aqueous-type standards, using relatively short total measurement times (typically 5 min). For that reason, if the test is to be meaningful, it should be optimized to reflect your real-world analytical situation.

Accuracy

Accuracy is a very difficult aspect of instrument performance to evaluate because it often reflects the skill of the person developing the method and analyzing the samples, instead of the capabilities of the instrument itself. If handled correctly, it is a very useful exercise to go through, particularly if you can get hold of reference material (ideally of similar matrices to your own) whose values are well defined. However, when attempting to compare the accuracy of different instruments, it is essential that you prepare every sample yourself, including the calibration standards, blanks, unknown samples, QC standards, or certified reference material (CRM). I suggest that you make up enough of each solution to give to each vendor for analysis. By doing this, you eliminate the uncertainty and errors associated with different people making up different solutions. It then becomes more of an assessment of the capability of the instrument, including its sample introduction system, interface region, ion optics, mass analyzer, detector, and measurement circuitry, to handle the unknown samples, minimize interferences, and get the correct results.

A word of caution should be expressed at this point. Having worked in ICP-MS for almost 20 years, I know that the experience of the person developing the method, running the samples, and performing the demonstration has a direct impact on the quality of the data generated in ICP-MS. There is no question in my mind that the analyst with the most application expertise has a much better chance of getting the right answer than someone who is either inexperienced or is not familiar with a particular type of sample. I think it is quite valid to compare the ability of the application specialist because this might be the person who is supporting you. However, if you want to assess the capabilities of the instrument alone, it is essential to take the skill of the operator out of the equation. This is not as straightforward as it sounds, but I have found that the best way to "level the playing field" is to send some of your sample matrices to each vendor before the actual demonstration. This allows the application person to spend time developing the method and become familiar with the samples. You can certainly hold back on your CRM or QC standards until you get to the demonstration, but at least it gives each vendor some uninterrupted time with your samples. This also allows you to spend most of the time at the demonstration evaluating the instrument, assessing hardware components, comparing features, and getting a good look at the software. It is my opinion that most instruments on the market should get the right answer—at least for the majority of routine applications. So, even though the accuracy of different instruments should be compared, it is more important to understand how the result was generated, especially when it comes to the analysis of very difficult samples.

Dynamic Range

When ICP-MS was first commercialized, it was primarily used to determine very low analyte concentrations. As a result, detection systems were only asked to measure concentration levels up to approximately five orders of magnitude. However, as the demand for greater flexibility grew, such systems were being called upon to extend their dynamic range to determine higher and higher concentrations. Today, the majority of commercial systems come standard with detectors that can measure signals up to eight orders of magnitude.

As mentioned in Chapter 11, there are subtle differences between how various detectors and detection systems achieve this, so it is important to understand how different instruments extend the dynamic range. The majority of quadrupole-based systems on the market extend the dynamic range by using a discrete dynode detector operated either in pulse-only mode or a combination of pulse and analog mode. When evaluating this feature, it is important to know whether this is done in one or two scans because it will have an impact on the types of samples you can analyze. The different approaches have been described earlier, but it is worth briefly going through them again:

Two-scan approach: Basically, two types of two-scan or prescan approaches have been used to extend the dynamic range:
  - In the first one, a survey or prescan is used to determine what masses are at high concentrations and what masses are at trace levels. Then, the second scan actually measures the signals by switching rapidly between analog and pulse counting.
  - In the second two-scan approach, the detector is first run in the analog mode to measure the high signals and then rescanned in pulse-counting mode to measure the trace levels.

One-scan approach: This approach is used to measure both the high levels and trace concentrations simultaneously in one scan. This is typically achieved by measuring the ion flux as an analog signal at some midpoint on the detector. When more than a threshold number of ions are detected, the ions are processed through the analog circuitry. When fewer than a threshold number of ions are detected, the ions cascade through the rest of the detector and are measured as a pulse signal in the conventional way.

Using pulse-only mode: The most recent development in extending the dynamic range is to use the pulse-only signal. This is achieved by monitoring the ion flux at one of the first few dynodes of the detector (before extensive electron multiplication has taken place) and then attenuating the signal by applying a control voltage. Electron pulses passed by the attenuation section are then amplified to yield pulse heights that are typical in normal pulse-counting applications. Under normal circumstances, this approach requires only one scan, but if the samples are complete unknowns, dynamic attenuation might need to be performed, where an additional premeasurement time is built into the settling time to determine the optimum detector attenuation for the selected dwell times used.

The methods that use a prescan or premeasurement time work very well, but they do have certain limitations for some applications, compared to the one-scan approach. Some of these include the following:

- The additional scan/measurement time means it will use more of the sample. Ordinarily this will not pose a problem, but if sample volume is limited to a few hundred microliters, it might be an issue.
- If concentrations of analytes are vastly different, the measurement circuitry reaction time of a prescan system might struggle to switch quickly enough between high and low concentration elements. This is not such a major problem, unless the measurement circuitry has to switch rapidly between consecutive masses in a multielement run, or there are large differences in the concentrations of two isotopes of the same element when carrying out ratio studies. In both these situations, there is a possibility that the detection system will miss counting some of the ions and produce erroneous data.
- The other advantage of the one-scan approach is that more time can be spent measuring the peaks of interest in a transient peak generated by a flow injection or laser ablation system. With a detector that uses two scans or a prescan approach, much time will be used just to characterize the sample. It is exaggerated even more with a transient peak, especially if the analyst has no prior knowledge of elemental concentrations in the sample.

This final point is exemplified in Figure 21.5, which shows the measurement of a flow injection peak of NIST 1643C potable water CRM, using an automated simultaneous pulse/analog EDR system. It can be seen that the K and Ca are at ppm levels, which requires the use of the analog counting circuitry, whereas the Pb and Cd are at ppb

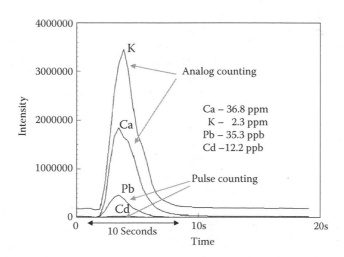

FIGURE 21.5 A one-scan approach to extending the dynamic range is more advantageous for handling a fast transient flow injection peak, in this example, generated by NIST 1643C. (From E. R. Denoyer, Q. H. Lu, *Atomic Spectroscopy,* **14**[6], 162–169, 1993.)

levels, which requires pulse counting.[8] This would not be such a difficult analysis for a detector, except that the transient peak has only lasted 10 s. This means that to get the highest-quality data, you want to be spending all the available time quantifying the peak. In other words, you cannot afford the luxury of doing a premeasurement, especially if you have no prior knowledge of the analyte concentrations.

For these reasons, it is important to understand how the detector handles high concentrations to evaluate them correctly. If you are truly interested in using ICP-MS to determine higher concentrations, you should check out the linearity of different masses across the mass range by measuring high-ppt (~500 ppt), low-ppb (~50 ppb), and ppm (10–100 ppm) levels. Do not be afraid to analyze a standard reference material (SRM) sample such as one of the NIST 1643 series of drinking water reference standards, which has both high (ppm) and low (ppb) levels. Finally, if you know there are large concentration differences between the same analytes, make sure the detector is able to determine them with good accuracy and precision. On the other hand, if your instrument is only going to be used to carry out ultratrace analysis, it probably is not worth spending the time to evaluate the capability of the extended dynamic range feature.

Interference Reduction

As mentioned in Chapter 14, there are two major types of interferences that have to be compensated for: spectral and matrix (space charge and physical). Although most instruments approach the principles of interference reduction in a similar way, the practical aspect of compensating for them will be different, depending on the differences in hardware components and instrument design. Let us look at interference reduction in greater detail and compare the different approaches used.

Reducing Spectral Interferences

The majority of spectral interferences seen in ICP-MS are produced by either the sample matrix, solvent, plasma gas, or various combinations of them. If the interference is caused by the sample, the best approach might be to remove the matrix by some kind of ion exchange column. However, this can be cumbersome and time consuming to do on a routine basis. If the interference is caused by solvent ions, simply desolvating the sample will have a positive effect on reducing the interference. For that reason, systems that come standard with chilled spray chambers to remove much of the solvent usually generate less sample-based oxide-, hydroxide-, and hydride-induced spectral interferences. There are alternative ways to reduce these types of interferences, but cooling the spray chamber or using a membrane desolvation system can be a very effective way of reducing the intensity of the solvent-based ionic species.

Spectral interferences are an unfortunate reality in ICP-MS, and it is now generally accepted that instead of trying to reduce or minimize them, the best way is to resolve the problem away using high-resolution technology such as a double-focusing magnetic sector mass analyzer.[9] Even though they are not considered ideal for a routine, high-throughput laboratory, they offer the ultimate in resolving power and have found a niche in applications that require ultratrace detection and a high degree of flexibility for the analysis of complex sample matrices. If you use a quadrupole-based instrument and are looking to purchase a second system to enhance the flexibility of your laboratory, it might be worth taking a serious look at magnetic sector technology. The full benefits of this type of mass analyzer for ICP-MS have been described in Chapter 8.

Let us now turn our attention to the different approaches used to reduce spectral interferences using quadrupole-based technology. Each approach should be evaluated on the basis of its suitability for the demands of your particular application.

Resolution Improvement

As described in Chapter 7, there are two very important performance specifications of a quadrupole—resolution and abundance sensitivity.[10] Although they both define the ability of a quadrupole to separate an analyte peak from a spectral interference, they are measured differently. Resolution reflects the shape of the top of the peak and is normally defined as the width of a peak at 10% of its height. Most instruments on the market have similar resolution specifications of 0.3–3.0 amu and typically use a nominal setting of 0.7–1.0 amu for all masses in a multielement run. For this reason, it is unlikely you will see any measurable difference when you make your comparison.

However, some systems allow you to change resolution settings on the fly on individual masses during a multielement analysis. Under normal analytical scenarios, this is rarely required, but at times it can be advantageous to improve the resolution for an analyte mass, particularly if it is close to a large interference and there is no other mass or isotope available for quantitation. This can be seen in Figure 21.6, which shows a spectral scan of 10 ppb $^{55}Mn^+$, which is monoisotopic, and 100 ppm

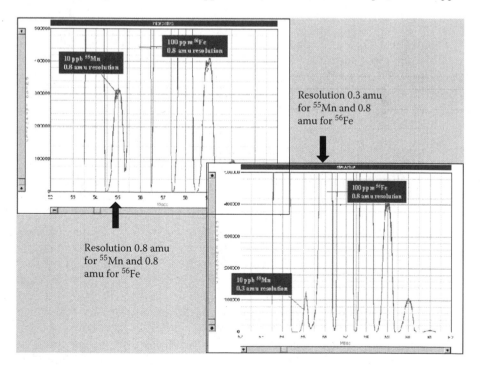

FIGURE 21.6 A resolution setting of 0.3 amu will improve the DL for $^{55}Mn^+$ in the presence of high concentrations of $^{56}Fe^+$ (courtesy of PerkinElmer Life and Analytical Sciences).

of $^{56}Fe^+$. The left-hand spectra shows the scan using a resolution setting of 0.8 amu for both $^{55}Mn^+$ and $^{56}Fe^+$, whereas the right-hand spectra shows the same scan, but using a resolution setting of 0.3 amu for $^{55}Mn^+$ and 0.8 amu for $^{56}Fe^+$. Even though the $^{55}Mn^+$ peak intensity is about 3 times lower at 0.3 amu resolution, the background from the tail of the large $^{56}Fe^+$ is about 7 times less, which translates into a fivefold improvement in the $^{55}Mn^+$ DL at a resolution of 0.3 amu, compared to 0.8 amu.

Higher Abundance Sensitivity Specifications

The second important specification of a mass analyzer is abundance sensitivity, which is a measure of the width of a peak at its base. It is defined as the signal contribution of the tail of a peak at one mass lower and one mass higher than the analyte peak, and generally speaking, the lower the specification, the better the performance of the mass analyzer. The abundance sensitivity of a quadrupole is determined by a combination of factors, including shape, diameter, and length of the rods; frequency of quadrupole power supply; and the slope of the applied RF/DC voltages. Even though there are differences between designs of quadrupoles in commercial ICP-MS systems, there appears to be very little difference in their practical performance.

When comparing abundance sensitivity, it is important to understand what the numbers mean. The trajectory of an ion through the analyzer means that the shape of the peak at one mass lower than the mass M, i.e., (M − 1), is slightly different from the other side of the peak at one mass higher, i.e., (M + 1). For this reason, the abundance sensitivity specification for all quadrupoles is always worse on the low-mass side than the high-mass side, and is typically 1×10^{-6} at M − 1 and 1×10^{-7} at M + 1. In other words, an interfering peak of 1 million cps at M − 1 would produce a background of 1 cps at M, whereas it would take an interference of 10 million cps at M + 1 to produce a background of 1 cps at M. In theory, hyperbolic rods will demonstrate better abundance sensitivity than round ones, as will a quadrupole with longer rods and a power supply with higher frequency. However, you have to evaluate whether this produces any tangible benefits when it comes to the analysis of your real-world samples.

Use of Cool Plasma Technology

All of the instruments on the market can be set up to operate under cool or cold plasma to achieve very low DLs for elements such as K, Ca, and Fe. Cool plasma conditions are achieved when the temperature of the plasma is cooled sufficiently low enough to reduce the formation of argon-induced polyatomic species.[11] This is typically achieved with a decrease in the RF power, an increase in the nebulizer gas flow, and sometimes a change in the sampling position of the plasma torch. Under these conditions, the formation of species such as $^{40}Ar^+$, $^{38}ArH^+$, and $^{40}Ar^{16}O^+$ is dramatically reduced, which allows the determination of low levels of $^{40}Ca^+$, $^{39}K^+$, and $^{56}Fe^+$, respectively.[12]

Under normal hot plasma conditions (typically, RF power of 1200–1600 W and a nebulizer gas flow of 0.8–1.0 L/min), these isotopes would not be available for quantitation because of the argon-based interferences. Under cool plasma conditions (typically, RF power of 600–800 W and a nebulizer gas flow of 1.2–1.6 L/min), the most sensitive isotopes can be used, offering low ppt detection in aqueous matrices. However, not all instruments offer the same level of cool plasma performance, so if these elements are important to you, it is critical to understand what kind of detection capability is achievable. A simple way to test cool plasma performance is to look at the BEC for iron at mass 56 with respect to cobalt at mass 59. This enables the background at mass 56 to be compared to a surrogate element such as Co, which has a similar ionization potential to Fe, without actually introducing Fe into the system and contributing to the ArO^+ background signal. When carrying out this test, it is important to use the cleanest deionized water to guarantee that there is no Fe in the blank. First, measure the background in counts/second at mass 56 aspirating deionized water. Then, record the analyte intensity of a 1 ppb Co solution at mass 59. The ArO^+ BEC can be calculated as follows:

$$BEC\ (ArO^+) = \frac{\text{Intensity of deionized water background at mass } 56 \times 1 \text{ ppb}}{\text{Intensity of 1 ppb of Co at mass } 59 - \text{background at mass } 56}$$

The ArO^+ BEC at mass 56 will be a good indication of the DL for $^{56}Fe^+$ under cool plasma conditions. The BEC value will typically be about an order of magnitude greater than the DL.

Although most instruments offer cool plasma capability, there are subtle differences in the way it is implemented. It is therefore very important to evaluate the ease of setup and how easy it is to switch from cool to normal plasma conditions and back in an automated multielement run. Also, remember that there will be an equilibrium time in switching from normal to cool plasma conditions. Make sure you know what this is, because an equivalent read-delay will have to be built into the method, which could be an issue if speed of analysis is important to you. If in doubt, set up a test to determine the equilibrium time by carrying out a short stability run while switching back and forth between normal and cool plasma conditions.

It is also critical to be aware that the electrical characteristics of a cool plasma are different from normal plasma. This means that unless there is a good grounding mechanism between the plasma and the RF coil, a secondary discharge can easily occur between the plasma and sampler cone. The result is an increased spread in kinetic energy of the ions entering the mass spectrometer, making them more difficult to control and steer through the ion optics into the mass analyzer. So, understand how this grounding mechanism is implemented and whether any hardware changes need to be made when going from cool to normal plasma conditions and vice versa (testing for a secondary discharge will be discussed later).

It should be noted that one of the disadvantages of the cool plasma approach is that cool plasma contains much less energy than a normal, high-temperature plasma. As a result, elemental sensitivity for the majority of elements is severely affected by the matrix, which basically precludes its use for the analysis of samples with a real matrix, unless the necessary steps are taken. This is shown in Figure 21.7, which shows cool plasma sensitivity for a selected group of elements in varying concentrations of nitric acid, and Figure 21.8, which shows the same group of elements under normal plasma conditions. It can be seen clearly that analyte sensitivity is dramatically reduced in a cool plasma as the acid concentration is increased, whereas,

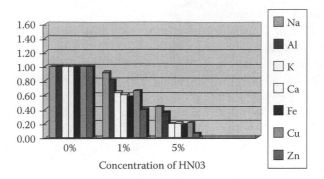

FIGURE 21.7 Sensitivity for a selected group of elements in varying concentrations of nitric acid, using cool plasma conditions (RF power–800 W, nebulizer gas–1.5 L/min). (From J. M. Collard, K. Kawabata, Y. Kishi, R. Thomas, *Micro*, January 2002.)

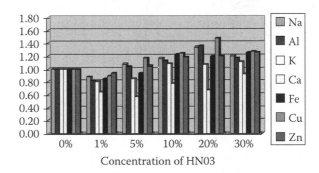

FIGURE 21.8 Sensitivity for a selected group of elements in varying concentrations of nitric acid, using normal plasma conditions (RF power–1600 W, nebulizer gas–1.0 L/min). (From J. M. Collard, K. Kawabata, Y. Kishi, R. Thomas, *Micro*, January 2002.)

under normal plasma conditions, the sensitivity for most of the elements varies only slightly with increasing acid concentration.[13]

In addition, because a cool plasma contains much less energy than a normal plasma, chemical matrices and acids with a high boiling point are often difficult to decompose in the plasma, which has the potential to cause corrosion problems on the interface of the mass spectrometer. This is the inherent weakness of the cool plasma approach—instrument performance is highly dependent on the sample being analyzed. As a result, unless simple aqueous-type samples are being analyzed, cool plasma operation often requires the use of standard additions or matrix matching to achieve satisfactory results. Additionally, to obtain the best performance for a full suite of elements, a multielement analysis often necessitates the use of two sets of operating conditions—one run for the cool plasma elements and another for normal plasma elements—which can be both time and sample consuming.

In fact, these application limitations have led some vendors to reject the cool plasma approach in favor of collision/reaction cell technology. So, it could be that the cool plasma capability of an instrument may not be that important if the equivalent elements are superior using the collision/reaction cell option. However, you should proceed with caution in this area, because on the current evidence not all collision/reaction cell instruments offer the same kind of performance. For some instruments, cool plasma DLs are superior to the same group of elements determined in the collision cell mode. For that reason, an assessment of the suitability of using cool plasma conditions or collision/reaction cell technology for a particular application problem has to be made based on the vendor's recommendations.

Using Collision/Reaction Cell and Interface Technology

Collision and reaction cells and interfaces are standard on most quadrupole-based instruments today and are used to reduce the formation of harmful polyatomic spectral interferences, such as $^{38}ArH^+$, $^{40}Ar^+$, $^{40}Ar^{12}C^+$, $^{40}Ar^{16}O^+$, and $^{40}Ar_2^+$, to improve detection capability for elements such as K, Ca, Cr, Fe, and Se. However, when comparing systems, it is important to understand how the interference reduction is carried out, what types of collision/reaction gases are used, and how the collision/reaction cell or interface deals with the many complex side reactions that take place—reactions that can potentially generate brand new interfering species and cause significant problems at other mass regions. The difference between collision cells and reaction cells and interfaces has been described in detail in Chapter 10. Two different approaches are used to reject these undesirable species. It can be done either by KED or by mass discrimination, depending on the type of multipole and the reaction gas used in the cell.

Unfortunately, the higher-order multipoles such as hexapoles or octapoles have less defined mass stability boundaries than lower-order multipoles, making them less than ideal to intercept these side reactions by mass discrimination. This means that some other mechanism has to be used to reject these unwanted species. The approach that has been traditionally used is to discriminate between them by kinetic energy. This is a well-accepted technique that is typically achieved by setting the collision cell potential slightly more negative than the mass filter potential. This means that the collision-product ions generated in the cell, which have a lower kinetic energy as a result of the collision process, are rejected; whereas the analyte ions, which have a higher kinetic energy, are transmitted. This method works very well but restricts their use to inert gases such as helium and less reactive gases such as hydrogen because of the limitations of higher-order multipoles in efficiently controlling the multitude of side reactions.

However, the use of highly reactive gases such as ammonia and methane can lead to more side reactions and potentially more interferences unless the by-products from these side reactions are rejected. The way around this problem is to utilize a lower-order multipole, such as a quadrupole, inside the reaction/collision cell and use it as a mass discrimination device. The advantages of using a quadrupole are that the stability boundaries are much better defined than a hexapole or an octapole, so it is relatively straightforward to operate the quadrupole inside the reaction cell as a mass or bandpass filter. Therefore, by careful optimization of the quadrupole electrical fields, unwanted reactions between the gas and the sample matrix or solvent, which could potentially lead to new interferences, are prevented. This means that every time an analyte and interfering ions enter the reaction cell, the bandpass of the quadrupole can be optimized for that specific problem and then changed on the fly for the next one.[14]

When assessing the capabilities of collision and reaction cells, it is important to understand the level of interference rejection that is achievable, which will be reflected in the instrument's DL and BEC values for the particular analytes being determined. This has been described in greater detail in Chapter 10, but depending on the nature of interference being reduced, there will be differences between the collision/reaction cell methods as well as with the collision/reaction interface approach. It is therefore critical to evaluate the capabilities of commercial instrumentation on the basis of your sample matrices and particular analytes of interest.

On the evidence published to date, it seems that the use of highly reactive gases and discrimination by mass appears to offer a more efficient way of reducing interfering ion background levels because the optimum reaction gas can be selected to create the most favorable ion–molecule reaction conditions for each analyte. In other words, the choice and flow of the reaction gas can be optimized for each and every application problem. However, they are not as straightforward when it comes to developing methods, especially when new samples are encountered. In addition, if more than one reaction gas needs to be used, they might not be ideally suited for a high-sample-throughput environment because of the lengthy analysis times involved.

On the other hand, the use of inert or low-reactivity gases and KED appears to offer a much simpler approach to reducing polyatomic spectral interferences. Normally, only one gas is used for a particular application problem, which is much better suited to routine analysis. It is possible to use other low gases such as hydrogen when helium does not work, but for the majority of elements one gas is sufficient. For some applications, the collision gas is kept flowing all the time, even for elements that do not need a collision cell. However, its major analytical disadvantage is that its interference reduction capabilities are generally not as good as a system that uses highly reactive gases. Because more collisions are required with an inert gas to suppress the interfering ions, the analyte will also undergo more collisions, and as a result, fewer of the analyte ions will make it through the kinetic energy barrier at the exit of the cell. For that reason, DLs for the majority of the elements that benefit from a collision/reaction cell are generally poorer using inert gases and KED than with a system that uses highly reactive gases and selective bandpass (mass) tuning.

However, it should be emphasized that when you are comparing systems, it should be done with your particular analytical problem in mind. In other words, evaluate the interference suppression capabilities of the different collision and reaction cell interface approaches by measuring BEC and DL performance for the suite of elements and sample matrices you are interested in. In other words, make sure it works for your application problem. This is even more important with the collision/reaction interface because it works on a slightly different principle, and as it does not use a traditional cell, it is unclear how it discriminates between all the generated by-product interfering ions from the analyte ions.

Every laboratory's analytical scenario is different, so it is almost impossible to determine which approach is better for a particular application problem. If you are not pushing DLs but are looking for a simplified approach to running samples on a routine basis, then maybe a collision cell using KED best suits your needs. However, if your samples are spectrally more complex and you are looking for more performance and flexibility because your DL requirements are more challenging, then the dynamic reaction cell using bandpass tuning is probably the best way to go. Also, be aware that systems that use collisional mechanisms and KED will either have to use higher-purity gases or a gas purifier (getter) because of the potential for impurities in the gas generating unexpected reaction by-product ions, which could potentially interfere with other analyte ions (refer to Chapter 10).[15] This not only has the potential to affect the detection capability, but ultra-high-purity gases are typically 2–3 times more expensive than industrial- or laboratory-grade gases.

Reduction of Matrix-Induced Interferences

As discussed in Chapter 14, there are three major sources of matrix-induced problems in ICP-MS. The first and simplest to overcome is often called a *sample transport* or *viscosity effect* and is a physical suppression of the analyte signal brought about by the matrix components. It is caused by the sample matrices' impact on droplet formation in the nebulizer or droplet size selection in the spray chamber. In some samples, it can also be caused by the variation in sample flow through the peristaltic pump. The second type of signal suppression is caused by the impact of the sample matrix on the ionization temperature of the plasma discharge. This typically occurs when different levels of matrix components or acids are aspirated into a cool plasma. The ionization conditions in the low-temperature plasma are so fragile that higher concentrations of matrix components result in severe suppression of the analyte signal. The third major cause of matrix suppression is the result of poor transmission of ions through the ion optics owing to matrix-induced space charge effects.[16] This has the effect of defocusing the ion beam, which leads to poor sensitivity and DLs, especially when trace levels of low-mass elements are being determined in the presence of large concentrations of high-mass matrix elements. Unless an electrostatic compensation is made in the ion optic region, the high-mass element will dominate the ion beam, resulting in severe matrix suppression on the lighter ones. All these types of matrix interferences are compensated to varying degrees by the use of internal standardization, where the intensity of a spiked element that is not present in the sample is monitored in samples, standards, and blank.

The single biggest difference in commercial instrumentation to focus the analyte ions into the mass analyzer is in the design of the ion lens system. Although they all basically do the same job of transporting the maximum number of analyte ions through the system, there are many different ways of implementing this fundamental process, including the use of an extraction lens, multicomponent lens systems, dynamically-scanned single ion lens, right-angled reflectors, or multipole ion guide systems. First, it is important to know how many lens voltages have to be optimized. If a system has many lens components, it is probably going to be more complex to carry out optimization on a routine basis. In addition, the cleaning and maintenance of a multicomponent lens system might be a little more time consuming. All of these are possible concerns, especially in a routine environment where maybe the skill level of the operator is not so high.

However, the design of the ion-focusing system or the number of lens components used is not as important as its ability to handle real-world matrices.[17] Most lens systems can operate in a simple aqueous sample because there are relatively few matrix ions to suppress the analyte ions. The test of the ion optics comes when samples with a real matrix are encountered. When a large number of matrix ions are present in the system, they can physically "knock" the analyte ions out of the ion beam. This shows itself as a suppression of the analyte ions, which means that less analyte ions are transmitted to the detector in the presence of a matrix. For this reason, it is important to measure the degree of matrix suppression of the instrument being evaluated across the full mass range. The best way to do this is to choose three or four of your typical analyte elements spread across the mass range (e.g., $^7Li^+$, $^{63}Cu^+$, $^{103}Rh^+$, and $^{138}Ba^+$). Run a calibration of a 20 ppb multielement standard in 1% HNO_3. Then make up a synthetic sample of 20 ppb of the same elements in one of your typical matrices. Read off this sample against the original calibration.

The percentage matrix suppression at each mass can then be calculated as follows:

$$\frac{20 \text{ ppb} - \text{Apparent concentration of 20 ppb analytes in your matrix}}{20 \text{ ppb}} \times 100$$

There is a strong possibility that your own samples will not really test the matrix suppression performance of the instrument, particularly if they are simple aqueous-type samples. If this is the case and you really would like to understand the matrix capabilities of your instrument, then make up a synthetic sample of your analytes in 500 ppm of a high-mass element such as thallium, lead, or uranium. For this test to be meaningful, you should tell the manufacturers to set up the ion optic voltages that are best suited for multielement analysis across the full mass range. If the ion optics are designed correctly for minimum matrix interferences, it should not matter if it incorporates an extraction lens, uses a photon stop, has an off-axis mass analyzer, or even whether it utilizes a single, multicomponent, or right-angled ion lens system.

It is also important to understand that an additional role of the ion optic system is to stop particulates and neutral species from making it through to the detector, which would increase the noise of the background signal. This will certainly impact the instrument's detection capability in the presence of complex matrices. Therefore, it is definitely worth carrying out a DL test in a difficult matrix such as lead or uranium, which tests the ability of the ion optics to transport the maximum number of analyte ions while rejecting the maximum number of matrix ions, neutral species, and particulates.

Another aspect of an instrument's matrix capability is its ability to aspirate lots of different types of samples, using both conventional nebulization and sampling accessories that generate a dry aerosol, such as laser ablation or ETV sampling. When changing sample types similar to this on a regular basis, parameters such as RF power, nebulizer gas flow, and sampling depth usually have to be changed. When this is done, there is an increased chance of altering the electrical characteristics of the plasma and producing a secondary discharge at the interface. All instruments should be able to handle this to some extent, but depending on how they compensate for the increase in plasma potential, parameters might need to be reoptimized because of the change in the spread of kinetic energy of the ions entering the mass spectrometer.[18] This may not be such a serious problem, but once again, it is important for you to be aware of this, especially if the instrument is running many different sample matrices on a routine basis.

Some of the repercussions of a secondary discharge, including increased doubly charged species, erosion of material from the skimmer cone, shorter lifetime of the sampler cone, a significantly different full-mass range response curve with laser ablation, and the occurrence of two signal maximums when optimizing nebulizer gas flow, have been well reported in the literature.[19–21] On the other hand, systems that do not show signs of this phenomena have reported an absence of these deleterious effects.[22]

A simple way of testing for the possibility of a secondary discharge is to aspirate one of your typical matrices containing approximately 1 ppb of a small group of elements across the mass range (such as $^7Li^+$, $^{115}In^+$, and $^{208}Pb^+$) and continuously monitor the signals while changing the nebulizer gas flow. In the absence of a secondary discharge, all three elements, which have widely different masses and ion energies, should track each other and have the same optimum nebulizer gas flow. This can be seen in Figure 21.9, which shows the signals for $^7Li^+$, $^{115}In^+$, and $^{208}Pb^+$ changing as the nebulizer gas flow is first increased and then decreased.

If the signals do not track each other or there is an erratic behavior in the signals, it could indicate that the normal kinetic energy of the ions has been altered by the change in the nebulizer gas flow. There are many reasons for this kind of behavior, but it could point to a possible secondary discharge at the interface or that the RF coil-grounding mechanism is not working correctly.[23] It should be emphasized that Figure 21.9 is just a graphical representation of what the relative signals might look like and might not exactly be the same for all instruments.

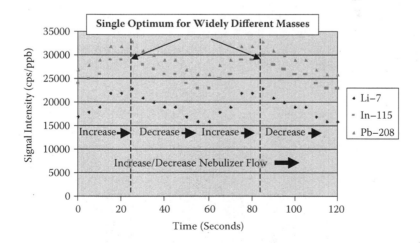

FIGURE 21.9 If the interface is grounded correctly, signals for 1 ppb $^7Li^+$, $^{115}In^+$, and $^{208}Pb^+$ should all track each other and have similar optimum values as the nebulizer gas flow is changed.

Sample Throughput

In laboratories where high sample throughput is a requirement, the overall cost of analysis is a significant driving force determining what type of instrument is purchased. However, in a high-workload laboratory there sometimes has to be a compromise between the number of samples analyzed and the DL performance required. For example, if the laboratory wants to analyze as many samples as possible, relatively short integration times have to be used for the suite of elements being determined. On the other hand, if DL performance is the driving force, longer integration times need to be used, which will significantly impact the total number of samples that can be analyzed in a given time. This was described in detail in Chapter 12, but it is worth revisiting to understand the full implications of achieving high sample throughput.

It is generally accepted that for a fixed integration time, peak hopping will always give the best DLs. As discussed earlier, measurement time is a combination of time spent on the peak taking measurements (dwell time) and the time taken to settle (settling time) before the measurement is taken. The ratio of the dwell time to the overall measurement time is often called the *measurement efficiency*. The settling time, as we now know, does not contribute to the analytical signal but definitely contributes to the analysis time. This means that every time the quadrupole sweeps to a mass and sits on the mass for the selected dwell time, there is an associated settling time. The greater the number of points that have been selected to quantitate the mass, the longer the total settling time and the worse the overall measurement efficiency.

For example, let us take a scenario where 20 elements need to be determined in duplicate. For argument's sake, let us use an integration time of 1 s per mass, comprising 20 sweeps at 50 ms per sweep. The total integration time that contributes to the analytical signal and the DL is therefore 20 s per replicate. However, every time the analyzer is swept to a mass, the associated scanning and settling time must be added to the dwell time. The greater the number of points that are taken to quantify the peak, the greater the magnitude of the settling time that must be added. For this scenario, let us assume that three points/peak are being used to quantify the peaks. Let us also assume for this case that the quadrupole and detector has a settling time of 5 ms. This means that a 15 ms settling time will be associated with every sweep of each individual mass. So, for 20 sweeps of 20 masses, this is equivalent to 6 s of nonanalytical time every replicate, which translates into 12 s (plus 40 s of actual measurement time) for every duplicate analysis. This is equivalent to a $40/(12 + 40) \times 100\%$ or a 77% measurement efficiency cycle. It does not take long to realize that the fewer the number of points taken per peak and the shorter the settling time, the better the measurement cycle. Just by reducing the number of points to one per peak and cutting the detector settling time by half, the nonanalytical time is reduced to 4 s, which is a $40/(2 + 40) \times 100\%$ or a 95% measurement efficiency per duplicate analysis. It is therefore very clear that the measurement protocol has a big impact on the speed of analysis and the number of the samples that can be analyzed in a given time. So, if sample throughput is important, you should understand how peak quantitation is carried out on each instrument.

The other aspect of sample throughput is the time taken for the sample to be aspirated through the sample introduction system into the mass spectrometer, reach a steady-state signal, and then be washed out when the analysis is complete. The wash-in and wash-out characteristics of the instrument will most definitely impact its sample throughput capabilities. Therefore, it is important for you to know what these times are for the system you are evaluating. You should also be aware that if the instrument uses a computer-controlled peristaltic pump to deliver the sample to the nebulizer and spray chamber, it can be speeded up to reduce the wash-in and wash-out times. So, this should also be taken into account when evaluating the memory characteristics of the sample introduction system.

Therefore, if speed of analysis is important to your evaluation criteria, it is worth carrying out a sample throughput test. Choose a suite of elements that represents your analytical challenge. Assuming you are also interested in achieving good detection capability, let the manufacturer set the measurement protocol (integration time, dwell time, settling time, number of sweeps, points/peak, sample introduction wash-in/wash-out times, etc.) to get their best DLs. If you are interested in measuring high and low concentrations, also make sure that the extended dynamic range feature is implemented. Then time how long it takes to achieve DL levels in duplicate from the time the sample probe goes into the sample to the time a result comes out on the screen or printer. If you have time, it might also be worth carrying out this test in an autosampler with a small number of your typical samples. It is important that the DL measurement protocol be used because factors such as integration times and wash-out times can be compromised to reduce the analysis time. All the measurement time issues discussed in this section and the memory characteristics of the sample introduction system will be fully evaluated with this kind of test. (Note: If sample throughput is an issue, there are automated sample delivery systems on the market that reduce the pre- and postmeasurement times to such a degree that they are realizing a 30–40% improvement in multielement analysis times.)

Transient Signal Capability

The demands on an instrument to handle transient signals generated by sampling accessories, such as laser ablation, chromatography separation devices, flow injection, or electrothermal vaporization systems, are very different from conventional multielement analysis using solution nebulization. Because the duration of a sampling accessory signal is short compared to a continuous signal generated by a pneumatic nebulizer, it is critical to optimize the measurement time to achieve the best multielement signal-to-noise ratio in the sampling time available. This was addressed in greater detail in Chapter 17, but basically the optimum design to capture the maximum amount of multielement data in a transient peak is to carry out the measurement in a simultaneous manner with a multicollector magnetic sector instrument or a TOF mass spectrometer.

However, a scanning system such as a quadrupole instrument can achieve good performance on a transient peak if the measurement time is maximized to get the best multielement signal-to-noise ratio. Therefore, instruments that utilize short settling times are more advantageous, because they achieve a higher measurement efficiency cycle. In addition, if the extended dynamic range is used to determine higher concentrations, the scanning and settling time of the detector will also have an impact on the quality of the signal. So, detectors that require two scans to characterize an unknown sample will use up valuable time in the quantitation process. For example, if a transient peak generated by a laser ablation device only lasts 10 s, a survey or prescan of 2 s will use up 20% of the available measurement time. This, of course, is a disadvantage when doing multielement analysis on a transient signal, especially if you have limited knowledge of the analyte concentration levels in your samples.

USABILITY ASPECTS

In most applications, analytical performance is a very important consideration when deciding what instrument to purchase. However, the vast majority of instruments being used today are being operated by technician-level chemists. They usually have had some experience in the use of trace element techniques such as AA or ICP-OES, but in no way could be considered experts in ICP-MS. Therefore, the usability aspects might be competing with analytical performance as the most important selection criteria, particularly if the application does not demand the ultimate in detection capability. Even though usability is in the eye of the user, there are some general issues that need to be addressed. They include, but are not limited to, the following:

- Ease of use
- Routine maintenance
- Compatibility with sampling accessories
- Installation requirements
- Technical support
- Training

Ease of Use

First of all, you need to determine the skill level of the operator who is going to run the instrument. If the operator is a PhD-type chemist, then maybe it is not critical that the instrument be easy to use. However, if the instrument is going to be used in a high-workload environment and possibly operated around the clock, there is a strong possibility that the operators will not be highly skilled. Therefore, you should be looking at how easy the software is to use and how similar it is to other trace element techniques that are used in your laboratory. This will definitely have an impact on the time it takes to get a person fully trained on the instrument. Another issue to consider is whether the person who runs the instrument on a routine basis is the same person who will be developing the methods. Correct method development is critical because it impacts the quality of your data, and therefore usually requires more expertise than just running routine methods. This is most definitely the case with collision/reaction cells and interfaces, especially if the method has never been done before. It can take a great deal of time and effort to select the best gases, gas flows, and optimization of the cell parameters to maximize the reduction of interferences for certain analytes in a new sample matrix.

I am not going to get into different software features or operating systems, because it is a complicated criterion to evaluate and decisions tend to be made more on a personal preference or comfort level than on the actual functionality of ICP-MS software features. It is also a moving target, as instrument software is continually being modified and updated. However, there are differences in the way software feels. For example, if you have come from an mass spectrometry background, you are probably comfortable with fairly complex research-type software. Alternatively, if you have come from a trace element background and have used AA or ICP-OES, you are probably used to more routine software that is relatively easy to use. You will find that different vendors have come to ICP-MS from a variety of different analytical chemistry backgrounds, which is often reflected in the way they design their software. Depending on the way the instrument will be used, an appropriate amount of time should be spent looking at software features that are specific to your application needs. For example, if you are working in a high-throughput environmental laboratory, you might be interested in turnkey methods that are used to run a particular EPA methodology, such as Method 200.8 for the determination of elements in water and wastes. In addition, maybe you should also be looking very closely at all the features of the automated "Quality Control" (QC) software, or if you do not have the time to export your data to an external spreadsheet to create reports, you might be more interested in software with comprehensive reporting capabilities. Alternatively, if your laboratory needs to characterize lots of unknown samples, you should carefully examine the "Semiquant" software and fully understand the kind of accuracy you can expect to achieve.

Routine Maintenance

ICP mass spectrometers are complex pieces of equipment that, if not maintained correctly, have the potential to fail when you least expect them to. For that reason, a major aspect of instrument usability is how often routine maintenance has to carried out, especially if complex sample matrices are being analyzed. You must not lose sight of the fact that your samples are being aspirated into the sample introduction system and the resulting ions generated in the plasma are steered into the mass analyzer via the interface and ion optics. In other words, the sample, in one form or another, is in contact with many components inside the instrument. So, it is essential to find out what components need to be changed and at what frequency to keep the instrument in good working order. Routine maintenance has been covered in great depth in Chapter 16, but you should be asking the vendor what needs to be changed or inspected on a regular basis and what type of maintenance should be done on daily, weekly, monthly, or yearly intervals. Some typical questions might include the following:

- If a peristaltic pump is being used to deliver the sample, how often should the tubing be changed?
- How often should the spray chamber drain system be checked?
- Can components be changed if a nebulizer gets damaged or blocked?
- How long does the plasma torch last?
- Can the torch sample injector be changed without discarding the torch?
- How is a neutral plasma maintained, and if an external shield or sleeves are used for grounding purposes, how often do they last?

- Is the RF generator solid state or does it use a power amplifier (PA) tube? This is important because PA tubes are expensive consumable items that typically need replacing every 1–2 years.
- How often do you need to clean the interface cones, and what is involved in cleaning them and keeping the cone orifices free of deposits?
- How long do the cones last?
- Do you have a platinum cone trade-in service, and what is their trade-in value?
- Which type of pump is used on the interface, and if it is a rotary-type pump, how often should the oil be changed?
- What mechanism is used to keep the ion optics free of sample particulates or deposits?
- How often should the ion optics be cleaned?
- What is the cleaning procedure for the ion optics?
- Do the turbomolecular pumps require any maintenance?
- How long do the turbomolecular pumps last?
- Does the mass analyzer require any cleaning or maintenance?
- How long does the detector last and how easy is it to change?
- What spare parts do you recommend to keep on hand? This can often indicate the components that are likely to fail most frequently.
- What maintenance needs to be carried out by a qualified service engineer, and how long does it take?

This is not an exhaustive list, but it should give you a good idea about what is involved in keeping an instrument in good working order. I also encourage you to talk to real-world users of the equipment to make sure you get their perspective of these maintenance issues.

Compatibility with Alternative Sampling Accessories

Alternative sample introduction techniques are becoming more necessary as ICP-MS is being utilized to analyze more complex sample types. Therefore, it is important to know if the sampling accessory is made by the ICP-MS instrument company or by a third-party vendor. Obviously, if it has been made by the same company, compatibility should not be an issue. However, if it is made by a third party, you will find that some sampling accessories work much better with some instruments than with others. It might be that the physical connection of coupling the accessory to the ICP-MS torch has been better thought out, or that the software "talks" to one system better than another. You should refer to Chapter 17 for more details on their suitability for your application, but if they are required, software/hardware compatibility should be one of your evaluation objectives. You should also read Chapter 18 if you are thinking of interfacing an LC system to your ICP-MS for trace element speciation studies.

Installation of Instrument

Installation of the instrument and where it is going to be located does not seem to be an obvious evaluation objective at first, but it could be important, particularly if space is limited. For example, is the instrument free-standing or bench-mounted? Maybe you have a bench available but no floor space, or vice versa? It could be that the instrument requires a temperature-controlled room to ensure good stability and mass calibration. If this is the case, have you budgeted for this kind of expense? If the instrument is being used for ultratrace detection levels, does it need to go into a class 1, 10, or 100 clean room? If it does, what is the size of the room and do the roughing pumps need to be placed in another room? In other words, it is important to fully understand the installation requirements for each instrument being evaluated and where it will be located. Refer to Chapter 16 for more information on instrument installation.

Technical Support

Technical and application support is a very important consideration, especially if you have had no previous experience with ICP-MS. You want to know that you are not going to be left on your own after you have made the purchase. Therefore, it is important to know not only the level of expertise of the specialist who is supporting you, but also whether they are local to you or located in the manufacturer's corporate headquarters. In other words, can the vendor give technical help whenever you need it? Another important aspect related to application support is the availability of application literature. Is there a wide selection of material available for you to read, either in the form of Web-based application reports or references in the open literature, to help you develop your methods? Also, find out if there are active user or Internet-based discussion groups, because they will be an invaluable source of technical and application help. One such source of help in this area can be found on the PlasmaChem Listserver, a plasma spectrochemistry discussion group out of Syracuse University.[24]

Training

Find out what kind of training course comes with the purchase of the instrument and how often it is run. Most instruments come with 2–3 day training course for one person, but most vendors should be flexible regarding the number of people who can attend. Some manufacturers also offer application training, where they teach you how to optimize methods for major application areas such as environmental, clinical, and semiconductor analysis. Talk to other users about the quality of the training they received when they purchased their instruments and also ask them what they thought of the operator's manuals. You will often find that this is a good indication of how important a manufacturer thinks customer training is.

RELIABILITY ISSUES

To a certain degree, instrument reliability is impacted by routine maintenance issues and the types of samples being analyzed, but it is generally considered more of a reflection of the design of an instrument. Most manufacturers will guarantee a minimum percentage uptime for their instrument, but this number (which is typically ~95%) is almost meaningless unless you really understand how it is calculated. Even when you know how it is calculated, it is still difficult to make the comparison, but at least you should understand the implications if the vendor fails to deliver. Good instrument reliability is taken for granted nowadays, but it has not always been the case. When ICP-MS was first commercialized, the early instruments were a little unpredictable, to say the least, and quite prone to frequent breakdowns. However, as the technique became more mature, the quality of instrument components improved, and therefore, the reliability improved. However, you should be aware that some components of the instrument are more problematic than others. This is particularly true when the design of an instrument is new or a model has had a major redesign. You will therefore find that in the life cycle of a newly designed instrument, the early years will be more susceptible to reliability problems than when the instrument is of an older design.

When we talk about instrument reliability, it is important to understand whether it is related to the samples being analyzed, the lack of expertise of the person operating the instrument, an unreliable component, or an inherent weakness in the design of the instrument. For example, how does the instrument handle highly corrosive chemicals such as concentrated mineral acids? Some sample introduction systems and interfaces will be more rugged than others and require less maintenance in this area. On the other hand, if the operator is not aware of the dissolved solids limitation of the instrument, they might attempt to aspirate a sample, which will slowly block the interface cones, causing signal drift and, in the long term, possible instrument failure. Or, it could be something as unfortunate as a major component, such as the RF generator power amplifier tube, discrete dynode detector, or turbomolecular pump (which all have a finite lifetime) failing in the first year of use.

Service Support

Instrument reliability is very difficult to assess at the evaluation stage, so you have to look very carefully at the kind of service support offered by the manufacturer. For example, how close is a qualified support engineer to you or what is the maximum amount of time you will have to wait to get a support engineer at your laboratory or at least to call you back to discuss the problem. Ask the vendor if they have the capability for remote diagnostics, where a service engineer can remotely run the instrument or check the status of a component by "talking to" your system computer via a modem. Even if this approach does not fix the problem, at least the service engineer can come to your laboratory with a very good indication of what the problem could be.

You should know up front what a service visit is going to cost you, irrespective of what component has failed. Also find out what routine maintenance jobs you can do and what requires an experienced service engineer. If it does require a service engineer, how long will it take them, because their time is not inexpensive. Most companies charge an hourly rate for a service engineer (which typically includes travel time as well), but if an overnight stay is required, fully understand what you are paying for (accommodations, meals, gas, etc.). Some companies might even charge for mileage between the service engineer's base and your laboratory. If you work in a commercial laboratory and cannot afford the instrument to be down for any length of time, find out what it is going to cost for 24/7 service coverage.

You can take a chance and just pay for each service visit, or you might want to budget for an annual preventative maintenance contract, where the service engineer checks out all the important instrumental components and systems frequently to make sure they are all working correctly. This might not be as critical if you work in an academic environment, where the instrument might be down for extended periods, but in my opinion, it is absolutely critical if you work in a commercial laboratory, which is using the instrument to generate revenue. Also find out what is included in the contract, because some also cover the cost of consumables or replacement parts, whereas others just cover the service visits. These annual preventative maintenance contracts typically make up about 10–15% of the cost of the instrument, but are well worth it if you do not have the expertise in-house, or if you just feel more comfortable having an insurance policy to cover instrument breakdowns.

Once again, talking to existing users will give you a very good perspective of the quality of the instrument and the service support offered by the manufacturer. There is no absolute guarantee that the instrument of choice is going to perform to your satisfaction 100% of the time, but if you work in a high-throughput, routine laboratory, make sure it will be down for the minimum amount of time. In other words, fully understand what it is going to cost you to maximize the uptime of all the instruments being evaluated.

FINANCIAL CONSIDERATIONS

The financial side of choosing an ICP mass spectrometer can often dominate the selection process. You may or may not have budgeted quite enough money to buy a top-of-the-line instrument or perhaps you had originally planned to buy another lower-cost trace element technique, or you could be using funds left over at the end of your financial year. All these scenarios dictate how much money you have available and what kind of instrument you can purchase. In my experience, you should proceed with caution in this kind of situation, because if only one manufacturer is willing to do a deal with you, the evaluation process will be a waste of time. Therefore, you should budget at least 12 months before you are going to make a purchase and add another 10–15% for inflation and any unforeseen price increases. In other words, if you want to get the right instrument for your application, never let price be the overriding factor in your decision. Always be wary of the vendor who will undercut everyone else to get your business. There could be a very good reason why they are doing this; for example, the instrument is being discontinued for a new model or it could be having some reliability problems that are affecting its sales.

This is not to say that price is unimportant, but what might appear to be the most expensive instrument to purchase, might be the least expensive to run. Therefore, you must never forget the cost of ownership in the overall financial analysis of your purchase. So, by all means compare the price of the instrument, computer, and any accessories you buy, but also factor in the cost of consumables, gases, and electricity based on your usage. Maybe instrument consumables from one vendor are much less expensive than from another vendor. This is particularly the case with interface cones and plasma torches. Or maybe the purity of collision/reaction gases is more critical with one cell-based instrument than another. For example, there is a factor-of 4 difference between the cost of high-purity (99.999%) hydrogen gas and ultra-high-purity (99.9999%) grade. So, be diligent when you compare prices. Look at the overall picture, and not just the cost of the instrument. Also, be aware of differences in sample throughput. Perhaps you can analyze more samples with one instrument because its measurement protocol is faster or it does not need recalibrating as often (less drift). It therefore follows that if you can get through your daily allocation of samples much faster with one instrument than another, then your argon consumption will be reduced.

Another aspect that should be taken into consideration is the salary of the operator. Even though you might think that this is a constant, irrespective of the instrument, you must assess the expertise required to run it. For example, if you are thinking of purchasing more complex technology such as a magnetic sector instrument for a research-type application, the operator needs to be of a much higher skill level than, say, someone who is being asked to run a routine application with a quadrupole-based instrument. As a result, the salary of that person will probably be higher.

Finally, if one instrument has to be installed in a temperature-controlled, air-conditioned, environment for stability purposes, the cost of preparing or building this kind of specialized room must be taken into consideration when doing your financial analysis. In other words, when comparing systems, never automatically reject the most expensive instrument. You will find that over the 10 years that you own the instrument, the cost of doing analysis and the overall cost of ownership are more important evaluation criteria.

THE EVALUATION PROCESS: A SUMMARY

As mentioned earlier in this chapter, it is not my intention to compare instrument designs and features, but to give you some general guidelines as to what are the most important evaluation criteria. Besides being a framework for your evaluation process, these guidelines should also be used in conjunction with the other chapters in this book and the cited referenced information.

However, if you want to find the best instrument for your application needs, be prepared to spend a few months evaluating the marketplace. Do not forget to prioritize your objectives and give each of them a weighting factor based on their degree of importance for the types of samples you analyze. Be careful to take the evaluation in the direction you want to go and not where the vendor wants to take it. In other words, it is important to compare apples with apples and not to be talked into comparing an apple with an orange that looks like an apple ... if you know what I mean. However, be prepared that there might not be a clear-cut winner at the end of the evaluation. If this is the case, then decide what aspects of the evaluation are most important and ask the manufacturer to put them in writing. Some vendors might be hesitant to do this, especially if it is guaranteeing instrument performance with your samples.

Talk to as many users in your field as you possible can—not only ones given to you as references by the vendor, but ones chosen by yourself also. This will give you a very good indication of the real-world capabilities of the instrument, which can often be overlooked at a demonstration. You might find from talking to "typical" users that it becomes obvious which instrument to purchase. If that is the case and your organization allows it, ask the vendor what your options are if you do not have samples to run and you do not want a demonstration. I guarantee you will be in a much better position to negotiate a lower price.

Never forget that it is a very competitive marketplace, and your business is extremely important to each of the ICP-MS manufacturers. Hopefully, this book has not only helped you understand the fundamentals of the technique a little better but has also given you some thoughts and ideas on how to find the best instrument for your needs. Refer to Chapter 23 for details on how to contact all the instrument vendors and consumables and accessories companies. Good luck with your evaluation.

REFERENCES

1. K. Nottingham, ICP-MS: It's Elemental, *Analytical Chemistry,* 35-38A, 2004, http://pubs.acs.org/subscribe/journals/ancham-a/76/i01/toc/toc_i01.html.
2. Royal Society of Chemistry, Report by the Analytical Methods Committee: Evaluation of Analytical Instrumentation—Part X Inductively Coupled Plasma Mass Spectrometers, *The Analyst,* **122**, 393–408, 1997.
3. Analytical figures of merit for ICP-MS, *Inductively Coupled Plasma Mass Spectrometry: An Introduction to ICP Spectrometries for Elemental Analysis*, A. Montasser, Ed., Wiley-VCH, 1998, pp. 16–28, chap. 1.4.
4. E. R. Denoyer, *Atomic Spectroscopy,* **13**(3), 93–98, 1992.
5. M. A. Thomsen, *Atomic Spectroscopy,* **13**(3), 93–98, 2000.
6. L. Halicz, Y. Erel, and A. Veron, *Atomic Spectroscopy,* **17**(5), 186–189, 1996.
7. R. Thomas, *Spectroscopy,* **17**(7), 44–48, 2002.
8. E. R. Denoyer, Q. H. Lu, *Atomic Spectroscopy,* **14**(6), 162–169, 1993.
9. R. Hutton, A. Walsh, D. Milton, and J. Cantle, *ChemSA,* **17**, 213–215, 1991.
10. *Quadrupole Mass Spectrometry and Its Applications*: P. H. Dawson, Ed., Elsevier, Amsterdam, 1976; reissued by AIP Press, Woodbury, NY, 1995.
11. S. J. Jiang, R. S. Houk, and M. A. Stevens, *Analytical Chemistry,* **60**, 217, 1988.
12. K. Sakata and K. Kawabata, *Spectrochimica Acta,* **49B**, 1027, 1994.
13. J. M. Collard, K. Kawabata, Y. Kishi, R. Thomas, *Micro,* January 2002.
14. S. D. Tanner and V. I. Baranov, *Atomic Spectroscopy,* **20**(2), 45–52, 1999.
15. B. Hattendorf and D. Günther, *Journal of Analytical Atomic Spectrometry,* **19**, 600–606, 2004.
16. S. D. Tanner, D. J. Douglas, and J. B. French, *Applied Spectroscopy,* **48**, 1373, 1994.
17. E. R. Denoyer, D. Jacques, E. Debrah, and S. D. Tanner, *Atomic Spectroscopy,* **16**(1), 1, 1995.
18. R. C. Hutton and A. N. Eaton, *Journal of Analytical Atomic Spectrometry,* **5**, 595, 1987.
19. A. L. Gray and A. Date, *Analyst,* **106**, 1255, 1981.
20. E. J. Wyse, D. W. Koppenal, M. R. Smith, and D. R. Fisher, 18th FACSS Meeting, Anaheim, CA, October 1991, Paper No. 409.
21. W. G. Diegor and H. P. Longerich, *Atomic Spectroscopy,* **21**(3), 111, 2000.
22. D. J. Douglas and J. B. French, *Spectrochimica Acta,* **41B**(3), 197, 1986.
23. E. R. Denoyer, *Atomic Spectroscopy,* **12**, 215–224, 1991.
24. PlasmaChem Listserver: A discussion group for plasma spectrochemists worldwide, Syracuse University, NY, http://www.lsoft.com/scripts/wl.exe?SL1=PLASMACHEM-L&H=LISTSERV.SYR.EDU.

22 Glossary of ICP-MS Terms

In all my years of working with inductively coupled plasma mass spectrometry (ICP-MS), I have never come across any written material that included a basic dictionary of terms, primarily aimed at someone new to the technique. When I first became involved in ICP-MS, most of the literature I read tended to give complicated descriptions of instrument components and explanations of fundamental principles that more often than not sailed over my head. It was not until I became more familiar with the technique that I began to get a better understanding of the complex ICP-MS jargon used in technical journals and presentations at scientific conferences. So, I knew that a glossary of ICP-MS terms was an absolute necessity in the next edition of my textbook. Even though the glossary is not exhaustive, it contains explanations and definitions of the most common ICP-MS words, expressions, and terms used in this book. It should mainly be used as a quick reference guide. If you want more information about the subject matter, you should use the index to find a more detailed explanation of the topic in the appropriate book chapter.

A

AA An abbreviation for **atomic absorption**.

abundance sensitivity A way of assessing the ability of a mass separation device, such as a quadrupole, to identify and measure a small analyte peak adjacent to a much larger interfering peak. An abundance sensitivity specification is a combination of two measurements. The first is expressed as the ratio of the intensity of the peak at 1 amu (atomic mass unit) below the analyte peak to the intensity of the analyte peak, and the second is the ratio of the peak intensity 1 amu above the analyte mass to the intensity of the analyte peak. Because of the motion of the ion through the mass filter, the abundance sensitivity specification of a mass-filtering device is always worse on the low-mass side compared to the high-mass side.

active film multipliers. Another name for **discrete dynode multipliers**, which are used to detect, measure, and convert ions into electrical pulses in ICP-MS. *Also refer to* **channel electron multiplier (CEM)** *and* **discrete dynode detector (DDD)**.

addition calibration A method of calibration in ICP-MS using standard additions. All samples are assumed to have a similar matrix, so spiking is only carried out on one representative sample and not the entire batch of samples, as per conventional standard additions used in graphite furnace AA analysis.

AE An abbreviation for **atomic emission**.

aerosol The result of breaking up a liquid sample into small droplets by the nebulization process in the sample introduction system. *Also refer to* **nebulizer** *and* **sample introduction system**.

alkylated metals A metal complex containing an alkyl group. Typically detected by coupling liquid chromatography with ICP-MS. *Also refer to* **speciation analysis**.

alpha-counting spectrometry A particle-counting technique that uses the measurement of the radioactive decay of alpha particles. *Also refer to* **particle-counting techniques**.

alternative sample introduction accessories Alternative ways of introducing samples into an ICP mass spectrometer other than conventional nebulization. Also known as **alternative sample introduction devices**. Often used to describe desolvation techniques or laser ablation.

analog counting A way of measuring high signals by changing the gain or voltage of the detector. *Also refer to* **pulse counting**.

argon The gas used to generate the plasma in an ICP.

argon-based interferences A polyatomic spectral interference generated by argon ions combining with ions from the matrix, solvent, or any elements present in the sample.

ashing A sample preparation technique that involves heating the sample (typically in a muffle furnace) until the volatile material is driven off and an ash-like substance is left.

atom A unit of matter. The smallest part of an element having all the characteristics of that element and consisting of a dense, central, positively charged nucleus surrounded by orbiting electrons. The entire structure has an approximate diameter of 10^{-8} cm and characteristically remains undivided in chemical reactions except for limited removal, transfer, or exchange of certain electrons.

atom-counting techniques A generic name given to techniques that use atom or ion counting to carry out elemental quantitation. Some common ones, besides ICP-MS, include secondary ionization mass spectrometry (SIMS), thermal ionization mass spectrometry (TIMS), accelerator mass spectrometry (AMS), and fission track analysis (FTA). *Also refer to* **ionizing radiation counting techniques**.

atomic absorption (AA) An analytical technique for the measurement of trace elements that uses the principle of generating free atoms (of the element of interest) in a flame or electrothermal atomizer (ETA) and measuring the amount of light absorbed from a wavelength-specific light source, such as a hollow cathode lamp (HCL) or electrode discharge lamp (EDL).

atomic emission (AE) A trace element analytical technique that uses the principle of exciting atoms in a high-temperature source such as a plasma discharge and measuring the amount of light the atoms emit when electrons fall back down to a ground (stable) state.

atomic mass or weight The average mass or weight of an atom of an element, usually expressed relative to the mass of carbon 12, which is assigned 12 atomic mass units.

atomic number The number of protons in an atomic nucleus.

atomic structure Describes the structural makeup of an atom. *Also refer to* **neutron,** **proton,** *and* **electron.**

attenuation (of the detector) Reduces the amplitude of the electrical signal generated by the detector, with little or no distortion. Usually carried out by applying a control voltage to extend the dynamic range of the detector. *Also refer to* **extended dynamic range.**

autodiluter A device to automatically dilute large numbers of samples with no manual intervention by the operator.

autosampler A device to automatically introduce large numbers of samples into the ICP-MS system with no manual intervention by the operator.

axial view An ICP-OES system in which the plasma torch is positioned horizontally (end-on) to the optical system as opposed to the conventional vertical (radial) configuration. It is generally accepted that viewing the end of the plasma, improves emission intensity by a factor of approximately 5- to 10-fold.

B

background equivalent concentration (BEC) Defined as the apparent concentration of the background signal based on the sensitivity of the element at a specified mass. The lower the BEC value, the more easily a signal generated by an element can be discerned from the background. Many analysts believe BEC is a more accurate indicator of the performance of an ICP-MS system than detection limit, especially when making comparisons of background reduction techniques, such as cool-plasma or collision/reaction cell and interface technology.

background noise The square root of the intensity of the blank in counts per second (cps) anywhere of analytical interest on the mass range. Detection limit (DL) is a ratio of the analyte signal to the background noise at the analyte mass. Background noise as an instrumental specification is usually measured at mass 220 amu (where there are no spectral features), while aspirating deionized water. *Also refer to* **background signal, instrument background noise,** *and* **detection limit.**

background signal The signal intensity of the blank in counts per second (cps) anywhere of analytical interest on the mass range. Detection Limit (DL) is a ratio of the analyte signal to the noise of the background at the analyte mass. Background as an instrumental specification, is usually measured at mass 220 amu (where there are no spectral features), while aspirating deionized water. *Also refer to* **background noise, instrument background signal,** *and* **detection limit.**

bandpass tuning/filtering A mechanism used in a dynamic reaction cell (DRC) to reject the by-products generated through secondary reactions utilizing the principle of mass discrimination. Achieved by optimizing the electrical fields of the reaction cell multipole (typically a quadrupole) to allow transmission of the analyte ion, while rejecting the polyatomic interfering ion.

BEC An abbreviation for **background equivalent concentration**.

by-product ions Ionic species formed as a result of secondary reactions that take place in a reaction/collision cell. *Also refer to* **secondary (side) reactions**.

C

calibration A plot, function, or equation generated using calibration standards and a blank, which describes the relationship between the concentration of an element and the signal intensity produced at the analyte mass of interest. Once determined, this relationship can be used to determine the analyte concentration in an unknown sample.

calibration standard A reference solution containing accurate and known concentrations of analytes for the purpose of generating a calibration curve or plot.

capacitive coupling An undesired electrostatic (or capacitive) coupling between the voltage on the load coil and the plasma discharge, which produces a potential difference of a few hundred volts. This creates an electrical discharge or arcing between the plasma and sampler cone of the interface, commonly known as a "secondary discharge" or "pinch effect."

capillary electrophoresis (CE) *Refer to* **capillary-zone electrophoresis (CZE)**.

capillary zone electrophoresis (CZE or CE) A chromatographic separation technique used to separate ionic species according to their charge and frictional forces. In traditional electrophoresis, electrically charged analytes move in a conductive liquid medium under the influence of an electric field. In capillary (zone) electrophoresis, species are separated based on their size-to-charge ratio inside a small capillary filled with an electrolyte. Its applicability to ICP-MS is mainly in the field of separation and detection of large biomolecules.

CE *Refer to* **capillary-zone electrophoresis**.

cell In ICP-MS terminology, a cell usually refers to a collision or reaction cell.

certified reference material (CRM) Well-established reference matrix that comes with certified values and associated statistical data that have been analyzed by other complementary techniques. Its purpose is to check the validity of an analytical method, including sample preparation, instrument methodology, and calibration routines to achieve sample results that are as accurate and precise as possible and can be defended when subjected to intense scrutiny.

channel electron multiplier (CEM) A detector used in ICP-MS to convert ions into electrical pulses using the principle of multiplication of electrons via a potential gradient inside a sealed tube.

Channeltron® Another name for a channel electron multiplier detector.

charge transfer reaction Sometimes referred to as "charge exchange." This is one of the ion–molecule reaction mechanisms that take place in a collision/reaction cell. Involves the transfer of a positive charge from the interfering ion to the reaction gas molecule, forming a neutral atom that is not seen by the mass analyzer. An example of this kind of reaction:

$$H_2 + {}^{40}Ar^+ = Ar + H_2^+.$$

chemical modification The process of chemically modifying the sample in electrothermal vaporization (ETV) ICP-MS work to separate the analyte from the matrix. *Also refer to* **chemical modifier** *and* **electrothermal vaporization**.

chemical modifier A chemical or substance that is added to the sample in an electrothermal vaporizer to change the volatility of the analyte or matrix. Typically added at the ashing stage of the heating program to separate the vaporization of the analyte away from the potential interferences of the matrix components. *Also refer to* **electrothermal vaporization**.

chromatographic separation device Any device that separates analyte species according to their retention times or mobility through a stationary phase. When coupled with an ICP-MS system, it is used for the separation, detection, and quantitation of speciated forms of trace elements. Examples include liquid, ion, gas, size exclusion, and capillary electrophoresis chromatography. *Also refer to* **speciation analysis**.

chromatography terminology (as applied to trace element speciation) The following are some of the most important terms used in the chapter on trace element speciation. For easy access, they are contained in one section and not distributed throughout the glossary.

> **buffer** A mobile-phase solution that is resistant to extreme pH changes, even with additions of small amounts of acids or bases.
>
> **chromatogram** The graphical output of the chromatographic separation. It is usually a plot of peak intensity of the separated species over time.
>
> **column** The main component of the chromatographic separation. It is typically a tube containing the stationary-phase material that separates the species and an eluent that elutes the species off the column.
>
> **counterions** The mobile phase contains a large number of ions that have a charge opposite to that of the surface-bound ions. These are known as counterions, which establish equilibrium with the stationary phase.
>
> **dead volume** Usually refers to the volume of the mobile phase between the point of injection and the detector that is accessible to the sample species, minus the volume of mobile phase that is contained in any union or connecting tubing.
>
> **gradient elution** Involves variation of the mobile-phase composition over time through a number of steps such as changing the organic content, altering the pH, changing the concentration of the buffer, or using a completely different buffer.
>
> **ion exchange** A technique in which separation is based on the exchange of ions (anions or cations) between the mobile phase and the ionic sites on a stationary phase bound to a support material in the column.
>
> **ion pairing** A type of separation that typically uses a reversed-phase column in conjunction with a special type of chemical in the mobile phase called an "ion-pairing reagent." *Also refer to* **reverse phase**.
>
> **isochratic elution** An elution of the analytes or species using the same solvent throughout the analysis.

mobile phase   A combination of the sample or species being separated or analyzed and the solvent that moves the sample through the column.

retention time   The time taken for a particular analyte or species to be separated and pass through the column to the detector.

reverse phase   A type of separation that is typically combined with ion pairing, and essentially means that the column's stationary phase is less polar and more organic than the mobile-phase solvents.

stationary phase   A solid material, such as silica or a polymer, that is set in place and packed into the column for the chromatographic separation to take place.

clean room   The general description given to a dedicated room for the sample preparation and analysis of ultrapure materials. Usually associated with a number that describes the number of particulates per cubic foot of air (e.g., a class 100 clean room will contain 100 particles/ft^3 of air). It is commonly accepted that the semiconductor industry has the most stringent demands, which necessitates the use of class 10 and sometimes class 1 clean rooms.

cluster ions   Ions that are formed by two or more molecular ions combining together in a collision/reaction cell to form molecular clusters.

cold plasma technology   Cool or cold plasma technology uses low-temperature plasma to minimize the formation of certain argon-based polyatomic species. Under normal plasma conditions (approximately 1000 W RF power and 1.0 L/min nebulizer gas flow), argon ions combine with matrix and solvent components to generate problematic spectral interferences, such as $^{38}ArH^+$, $^{40}Ar^+$, and $^{40}Ar^{16}O^+$, which impact the detection limits of a small number of elements including K, Ca, and Fe. By using cool plasma conditions (approximately 600 W RF power and 1.6 L/min nebulizer gas flow), the ionization conditions in the plasma are changed so that many of these interferences are dramatically reduced and detection limits are improved.

cold vapor atomic absorption (CVAA)   An analytical approach to determine low levels of mercury by generating mercuric vapor in a quartz cell and measuring the number of mercury atoms produced, using the principle of atomic absorption. *Also refer to* **hydride generation atomic absorption**.

collision cell   Specifically, a cell that predominantly uses the principle of collisional fragmentation to break apart polyatomic interfering ions generated in the plasma discharge. Collision cells typically utilize higher-order multipoles (such as hexapoles or octapoles) with inert or low-reactive gases (such as helium and hydrogen) to first stimulate ion–molecule collisions, and then kinetic energy discrimination to reject any undesirable by-product ionic species formed.

collision-induced dissociation (CID)   A basic principle, first used for the study of organic molecules using tandem mass spectrometry, that relies on using a nonreactive gas in a collision cell to stimulate ion–molecule collisions. The more collision-induced daughter species that are generated, the better the chance of identifying the structure of the parent molecule.

collision/reaction cell (CRC) technology   A generic term applied to collision and reaction cells that use the principle of ion–molecule collisions and reactions to cleanse the ion beam of problematic polyatomic spectral interferences before they enter the mass analyzer. Both collision and reaction cells are positioned in the mass spectrometer vacuum chamber after the ion optics but prior to the mass analyzer. *Also refer to* **collision cell** *and* **reaction cell**.

collision/reaction interface (CRI) technology A collision/reaction mechanism approach, which instead of using a pressurized cell, injects a gas directly into the interface between the sampler and skimmer cones. The injection of the collision/reaction gas into this region of the ion beam produces high collision frequency between the argon gas and the injected gas molecules. This has the effect of removing argon-based polyatomic interferences before they are extracted into the ion optics.

collisional damping A mechanism that describes the temporal broadening of ion packets in a quadrupole-based dynamic reaction cell to dampen out fluctuations in ion energy. By optimizing cell conditions such as gas pressure, RF stability boundary (q parameter), entrance/exit lens potentials, and cell rod offsets, it has been shown that fluctuation in ion energies can be dampened sufficiently to carry out isotope ratio precision measurements near their statistical limit.

collisional focusing The mechanism of focusing ions toward the center of the ion beam in a collision/reaction cell. By using a neutral collision gas of lower molecular weight than the analyte, the analyte ions will lose kinetic energy and migrate toward the axis as a result of the collisions with the gaseous molecules. Therefore, the number of ions exiting the cell and reaching the detector will increase. *Also refer to* **collision cell** *and* **reaction cell**.

collisional fragmentation The mechanism of breaking apart (fragmenting) a polyatomic interfering ion in a collision/reaction cell using collisions with a gaseous molecule. The predominant mechanism used in a collision cell, as opposed to a reaction cell. *Also refer to* **collision cell** *and* **reaction cell**.

collisional mechanisms The mechanisms by which the interfering ion is reduced or minimized to allow the determination of the analyte ion. The most common collisional mechanisms seen in collision/reaction cells include collisional focusing, dissociation, and fragmentation, whereas the major reaction mechanisms include exothermic/endothermic associations, charge transfer, molecular associations, and proton transfer.

collisional retardation A mechanism in a collision/reaction cell where the gas atoms/molecules undergo multiple collisions with the polyatomic interfering ion, retarding or lowering its kinetic energy. Because the interfering ion has a larger cross-sectional area than the analyte ion, it undergoes more collisions, and as a result, can be separated or discriminated from the analyte ion based on their kinetic energy differences.

concentric nebulizer A nebulizer that uses two narrow concentric capillary tubes (one inside the other) to aspirate a liquid into the ICP-MS spray chamber. Argon gas is usually passed through the outer tube, which creates a venturi effect, and as a result, the liquid is sucked up through the inner capillary tube.

cones *Refer to* **interface cones**.

cool plasma technology *Refer to* **cold plasma technology**.

cooled spray chamber   A spray chamber that is cooled in order to reduce the amount of solvent entering the plasma discharge. Used for a variety of reasons, including reducing oxide species, minimizing solvent-based spectral interferences, and allowing the trouble-free aspiration of organic solvents.

correction equation   A mathematical approach used to compensate for isobaric and polyatomic spectral overlaps. It works on the principle of measuring the intensity of the interfering species at another mass, which is ideally free of any interference. A correction is then applied by knowing the ratio of the intensity of the interfering species at the analyte mass to its intensity at the alternate mass.

counts per second (cps)   Units of signal intensity used in ICP-MS. Number of detector electronic pulses counted per second.

cps   An abbreviation for **counts per second**.

CRC   An abbreviation for **collision/reaction cell technology**.

CRI   An abbreviation for **collision/reaction interface technology**.

CRM   An abbreviation for certified reference materials.

cross-calibration   A calibration method that is used to correlate both pulse (low levels) and analog (high levels) signals in a dual-mode detector. This is possible because the analog and pulse outputs can be defined in identical terms (of incoming pulse counts per second) based on knowing the voltage at the first analog stage, the output current, and a conversion factor defined by the detection circuitry electronics. By carrying out a cross-calibration across the mass range, a dual-mode detector is capable of achieving approximately eight to nine orders of dynamic range in one simultaneous scan.

cross-flow nebulizer   A nebulizer that is designed for samples that contain a heavier matrix or small amounts of undissolved solids. In this design, the argon gas flow is directed at right angles to the tip of a capillary tube through which the sample is drawn up with a peristaltic pump.

CVAA   An abbreviation for **cold vapor atomic absorption**.

cyclonic spray chamber   A spray chamber that operates using the principle of centrifugal force. Droplets are discriminated according to their size by means of a vortex produced by the tangential flow of the sample aerosol and argon gas inside the spray chamber. Smaller droplets are carried with the gas stream into the ICP-MS, while the larger droplets impinge on the walls and fall out through the drain.

CZE   An abbreviation for **capillary-zone electrophoresis**.

D

data-quality objectives   A term used to describe the quality goals of the analytical result. Typically achieved by optimizing the measurement protocol to achieve the desired accuracy/precision/sample throughput required for the analysis.

dead time correction   Sometimes ions hit the detector too fast for the measurement circuitry to handle in an efficient manner. This is caused by ions arriving at the detector during the output pulse of the preceding ion and not being detected by the counting system. This "dead time," as it is known, is a fundamental limitation of the multiplier detector and is typically 30–50 ns, depending on the detection system. A compensation or "dead time correction" has to be made in the measurement circuitry in order to count the maximum number of ions hitting the detector.

Debye length The distance over which ions exert an electrostatic influence over one another as they move from the interface region into the ion optics. In the ion-sampling process, this distance is small compared to the orifice diameter of the sampler or skimmer cone. As a result, there is little electrical interaction between the ion beam and the cones, and relatively little interaction between the individual ions within the ion beam. In this way, the compositional integrity of the ion beam is maintained throughout the interface region.

desolvating microconcentric nebulizer A microconcentric nebulizer that uses some type of desolvation system to remove the sample solvent. *Also refer to* **desolvation device** *and* **membrane desolvation**.

desolvating spray chamber A general name given to a spray chamber that removes or reduces the amount of solvent from a sample using the principle of desolvation. Some of the approaches that are typically used include conventional water cooling, heating with cooling condensers, Peltier (thermoelectric) cooling, or membrane-based desolvation techniques.

desolvation device A general name given to a device that removes or reduces the amount of solvent from a sample using the principle of desolvation. Some of the approaches that are typically used include conventional water cooling, heating/condensing units, Peltier (thermoelectric) cooling, or membrane-based desolvation techniques.

detection capability A generic term used to assess the overall detection performance of an ICP mass spectrometer. There are a number of different ways of evaluating detection capability, including instrument detection limit (IDL), method detection limit (MDL), element sensitivity, and background equivalent concentration (BEC).

detection limit Most often refers to the instrument detection limit (IDL) and is typically defined as a ratio of the analyte signal to the noise of the background at a particular mass. For a 99% confidence level, it is usually calculated as 3× standard deviation (SD) of 10 replicates (measurements) of the sample blank expressed as concentration units.

detector A generic name used for a device that converts ions into electrical pulses in ICP-MS.

detector dead time *Refer to* **dead time correction**.

digital counting Refers to the process of counting the number of pulses generated by the conversion of ions into an electrical signal by the detector measurement circuitry.

DIHEN An abbreviation for **direct injection high-efficiency nebulizer**.

DIN An abbreviation for **direct injection nebulizer**.

direct injection high-efficiency nebulizer (DIHEN) A more recent refinement of the direct injection nebulizer (DIN), which appears to have overcome many of the limitations of the original design.

direct injection nebulizer (DIN) A nebulizer that injects a liquid sample under high pressure directly into the base of the plasma torch. The benefit of this approach is that no spray chamber is required, which means that an extremely small volume of sample can be introduced directly into the ICP-MS with virtually no carryover or memory effects from the previous sample.

discrete dynode detector (DDD) The most common type of detector used in ICP-MS. As ions emerge from the quadrupole rods onto the detector, they strike the first dynode, liberating secondary electrons. The electron-optic design of the dynode produces acceleration of these secondary electrons to the next dynode, where they generate more electrons. This process is repeated at each dynode, generating a pulse of electrons that are finally captured by the multiplier anode. *Also refer to* **active film multipliers**.

double-focusing magnetic sector mass spectrometer (analyzer) A mass spectrometer that uses a very powerful magnet combined with an electrostatic analyzer (ESA) to produce a system with very high resolving power. This approach, known as "double focusing," samples the ions from the plasma. The ions are accelerated in the plasma to a few kilovolts into the ion-optic region before they enter the mass analyzer. The magnetic field, which is dispersive with respect to ion energy and mass, then focuses all the ions with diverging angles of motion from the entrance slit. The ESA, which is only dispersive with respect to ion energy, then focuses all the ions onto the exit slit, where the detector is positioned. If the energy dispersions of the magnet and ESA are equal in magnitude but opposite in direction, they will focus both ion angles (first focusing) and ion energies (second focusing) when combined together. *Also refer to* **electrostatic analyzer**.

double-pass spray chamber A spray chamber that comprises an inner (central) tube inside the main body of the spray chamber. The smaller droplets are selected by directing the aerosol from the nebulizer into the central tube. The aerosol emerges from the tube, where the larger droplets fall out (because of gravity) through a drain tube at the rear of the spray chamber. The smaller droplets then travel back between the outer wall and the central tube into the sample injector of the plasma torch. The most common type of double-pass spray chamber is the Scott design.

doubly charged ion A species that is formed when an ion is generated with a double positive charge as opposed to a normal single charge and produces an isotopic peak at half its mass. For example, the major isotope of barium at mass 138 amu also exhibits a doubly charged ion at mass 69 amu, which can potentially interfere with gallium at mass 69. Some elements such as the rare earths readily form doubly charged species, whereas others do not. Formation of doubly charged ions is also impacted by the ionization conditions (RF power, nebulizer gas flow, etc.) in the plasma discharge.

DRC An abbreviation for **dynamic reaction cell**.

droplet Refers to individual particles (either small or large) that make up an aerosol generated by the nebulizer.

duty cycle (%) Also known as the "measurement duty cycle." It refers to the actual peak measurement time and is expressed as a percentage of the overall integration time. It is calculated by dividing the total peak quantitation time (dwell time × number of sweeps × replicates × elements) by the total integration time ([dwell time + settling/scanning time] × number of sweeps × replicates × elements).

dwell time   The time spent sitting (dwelling) on top of the analytical peak (mass) and taking measurements.

dynamic reaction cell (DRC)   A type of collision/reaction cell. Unlike a simple collision cell, a quadrupole is used instead of a hexapole or octapole. A highly reactive gas such as ammonia or methane is bled into the cell, which is a catalyst for ion–molecule chemistry to take place. By a number of different reaction mechanisms, the gaseous molecules react with the interfering ions to convert them either into an innocuous species different from the analyte mass or a harmless neutral species. The analyte mass then emerges from the dynamic reaction cell, free of its interference, and is steered into the analyzer quadrupole for conventional mass separation. Through careful optimization of the quadrupole electrical fields, unwanted reactions between the gas and the sample matrix or solvent, which could potentially lead to new interferences, are prevented. Therefore, every time an analyte and interfering ions enter the dynamic reaction cell, the bandpass of the quadrupole can be optimized for that specific problem and then changed on the fly for the next one.

dynamically scanned ion lens   A commercial ion optic approach to focus the maximum number of ions into the mass analyzer. In this design, the voltage is dynamically ramped on the fly in concert with the mass scan of the analyzer. The benefit is that the optimum lens voltage is placed on every mass in a multi-element run to allow the maximum number of analyte ions through, while keeping the matrix ions down to an absolute minimum. This is typically used in conjunction with a grounded stop acting as a physical barrier to reduce particulates, neutral species, and photons from reaching the mass analyzer and detector.

E

EDR   An abbreviation for the term **extended dynamic range**, used in detector technology.

electron   A negatively charged fundamental particle orbiting the nucleus of an atom. It has a mass equal to 1/1836 of a proton's mass. Removal of an electron by excitation in the plasma discharge generates a positively charged ion.

electrostatic analyzer (ESA)   An ion-focusing device (utilizing a series of electrostatic lens components) that varies the electric field to allow the passage of ions of certain energy. In ICP-MS, it is typically used in combination with a conventional electromagnet to focus ions based on their angular motion and their kinetic energy to produce very high resolving power. *Also refer to* **double-focusing magnetic sector mass spectrometer (analyzer)**.

electrothermal atomization (ETA)   An atomic absorption (AA) analytical technique that uses a heated metal filament or graphite tube (in place of the normal flame) to generate ground-state analyte atoms. The sample is first injected into the filament or tube, which is heated up slowly to remove the matrix components. Further heating then generates ground-state atoms of the analyte, which absorb light of a particular wavelength from an element-specific, hollow cathode lamp source. The amount of light absorbed is measured by a monochromator (optical system) and detected by a photomultiplier or solid-state detector, which converts the photons into an electrical pulse. This absorbance signal is used to determine the concentration of that element in the sample. Typically used for ppb-level determinations.

electrothermal vaporization (ETV) A sample pretreatment technique used in ICP-MS. Based on the principle of electrothermal atomization (ETA) used in atomic absorption (AA), ETV is not used to generate ground-state atoms, but instead uses a carbon furnace (tube) or metal filament to thermally separate the analytes from the matrix components and then sweep them into the ICP mass spectrometer for analysis. This is achieved by injecting a small amount of the sample into a graphite tube or onto a metal filament. After the sample is introduced, drying, charring, and vaporization are achieved by slowly heating the graphite tube or metal filament. The sample material is vaporized into a flowing stream of carrier gas, which passes through the furnace or over the filament during the heating cycle. The analyte vapor recondenses in the carrier gas and is then swept into the plasma for ionization.

elemental fractionation A term used in laser ablation. It is typically defined as the variation in intensity of a particular element over time compared to the total amount of dry aerosol generated by the sample. It is generally sample and element specific, but there is evidence to suggest that the shorter-wavelength excimer lasers exhibit better elemental fractionation characteristics than the longer-wavelength Nd:YAG design because they produce smaller particles that are easier to volatilize.

endothermic reaction In thermodynamics, this describes a chemical reaction that absorbs energy in the form of heat. In ICP-MS, it generally refers to an ion–molecule reaction in a collision/reaction cell that is not allowed to proceed because the ionization potential of the analyte ion is significantly less than that of the reaction gas molecule. *Also refer to* **exothermic reaction**.

ESA An abbreviation for **electrostatic analyzer**.

ETA An abbreviation for **electrothermal atomization**.

ETV An abbreviation for **electrothermal vaporization**.

excimer laser A gas-filled laser in which a very short electrical pulse excites a mixture containing a halogen such as fluorine and a rare gas such as argon or krypton. It produces a brief, intense pulse of UV light. The output of an excimer laser is used for writing patterns on semiconductor chips because the short wavelength can write very fine lines. In ICP-MS, the most common excimer laser used is ArF at 193 nm and is typically used to ablate material with a very small size, such as inclusions on the surface of a geological sample.

exothermic reaction In thermodynamics, this describes a chemical reaction that releases energy in the form of heat. In ICP-MS, it generally refers to an ion–molecule reaction in a collision/reaction cell that is spontaneous because the ionization potential of the interfering ion is much greater than the reaction gas molecule. *Also refer to* **endothermic reaction**.

extended dynamic range (EDR) An approach used in ICP-MS to extend the linear dynamic range of the detector from five orders of magnitude up to eight or nine orders of magnitude. *Also refer to* **discrete dynode detector** *and* **Faraday cup detector**.

external standardization The normal mode of calibration used in ICP-MS by comparing the analyte intensity of unknown samples to the intensity of known calibration or reference standards.

extraction lens An ion lens used to electrostatically extract the ions out of the interface region.

F

FAA An abbreviation for **flame atomic absorption**.

Faraday collector Another name for a Faraday cup detector.

Faraday cup detector A simple metal electrode detector used to measure high ion counts. When the ion beam hits the metal electrode, it will be charged, whereas the ions are neutralized. The electrode is then discharged to measure a small current equivalent to the number of discharged ions. By measuring the ion current on the metal part of the circuit, the number of ions in the circuit can be determined. Unfortunately, with this approach, there is no control over the applied voltage (gain). So, it can only be used for high ion counts, and therefore is not suitable for ultratrace determinations.

FIA An abbreviation for **flow injection analysis**.

flame atomic absorption (FAA) An atomic absorption analytical technique that uses a flame (usually air–acetylene or nitrous oxide–acetylene) to generate ground-state atoms. The sample solution is aspirated into the flame via a nebulizer and a spray chamber. The ground-state atoms of the sample absorb light of a particular wavelength from an element-specific, hollow cathode lamp source. The amount of light absorbed is measured by a monochromator (optical system) and detected by a photomultiplier or solid-state detector, which converts the photons into an electrical pulse. This absorbance signal is used to determine the concentration of the element in the sample. Typically used for ppm-level determinations.

flight tube A generic name given to the housing that contains a series of optical components which focus ions onto the detector of a time-of-flight (TOF) mass analyzer. There are basically two different kinds of flight tubes that are used in commercial TOF mass analyzers. One is the orthogonal design, where the flight tube is positioned at right angles to the sampled ion beam, and the other, the axial design, where the flight tube is in the same axis as the ion beam. In both designs, all ions are sampled through the interface region, but instead of being focused into the mass filter in the conventional sequential way, packets (groups) of ions are electrostatically injected into the flight tube at exactly same time.

flow injection analysis (FIA) A powerful front-end sampling accessory for ICP-MS that can be used for preparation, pretreatment, and delivery of the sample. It involves the introduction of a discrete sample aliquot into a flowing carrier stream. Using a series of automated pumps and valves, procedures can be carried out on-line to physically or chemically change the sample or analyte before introduction into the mass spectrometer for detection.

fringe rods A set of four short rods operated in the RF-only mode, positioned at the entrance of a quadrupole mass analyzer. Their function is to minimize the effect of the fringing fields at the entrance of a quadrupole mass analyzer and thus improve the efficiency of transmission of ions into the mass analyzer. They are usually straight, but it has been suggested that curved fringe rods might reduce background levels.

fusion mixture A compound or mixture added to solid samples as an aid to get them into solution. Fusion mixtures are usually alkaline salts (e.g., lithium metaborate, sodium carbonate) that are mixed with the sample (in powdered form) and heated in a muffle furnace to create a chemical/thermal reaction between the sample and the salt. The fused mixture is then dissolved in a weak mineral acid to get the analytes into solution.

G

gamma-counting spectrometry A particle-counting technique that uses the measurement of the radioactive decay of gamma particles. *Also refer to* **particle-counting techniques**.

gas dynamics In ICP-MS, it refers to the flow and velocity of the plasma gas through the interface region. It dictates that the composition of the ion beam immediately behind the sampler cone be the same as the composition in front of the cone because the expansion of the gas at this stage is not controlled by electrodynamics. This happens because the distance over which ions exert influence on one another (the Debye length) is small compared to the orifice diameter of the sampler or skimmer cone. Consequently, there is little electrical interaction between the ion beam and the cone and relatively little interaction between the individual ions in the beam. In this way, gas dynamics ensures that the compositional integrity of the ion beam is maintained throughout the interface region.

getter (gas purifier) A device that "cleans up" inorganic and organic contaminants in pure gases. The getter usually refers to a metal that oxidizes quickly, and when heated to a high temperature (usually by means of RF induction), evaporates and absorbs/reacts with any residual impurities in the gas.

GFAA An abbreviation for **graphite furnace atomic absorption**.

graphite furnace atomic absorption (GFAA) An electrothermal atomization (ETA) analytical technique that specifically uses a graphite tube (in place of the normal flame) to generate ground-state analyte atoms. The sample is first injected into the tube, which is heated up slowly to remove the matrix components. Further heating then generates ground-state atoms of the analyte, which absorb light of a particular wavelength from an element-specific, hollow cathode lamp source. The amount of light absorbed is measured by a mono-chromator (optical system) and detected by a photomultiplier or solid-state detector, which converts the photons into an electrical pulse. This absorbance signal is used to determine the concentration of that element in the sample. Typically used for ppb-level determinations. *Also refer to* **electrothermal atomization (ETA)**.

grounding mechanism A way of eliminating the secondary discharge (pinch effect) produced by capacitive (RF) coupling of the load coil to the plasma. This undesired coupling between the RF voltage on the load coil and the plasma discharge produces a potential difference of a few hundred volts, which creates an electrical discharge (arcing) between the plasma and sampler cone of the interface. This mechanism varies with different instrument designs, but basically involves grounding the load coil to make sure the interface region is maintained at zero potential.

H

half-life The time required for half the atoms of a given amount of a radioactive substance to disintegrate. This principle is used in particle-counting measuring techniques.

heating zones The zones that describe the different temperature regions within a plasma discharge, where the sample passes through. The most common zones include the preheating zone (PHZ), where the sample is desolvated; the initial radiation zone (IRZ), where the sample is broken down into its molecular form; and the normal analytical zone (NAZ), where the sample is first atomized and then ionized.

HEN An abbreviation for **high-efficiency nebulizer**.

hexapole A multipole containing six rods, used in collision/reaction cell technology.

HGAA An abbreviation for **hydride generation atomic absorption**.

high-efficiency nebulizer (HEN) A generic name given to a nebulizer that is very efficient, with very little wastage. Usually used to describe direct injection or microconcentric-designed systems, which deliver all or a very high percentage of the sample aerosol into the plasma discharge.

high-resolution mass analyzer A generic name given to a mass spectrometer with very high resolving power. Commercial designs are usually based on the double-focusing magnetic sector design.

high-sensitivity interface (HSI) High-sensitivity interfaces (HSI) are offered as an option with most commercial ICP-MS systems. They all work slightly differently but share similar components. By using a slightly different cone geometry, higher vacuum at the interface, one or more extraction lenses, or modified ion optic design, they offer up to 10 times the sensitivity of a traditional interface. However, their limitations are that background levels are often elevated, particularly when analyzing samples with a heavy matrix. Therefore, they are more suited for the analysis of clean solutions.

high-solids nebulizers Nebulizers that are used to aspirate higher concentrations of dissolved solids into the ICP-MS. The most common types used are the

Babbington, V-groove, and cone-spray designs. Not widely used for ICP-MS because of the dissolved-solids limitations of the technique, but are sometimes used with flow injection sample introduction techniques.

hollow ion mirror   A more recent development in ion-focusing optics. The ion mirror, which has a hollow center, creates a parabolic electrostatic field to reflect and refocus the ion beam at right angles to the ion source. This allows photons, neutrals, and solid particles to pass through it, while allowing ions to be reflected at right angles into the mass analyzer. The major benefit of this design is the highly efficient way the ions are refocused, offering extremely high sensitivity and low background across the mass range.

homogenized sample beam   The laser beam in an excimer laser, which produces a much flatter beam profile and more precise control of the ablation process.

hydride generation atomic absorption (HGAA)   A very sensitive analytical technique for determining trace levels of volatile elements such as As, Bi, Sb, Se, and Te. Generation of the elemental hydride is carried out in a closed vessel by the addition of a reducing agent, such as sodium borohydride, to the acidic sample. The resulting gaseous hydride is swept into a special heated quartz cell (in place of the traditional flame burner head), where atomization occurs. Atomic absorption quantitation is then carried out in the conventional way, by comparing the absorbance of unknown samples against known calibration or reference standards.

hydrogen atom transfer   An ion–molecule reaction mechanism in a collision/reaction cell where a hydrogen atom is transferred to the interfering ion, which is converted to an ion at one mass higher.

hyperbolic fields   The four rods that make up a quadrupole are usually cylindrical or elliptical in shape. The electrical fields produced by these rods are typically hyperbolic in shape.

I

ICP   An abbreviation for **inductively coupled plasma**.

ICP-OES   An abbreviation for **inductively coupled plasma optical emission spectrometry**.

impact bead (nebulizer)   A type of spray chamber more commonly used in atomic absorption spectrometers. The aerosol from the nebulizer is directed onto a spherical bead, where the impact breaks the sample into large and small droplets. The large droplets fall out due to gravitational force and the smaller droplets are directed by the nebulizer gas flow into the atomization/excitation/ionization source.

inductively coupled plasma (ICP)   The high-temperature source used to generate ions in ICP-MS. It is formed when a tangential (spiral) flow of argon gas is directed between the outer and middle tube of a quartz torch. A load coil (usually copper) surrounds the top end of the torch and is connected to an RF generator. When RF power (typically, 750–1500 W) is applied to the load coil, an alternating current oscillates within the coil at a rate corresponding to the frequency of the generator. The RF oscillation of the current in the coil creates an intense electromagnetic field in the area at the top of the torch. With argon gas flowing through the torch, a high-voltage spark is applied to the gas, causing some electrons to be stripped from their argon atoms. These electrons, which are caught up and accelerated in the magnetic field, then collide with other argon atoms, stripping off still more electrons. This collision-induced ionization of the argon continues in a chain reaction, breaking down the gas into argon atoms, argon ions, and electrons, forming what is known as an "inductively coupled plasma (ICP) discharge" at the open end of the plasma torch.

inductively coupled plasma optical emission spectrometry (ICP-OES) A multi-element technique that uses an inductively coupled plasma to excite ground-state atoms to the point where they emit wavelength-specific photons of light, characteristic of a particular element. The number of photons produced at an element-specific wavelength is measured using high-resolving optical components to separate the analyte wavelengths and a photon-sensitive detection system to measure the intensity of the emission signal produced. This emission signal is directly related to the concentration of that element in the sample. Commercial instrumentation comes in two configurations: a traditional radial view, where the plasma is vertical and is viewed from the side (side-on viewing), and an axial view, where the plasma is positioned horizontally and is viewed from the end (end-on viewing).

infrared (IR) lasers Laser ablation systems that operate in the IR region of the electromagnetic spectrum, such as the Nd:YAG laser, which has its primary wavelength at 1064 nm.

instrument background noise Square root of the spectral background of the instrument (in cps), usually measured at mass 220 amu, where there are no spectral features. *Also refer to* **background signal**, **background noise**, *and* **detection limit**.

instrument background signal Spectral background of the instrument (in cps), usually measured at mass 220 amu, where there are no spectral features. *Also refer to* **background signal**, **background noise**, *and* **detection limit**.

integration time The total time spent measuring an analyte mass (peak). Comprising the time spent dwelling (sitting) on the peak multiplied by the number of points used for peak quantitation multiplied by the number of scans used in the measurement protocol. *Also refer to* **duty cycle**, **measurement duty cycle**, **peak measurement protocol**, **settling time**, *and* **dwell time**.

interface The plasma discharge is coupled to the mass spectrometer via the interface. The interface region comprises a water-cooled metal housing containing the sampler cone and the skimmer cone, which directs the ion beam from the central channel of the plasma into the ion optic region.

interface cones Refer to the sampler and skimmer cones housed in the interface region. *Also refer to* **interface** *and* **interface region**.

interface pressure The pressure between the sampler cone and skimmer cone. This region is maintained at a pressure of approximately 1–2 torr by a mechanical roughing pump.

interface region A region comprising a water-cooled metal housing containing the sampler cone and the skimmer cone, which directs and focuses the ion beam from the central channel of the plasma into the ion optic region.

interferences A generic term given to a nonanalyte component that enhances or suppresses the signal intensity of the analyte mass. The most common interferences in ICP-MS are spectral, matrix, or sample transport in nature.

internal standardization (IS) A quantitation technique used to correct for changes in analyte sensitivity caused by variations in the concentration and type of matrix components found in the sample. An internal standard is a nonanalyte isotope that is added to the blank solution, standards, and samples before analysis. It is typical to add three or four internal standard elements to the samples to cover all the analyte elements of interest across the mass range. The software adjusts the analyte concentration in the unknown samples by comparing the intensity values of the internal standard elements in the unknown sample to those in the calibration standards. Because ICP-MS is prone to many matrix- and sample-transport-based interferences, internal standardization is considered necessary to analyze most sample types.

ion An electrically charged atom or group of atoms formed by the loss or gain of one or more electrons. A cation (positively charged ion) is created by the loss of an electron, and an anion (negatively charged ion) is created by the gain of an electron. The valency of an ion is equal to the number of electrons lost or gained and is indicated by a plus sign for cations and a minus sign for anions. ICP-MS typically involves the detection and measurement of positively charged ions generated in a plasma discharge.

ion chromatography (IC) A chromatographic separation technique used for determination of anionic species such as nitrates, chlorides, and sulfates. When coupled with ICP-MS, it becomes a very sensitive hyphenated technique for the determination of a wide variety of elemental ionic species.

ion energy In ICP-MS, it refers to the kinetic energy of the ion, in electronvolts (eV). It is a function of both the mass and velocity of the ion ($KE = \frac{1}{2}MV^2$). It is generally accepted that the spread of kinetic energies of all the ions in the ion beam entering the mass spectrometer must be on the order of a few electronvolts to be efficiently focused by the ion optics and resolved by the mass analyzer.

ion energy spread The variation in kinetic energy of all the ions in the ion beam emerging from the ionization source (plasma discharge). It is generally accepted that this variation (spread) of kinetic energies must be on the order of a few electronvolts to be efficiently focused by the ion optics and resolved by the mass analyzer.

ion flow The flow of ions from the interface region through the ion optics into the mass analyzer.

ion-focusing guide An alternative name for the **ion optics**.

ion-focusing system An alternative name for the **ion optics**.

ion formation The transfer of energy from the plasma discharge to the sample aerosol to form an ion. By traveling through the different heating zones in the plasma, where the sample is first dried, vaporized and atomized, then finally converted to an ion.

ion kinetic energy *Refer to* **ion energy**.

ion lens Often referred to as a single-lens component in the ion optic system. *Also refer to* **ion optics**.

ion lens voltages The voltages put on one or more lens components in the ion optic system to electrostatically steer the ion beam into the mass analyzer. *Also refer to* **ion optics**.

ion mirror A more recent development in ion-focusing optics. With this design, a parabolic electrostatic field is created with a hollow ion mirror to reflect and refocus the ion beam at right angles to the ion source. The ion mirror is an electrostatically charged ring, which is hollow in the center. This allows photons, neutrals, and solid particles to pass through it, while allowing ions to be reflected at right angles into the mass analyzer. The major benefit of this design is the highly efficient way the ions are refocused, offering extremely high sensitivity across the mass range with very little compromise in oxide performance. In addition, there is very little contamination of the ion optics because a vacuum pump sits behind the ion mirror to immediately remove these particles before they have a chance to penetrate further into the mass spectrometer.

ion optics Comprises one or more electrostatically charged lens components that are positioned immediately after the skimmer cone. They are made up of a series of metallic plates, barrels, or cylinders, which have a voltage placed on them. The function of the ion optic system is to take ions after they emerge from the interface region and steer them into the mass analyzer. Another function of the ion optics is to reject the nonionic species such as particulates, neutral species, and photons and prevent them from reaching the detector. Depending on the design, this is achieved by using some kind of physical barrier, positioning the mass analyzer off axis relative to the ion beam, or electrostatically bending the ions by 90° into the mass analyzer.

ion packet A "slice of ions" that is sampled from the ion beam in a time-of-flight (TOF) mass analyzer. In the TOF design, all ions are sampled through the interface cones, but instead of being focused into the mass filter in the conventional way, packets (groups) of ions are electrostatically injected into the flight tube at exactly the same time. Whether the orthogonal (right-angle) or axial (straight-on) approach is used, an accelerating potential is applied to the continuous ion beam. The ion beam is then "chopped" by using a pulsed voltage supply to provide repetitive voltage "slices" at a frequency of a few kilohertz. The "sliced" packets of ions are then allowed to "drift" into the flight tube, where the individual ions are temporally resolved according to their differing velocities.

ion repulsion The degree to which positively charged ions repel each other as they enter the ion optics. The generation of a positively charged ion beam is the first stage in the charge separation process. Unfortunately, the net positive charge of the ion beam means that there is now a natural tendency for the ions to repel one another. If nothing is done to compensate for this repulsion, ions of higher mass-to-charge ratio will dominate the center of the ion beam and force the lighter ions to the outside. The degree of loss will depend on the kinetic energy of the ions—ions with high kinetic energy (high-mass elements) will be transmitted in preference to ions with medium (midmass elements) or low kinetic energy (low-mass elements).

ionization source In ICP-MS, the ionization source is the plasma discharge, which reaches temperatures of up to 10,000 K to ionize the liquid sample.

ionizing radiation counting techniques Particle-counting techniques such as alpha, gamma, and scintillation counters that are used to measure the isotopic composition of radioactive materials. However, the limitation of particle-counting techniques is that the half-life of the analyte isotope has a significant impact on the method detection limit. This implies that they are better suited for the determination of short-lived radioisotopes, because meaningful data can be obtained in a realistic amount of time. They have also been successfully applied to the quantitation of long-lived radionuclides, but unfortunately require a combination of extremely long counting times and large amounts of sample to achieve low levels of quantitation.

ion–molecule chemistry A chemical reaction between the analyte or interfering ion and molecules of the reaction gas in a collision/reaction cell. A reactive gas, such as hydrogen, ammonia, oxygen, methane, or gas mixtures, is bled into the cell, which is a catalyst for ion–molecule chemistry to take place. By a number of different reaction mechanisms, the gaseous molecules react with the interfering ions to convert them into either an innocuous species different from the analyte mass or a harmless neutral species. The analyte mass then emerges from the cell free of its interference and is steered into the analyzer quadrupole for conventional mass separation. In some cases, the chemistry can take place between the gaseous molecule and the analyte to form a new analyte ion free of the interfering species.

isobar (or isobaric) Used in the context of atomic principles, it refers to two or more atoms with the same atomic mass (same number of neutrons) but different atomic number (different number of protons). *Also refer to* **isobaric interferences**.

isobaric interferences The word "isobaric" is used in the context of atomic principles, and refers to two or more atoms with the same atomic mass but different atomic number. In ICP-MS, they are a classification of spectrally induced interferences produced mainly by different isotopes of other elements in the sample, creating spectral interferences at the same mass as the analyte.

isotope A different form of an element having the same number of protons in the nucleus (i.e., same atomic number) but a different number of neutrons (i.e., different atomic mass). There are 275 isotopes of the 81 stable elements in the periodic table, in addition to over 800 radioactive isotopes. Isotopes of a single element possess very similar properties.

isotope dilution An absolute means of quantitation in ICP-MS based on altering the natural abundance of two isotopes of an element by adding a known amount of one of the isotopes. The principle works by spiking the sample solution with a known weight of an enriched stable isotope. By knowing the natural abundance of the two isotopes being measured, the abundance of the spiked enriched isotope, the weight of the spike, and the weight of the sample, it is possible to determine the original trace element concentration. It is considered one of the most accurate and precise quantitation techniques for elemental analysis by ICP-MS.

isotope ratio The ability of ICP-MS to determine individual isotopes makes it suitable for an isotopic measurement technique called "isotope ratio analysis." The ratio of two or more isotopes in a sample can be used to generate very useful information, such as an indication of the age of a geological formation, a better understanding of animal metabolism, and the identification of sources of environmental contamination. Similar to isotope dilution, isotope ratio analysis uses the principle of measuring the exact ratio of two isotopes of an element in the sample. With this approach, the isotope of interest is typically compared to a reference isotope of the same element, but can also be referenced to an isotope of another element.

isotope ratio precision The reproducibility or precision of measurement of isotope ratios is very critical for some applications. For the highest-quality isotopic ratio precision measurements, it is generally acknowledged that either magnetic sector or time-of-flight (TOF) instrumentation offers the best approach over quadrupole ICP-MS.

isotopic abundance The percentage abundance of an isotope compared to the element's total abundance in nature. *Also refer to* **natural abundance** *and* **relative abundance of natural isotopes**.

K

KE An abbreviation for **kinetic energy**.

KED An abbreviation for **kinetic energy discrimination**.

kinetic energy (KE) The energy possessed by a moving body due to its motion. It is equal to one-half the mass of the body times the square of its speed (velocity): $KE = \frac{1}{2}MV^2$. For kinetic energy as applied to moving ions, *refer to* **ion energy**.

kinetic energy discrimination (KED) In collision/reaction cell technology, it is one way to separate the newly formed by-product ions from the analyte ions. It is typically achieved by setting the collision cell potential (voltage) slightly more negative than the mass filter potential. This means that the collision by-product ions generated in the cell, which have a lower kinetic energy as a result of the collision process, are rejected, whereas the analyte ions, which have a higher kinetic energy, are transmitted to the mass analyzer.

L

laser ablation A sample preparation technique that uses a high-powered laser beam to vaporize the surface of a solid sample and sweep it directly into the ICP-MS system for analysis. It is mainly used for samples that are extremely difficult to get into solution or for samples that require the analysis of small spots or inclusions on the surface.

laser absorption The "coupling" efficiency of the sample with the laser beam in laser ablation work. The more light the sample absorbs, the more efficient the ablation process becomes. It is generally accepted that the shorter-wavelength excimer lasers have better absorption characteristics than the longer-wavelength IR laser systems for UV-transparent/opaque materials such as calcites, fluorites, and silicates and, as a result, generate smaller particle size and higher flow of ablated material.

laser fluence A term used to describe the power density of a laser beam in laser ablation studies. It is defined as the laser pulse energy per focal spot area, measured in J/cm². It is related to laser irradiance, which is the ratio of the fluence to the width of the laser pulse.

laser irradience A term used to describe the power density of a laser beam in laser ablation studies. Laser irradiance is the ratio of the laser pulse energy per focal spot area (i.e., fluence) to the width of the laser pulse. *Also refer to* **laser fluence**.

laser sampling *Refer to* **laser ablation**.

laser vaporization *Refer to* **laser ablation**.

laser wavelength The primary wavelength of the optical components used in the design of a laser ablation system.

load coil Another name for the RF coil used to generate a plasma discharge. *Also refer to* **RF generator**.

low-temperature plasma An alternative name for cool or cold plasma.

M

magnetic field A region around a magnet, an electric current, or a moving charged particle that is characterized by the existence of a detectable magnetic force at every point in the region and by the existence of magnetic poles. In ICP-MS, it usually refers to the magnetic field around the RF coil of the plasma discharge or the magnetic field produced by a quadrupole or an electromagnet.

magnetic sector mass analyzer A design of mass spectrometer used in ICP-MS to generate very high resolving power as a way of reducing spectral interferences. Commercial designs typically utilize a very powerful magnet combined with an electrostatic analyzer (ESA). In this approach, known as the double-focusing design, the ions from the plasma are sampled. In the plasma, the ions are accelerated to a few kilovolts into the ion optic region before they enter the mass analyzer. The magnetic field, which is dispersive with respect to ion energy and mass, then focuses all the ions with diverging angles of motion from the entrance slit. The ESA, which is only dispersive with respect to ion energy, then focuses all the ions onto the exit slit, where the detector is positioned. If the energy dispersion of the magnet and ESA are equal in magnitude but opposite in direction, they will focus both ion angles (first focusing) and ion energies (second focusing) when combined together. *Also refer to* **electrostatic analyzer**.

mass analyzer The part of the mass spectrometer where the separation of ions (based on their mass-to-charge ratio) takes place. In ICP-MS, the most common type of mass analyzers are quadrupole, magnetic sector, and time-of-flight (TOF) systems.

mass calibration The ability of the mass spectrometer to repeatedly scan to the same mass position every time during a multielement analysis. Instrument manufacturers typically quote a mass calibration stability specification for their design of mass analyzer based on the drift or movement of the peak (in atomic mass units) position over a fixed period of time (usually 8 h).

mass calibration stability *Refer to* **mass calibration**.

mass discrimination Sometimes called "mass bias." In ICP-MS, it occurs when a higher-concentration isotope is suppressing the signal of the lower-concentration isotope, producing a biased result. The effect is not so obvious if the concentrations of the isotopes in the sample are similar, but can be quite significant if the concentrations of the two isotopes are vastly different. If that is the case, it is recommended to run a standard of known isotopic composition to compensate for the effects of the suppression.

mass filter Another name for a mass analyzer.

mass-filtering discrimination A way of discriminating between analyte ions and the unwanted by-product interference ions generated in a collision/reaction cell.

mass resolution A measure of a mass analyzer's ability to separate an analyte peak from a spectral interference. The resolution of a quadrupole is nominally 1 amu and is traditionally defined as the width of a peak at 10% of its height.

mass scanning The process of electronically scanning the mass separation device to the peak of interest and taking analytical measurements. Basically, two approaches are used: single-point peak hopping, in which a measurement is typically taken at the peak maximum, and the multipoint-scanning approach, in which a number of measurements are taken across the full width of the peak. *Also refer to* **ramp scanning**, **integration time**, **dwell time**, **settling time**, **peak measurement protocol**, *and* **peak hopping**.

mass separation The process of separating the analyte ions from the nonanalyte, matrix, solvent, and interfering ions with the mass analyzer.

mass separation device Another name for a mass analyzer.

mass spectrometer The mass spectrometer section of an ICP-MS system is generally considered to be everything in the vacuum chamber from the interface region to the detector, including the interface cones, ion optics, mass analyzer, detector, and vacuum pumps.

matching network (RF) The matching network of the RF generator compensates for changes in impedance (a material's resistance to the flow of an electric current) produced by the sample's matrix components or differences in solvent volatility. In crystal-controlled generators, this is usually done with mechanically driven servo-type capacitors. With free-running generators, the matching network is based on electronic tuning of small changes in the RF brought about by the sample, solvent, or matrix components.

mathematical correction equations Used to compensate or correct for spectral interference in ICP-MS. Similar to interelement corrections (IECs) used in ICP-OES, they work on the principle of measuring the intensity of the interfering isotope or interfering species at another mass, which is ideally free of any interferences. A correction is then applied, depending on the ratio of the intensity of the interfering species at the analyte mass to its intensity at the alternate mass.

Mathieu stability plot A graphical representation of the stability of an ion as it passes through the rods of a multipole mass separation device. It is a function of the ratio of the RF to the DC current placed on each pair of rods. A plot of these ratios of multiple ions traveling through the multipole shows which ions are stable and make it through the rods to the detector and which ions are unstable and get ejected from the multipole. The most well-defined stability boundaries are obtained with a quadrupole and become more diffuse with higher-order multipoles such as hexapoles and octapoles.

matrix interferences There are basically three types of matrix-induced interferences. The first, and simplest to overcome, is often called a "sample transport or viscosity effect" and is a physical suppression of the analyte signal brought on by the level of dissolved solids or acid concentration in the sample. The second type of matrix suppression is caused when the sample matrix affects the ionization conditions of the plasma discharge, which results in varying amounts of signal suppression depending on the concentration of the matrix components. The third type of matrix interference is often called "space-charge matrix suppression." This occurs mainly when low-mass analytes are being determined in the presence of larger concentrations of high-mass matrix components. It has the effect of defocusing the ion beam, and unless any compensation is made, the high-mass matrix element will dominate the ion beam, pushing the lighter elements out of the way, leading to low sensitivity and poor detection limits. The classical way to compensate for matrix interferences is to use internal standardization.

matrix separation Usually refers to some kind of chromatographic column technology to remove the matrix components from the sample before it is introduced into the ICP-MS system.

Mattauch-Herzog magnetic sector design One of the earliest designs of double-focusing magnetic sector mass spectrometers. In this design, which was named after the German scientists who invented it, two or more ions of different mass-to-charge ratios are deflected in opposite directions in the electrostatic and magnetic fields. The divergent monoenergetic ion beams are then brought together along the same focal plane.

measurement duty cycle Also known as the duty cycle, it refers to a percentage of actual quantitation time compared to total integration time. It is calculated by dividing the total quantitation time (dwell time × number of sweeps × replicates × elements) by total integration time ([dwell time + settling/scanning time] × number of sweeps × replicates × elements).

membrane desolvation Can be used with any sample introduction technique to remove solvent vapors. However, it is typically used with an ultrasonic or microconcentric nebulizer to remove the solvent from a liquid sample. In this design, the sample aerosol enters the membrane desolvator, where the solvent vapor passes through the walls of a tubular microporous PTFE or Nafion membrane. A flow of argon gas removes the volatile vapor from the exterior of the membrane, while the analyte aerosol remains inside the tube and is carried into the plasma for ionization.

microconcentric nebulizer Is based on the concentric nebulizer design, but operates at much lower flow rates. Conventional nebulizers have a sample uptake rate of about 1 mL/min with an argon gas pressure of 1 L/min, whereas microconcentric nebulizers typically run at less than 0.1 mL/min and typically operate at much higher gas pressure to accommodate the lower sample flow rates.

microflow nebulizer A generic name for nebulizers that operate at much lower flow rates than conventional concentric or cross-flow designs. *Also refer to* **microconcentric nebulizer**.

microporous membrane A tubular membrane made of an organic microporus material such as Teflon or Nafion, used in membrane desolvation. The sample aerosol enters the desolvation system, where the solvent vapor passes through the walls of the tubular membrane. A flow of argon gas then removes the volatile vapor from the exterior of the membrane, while the analyte aerosol remains inside the tube and is carried into the plasma for ionization

microsampling A generic name given to any front-end sampling device in atomic spectrometry that can be used for the preparation, pretreatment, and delivery of the sample to the spectrometric analyzer. The most common type of microsampling device used in ICP-MS is the flow injection technique, which involves the introduction of a discrete sample aliquot into a flowing carrier stream. Using a series of automated pumps and valves, procedures can be carried out on-line to physically or chemically change the sample or analyte, before introduction into the mass spectrometer for detection.

microwave digestion A method of digesting difficult-to-dissolve solid samples using microwave technology. Typically, a dissolution reagent such as a concentrated mineral acid is added to the sample in a closed acid-resistant vessel contained in a specially designed microwave oven. By optimizing the current, temperature, and pressure settings, difficult samples can be dissolved in a relatively short time compared to traditional hot plate sample digestion techniques.

microwave dissolution An alternative name for **microwave digestion**.

microwave-induced plasma (MIP) The most basic form of electrodeless plasma discharge. In this device, microwave energy (typically, 100–200 W) is supplied to the plasma gas from an excitation cavity around a glass/quartz tube. The plasma discharge in the form of a ring is generated inside the tube. Unfortunately, even though the discharge achieves a very high power density, the high excitation temperatures only exist along a central filament. The bulk of the MIP never goes above 2000–3000 K, which means it is prone to very severe matrix effects. In addition, it is easily extinguished when aspirating liquid samples, so it has a found a niche as a detection system for gas chromatography.

MIP An abbreviation for **microwave-induced plasma**.

molecular association reaction An ion–molecule reaction mechanism in a collision/reaction cell, where an interfering ion associates with a neutral species (atom or molecule) to form a molecular ion.

molecular cluster ions Species that are formed by two or more molecular ions combining together in a reaction cell to form molecular clusters.

molecular spectral interferences Another name for polyatomic spectral interferences, which are typically generated in the plasma by the combination of two or more atomic ions. They are caused by a variety of factors, but are usually associated with the argon plasma/nebulizer gas used, matrix components in the solvent/sample, other elements in the sample, or entrained oxygen/nitrogen from the surrounding air.

multicomponent ion lens An ion lens system consisting of several lens components, all of which have a specific role to play in the transmission of the analyte ions into the mass filter. To achieve the desired analyte specificity, the voltage can be optimized on every ion lens and is usually combined with an off-axis mass analyzer to reject unwanted photons and neutral species.

multichannel analyzer The data acquisition system that stores and counts the ions as they strike the detector. As the ions emerge from the end of the quadrupole rods, they are converted into electrical pulses by the detector and stored by the multichannel analyzer. This multichannel data acquisition system typically has 20 channels per mass, and as the electrical pulses are counted in each channel, a profile of the mass is built up over the 20 channels, corresponding to the spectral peaks of the analyte masses being determined.

multichannel data acquisition The process of storing and counting ions in ICP-MS. *Also refer to* **multichannel analyzer**.

multipole The generic name given to a mass filter that isolates an ion of interest by applying DC or RF currents to pairs of rods. The most common type of mass analyzer multipole used in ICP-MS is the quadrupole (four rods). However, other higher orders of multipoles used in collision/reaction cell technology include hexapoles (six rods) and octapoles (eight rods).

N

natural abundance The natural amount of an isotope occurring in nature. *Also refer to* **isotopic abundance**.

natural isotopes Different isotopic forms of an element that occur naturally on or beneath the earth's crust.

Nd:YAG laser Nd:YAG is an acronym for neodymium-doped yttrium aluminum garnet, a compound that is used as the lasing medium for certain solid-state lasers. In this design, the YAG host is typically doped with around 1% neodymium by weight. Nd:YAG lasers are optically pumped using a flashlamp or laser diodes and emit light with a wavelength of 1064 nm in the infrared region. However, for many applications, the infrared light is frequency-doubled, -tripled, -quadrupled, or -quintupled by using additional optical components to generate output wavelengths in the visible and UV regions.

Typical wavelengths used for laser ablation/ICP-MS work include 532 nm (doubled), 266 nm (quadrupled), and 213 nm (quintupled). Pulsed Nd:YAG lasers are usually operated in the so called "Q-switching" mode, where an optical switch is inserted in the laser cavity, waiting for a maximum population inversion in the neodymium ions before it opens. Then the light wave can run through the cavity, depopulating the excited laser medium at maximum population inversion. In this Q-switched mode, output powers of 20 MW and pulse durations of less than 10 ns are achieved.

nebulizer The component of the sample introduction system that takes the liquid sample and pneumatically breaks it down into an aerosol using the pressure created by a flow of argon gas. The concentric and cross-flow designs are the most common in ICP-MS.

neutral species Species generated in the plasma torch that have no positive or negative charge associated with them. If they are not eliminated, they can find their way into the detector and produce elevated background levels.

neutron A fundamental particle that is neutral in charge, found in the nucleus of an atom. It has a mass equal to that of a proton. The number of neutrons in the atomic nucleus defines the isotopic composition of that element.

Nier–Johnson magnetic sector design Nier–Johnson double-focusing magnetic sector instrumentation is the technology that all modern magnetic sector instrumentation is based on. Named after the scientists who developed it, Nier–Johnson geometry comes in two different designs, the "standard" and "reverse" Nier–Johnson geometry. Both these designs, which use the same basic principles, consist of two analyzers: a traditional electromagnet analyzer and an electrostatic analyzer (ESA). In the standard (sometimes called "forward") design, the ESA is positioned before the magnet, and in the reverse design, it is positioned after the magnet.

octapole A multipole mass-filtering device containing eight rods. In ICP-MS, octapoles are typically used in collision/reaction cell technology.

off-axis ion lens An ion lens system that is not on the same axis as the mass analyzer. Designed to stop particulates, neutral species, and photons from hitting the detector.

oxide ions Polyatomic ions that are formed between oxygen and other elemental components in the plasma gas, sample matrix, or solvent. They are generally not desirable because they can cause spectral overlaps on the analyte ions. Oxide formation is typically worse in the cooler zones of the plasma, and as a result, can be reduced by optimizing the RF power, nebulizer gas flow, and sampling position.

P

parabolic field Shape of the magnetic fields produced by a quadrupole.

particle-counting techniques Include alpha, gamma, and scintillation counters that are used to measure the isotopic composition of radioactive materials. However, the limitation of particle-counting techniques is that the half-life of the analyte isotope has a significant impact on the method's detection limit.

This means that to get meaningful data in a realistic amount of time, they are better suited for the determination of short-lived radioisotopes. They have been successfully applied to the quantitation of long-lived radionuclides, but unfortunately require a combination of extremely long counting times and large amounts of sample to achieve low levels of quantitation.

peak hopping A quantitation approach in which the quadrupole power supply is driven to a discrete position on the analyte mass (normally the maximum point), allowed to settle (settling time), and a measurement taken for a fixed amount of time (dwell time). The integration time for that peak is the dwell time multiplied by the number of scans (scan time). Multielement peak quantitation involves peak hopping to every mass in the multielement run. *Also refer to* **measurement duty cycle** *and* **peak measurement protocol**.

peak integration The process of integrating an analytical peak (mass). *Also refer to* **integration time**, **peak measurement protocol**, *and* **measurement duty cycle**.

peak measurement protocol The protocol of scanning the quadrupole and measuring a peak in ICP-MS. In multielement analysis, the quadrupole is scanned to the first mass. The electronics are allowed to settle (settling time), left to dwell for a fixed period of time at one or multiple points on the peak (dwell time), and signal intensity measurements are taken (based on the dwell time). The quadrupole is then scanned to the next mass and the measurement protocol repeated. The complete multielement measurement cycle (sweep) is repeated as many times as is needed to make up the total integration per peak and the number of required replicate measurements per sample analysis.

peak quantitation The process of quantifying the peak in ICP-MS using calibration standards. *Also refer to* **peak hopping**, **peak integration**, *and* **peak measurement protocol**.

Peltier cooler A thermoelectric cooler using the principle of generating a cold environment by creating a temperature gradient between two different materials. It uses electrical energy via a solid-state heat pump to transfer heat from a material on one side of the device to a different material on the other side, thus producing a temperature gradient across the device (similar to a household air conditioning system).

peristaltic pump A small pump in the sample introduction system that contains a set of minirollers (typically, 12) all rotating at the same speed. The constant motion and pressure of the rollers on the pump tubing feeds the sample through to the nebulizer. Peristaltic pumps are usually used with cross-flow nebulizers.

photon stop A grounded metal disk in the ion lens system that is used as a physical barrier to stop particulate matter, neutral species, and photons from getting to the detector.

physical interferences An alternative term used to describe sample transport- or viscosity-based suppression interferences.

pinch effect An effect caused by an undesired electrostatic (capacitive) coupling between the voltage on the load coil and the plasma discharge, which produces a potential difference of a few hundred volts. This capacitive coupling is commonly referred to as the pinch effect and shows itself as a secondary discharge (arcing) in the region where the plasma is in contact with the sampler cone.

plasma discharge Another name for an inductively coupled plasma (ICP).

plasma source Refers to the RF hardware components that create the plasma discharge, including the RF generator, matching network, plasma torch, and argon gas pneumatics.

plasma torch Another name for the quartz torch that is used to generate the plasma discharge. The plasma torch consists of three concentric tubes: an outer tube, middle tube, and sample injector. The torch can either be one piece, where all three tubes are connected, or it can have a demountable design, in which the tubes and the sample injector are separate. The gas (usually argon) that is used to form the plasma (plasma gas) is passed between the outer and middle tubes at a flow rate of 12–17 L/min. A second gas flow (auxiliary gas) passes between the middle tube and the sample injector at 1 L/min, and is used to change the position of the base of the plasma relative to the tube and the injector. A third gas flow (nebulizer gas), also at 1 L/min brings the sample, in the form of a fine-droplet aerosol, from the sample introduction system and physically punches a channel through the center of the plasma. The sample injector is often made from other materials besides quartz, such as alumina, platinum, and sapphire, if highly corrosive materials need to be analyzed.

polyatomic spectral interferences Another name for molecular-based spectral interferences, which are typically generated in the plasma by the combination of two or more atomic ions. They are caused by a variety of factors, but are usually associated with the argon plasma/nebulizer gas used, matrix components in the solvent/sample, other elements in the sample, or entrained oxygen/nitrogen from the surrounding air.

precursor ion Usually refers to a polyatomic or isobaric interfering ion that is formed in the plasma as opposed to a product (or by-product) ion that is formed in the collision/reaction cell.

product ion Usually refers to a product (or by-product) ion that is formed in the collision/reaction cell as opposed to a precursor interfering ion (polyatomic or isobaric) that is formed in the plasma.

proton A stable, positively charged fundamental particle that shares the atomic nucleus with a neutron. It has a mass 1836 times that of the electron.

proton transfer A reaction mechanism in a collision/reaction cell in which the interfering polyatomic species gives up a proton, which is then transferred to the reaction gas molecule to form a neutral atom.

pulse counting Refers to the conventional mode of counting ions with the detector measurement circuitry. Depending on the type of detection system that is used, an ion emerges from the quadrupole and strikes the ion-sensitive surface (discrete dynode, Channeltron, etc.) of the detector to generate electrons. These electrons move down the detector and generate more secondary electrons. This process is repeated at each stage of the detector, producing a pulse of electrons that is finally captured by the detector's collecting and counting circuitry.

Q

quadrupole The most common type of mass separation device used in commercial ICP-MS systems. It consists of four cylindrical or hyperbolic metallic rods of the same length (15–20 cm) and diameter (approximately 1 cm). The rods are typically made of stainless steel or molybdenum and sometimes coated with a ceramic coating for corrosion resistance. A quadrupole operates by placing both a DC field and a time-dependent AC of RF 2–3 MHz on opposite pairs of the four rods. By selecting the optimum AC/DC ratio on each pair of rods, ions of a selected mass are then allowed to pass through the rods to the detector, while the others are unstable and ejected from the quadrupole.

quadrupole power supply Another name for the electronic components that control the RF and DC voltages to change the mass-filtering characteristics.

quadrupole scan rate Scan rates of commercial quadrupole mass analyzers are on the order of 2500 amu/s. The quadrupole scan rate and the slope at which the RF and DC voltages of the quadrupole power supply are scanned will determine the desired resolution setting. A steeper slope translates to higher resolution, whereas a shallower slope means poorer resolution.

quadrupole stability regions The region of the Mathieu stability plot where the trajectory of an ion is stable and makes it through to the end of the quadrupole rods. All commercial ICP-MS systems that utilize quadrupole technology as the mass separation device operate in the 1st stability region, where resolving power is typically on the order of 500–600. If the quadrupole is operated in the 2nd or 3rd stability regions, resolving powers of 4000 and 9000, respectively, have been achieved. However, improving resolution using this approach has resulted in a significant loss of signal and higher background levels.

quantitative methods The different kinds of quantitative analyses available in ICP-MS, which include traditional quantitative analysis (using external calibration, standard additions, or addition calibration), semiquantitative routines (semiquant), isotope dilution (ID) methods, isotope ratio (IR) measurements, and classical internal standardization (IS).

quartz torch The standard plasma torch used in ICP-MS. *Also refer to* **plasma torch**.

R

radio frequency (RF) generator The power supply used to create the plasma discharge. Hardware includes the RF generator, matching network, plasma torch, and argon gas pneumatics.

radioactive isotope Sometimes known as "radioisotope," a radioactive isotope is a natural or artificially created isotope of an element having an unstable nucleus that decays, emitting alpha, beta, or gamma rays until stability is reached. The stable end product is typically a nonradioactive isotope of another element.

ramp scanning One of the two approaches for quantifying a peak in ICP-MS (peak hopping being the other). In the multichannel ramp scanning approach, a continuous smooth ramp of $1 - n$ channels (where n is typically 20) per mass is made across the peak profile. Mainly used for accumulating spectral and peak shape information when doing mass scans. It is normally used for doing mass calibration and resolution checks and as a classical qualitative method development tool to find out what elements are present in the sample and to assess their spectral implications on the masses of interest. Full-peak ramp scanning is not normally used for doing rapid quantitative analysis, because valuable analytical time is wasted taking data on the wings and valleys of the peak where the signal-to-noise ratio is poorest. For this kind of work, peak hopping is normally chosen.

reaction cell A collision/reaction cell that specifically uses ion–molecule reactions to eliminate the spectral interference. Often used to describe a dynamic reaction cell (DRC).

reaction mechanism The mechanism by which the interfering ion is reduced or minimized to allow the determination of the analyte ion. The most common collisional mechanisms seen in collision/reaction cells include collisional focusing, dissociation, and fragmentation, whereas the major reaction mechanisms include exothermic/endothermic associations, charge transfer, molecular associations, and proton transfer.

reactive gases In ICP-MS, the term refers to gases such as hydrogen, ammonia, oxygen, methane, or those used to stimulate ion–molecule reactions in a collision/reaction cell (CRC) or collision/reaction interface (CRI).

relative abundance of natural isotopes The isotopic composition expressed as a percentage of the total abundance of that element found in nature.

resolution A measure of the ability of a mass analyzer to separate an analyte peak from a spectral interference. The resolution of a quadrupole is nominally 1 amu and is traditionally defined as the width of a peak at 10% of its height.

resolving power Although resolving power and resolution are both a measure of a mass analyzer's ability to separate an analyte peak from a spectral interference, the term "resolving power" is normally associated with magnetic sector technology and is represented by the equation $R = m/\Delta m$, where m is the nominal mass at which the peak occurs and Δm is the mass difference between two resolved peaks. The resolving power of commercial double-focusing magnetic sector mass analyzers is on the order of 1000–10,000, depending on the resolution setting chosen.

response tables The intensity values for known concentrations of every elemental isotope stored in the instrument's calibration software. When semiquantitative analysis is carried out, the signal intensity of an unknown sample is compared against the stored response tables. By correcting for common spectral interferences and applying heuristic, knowledge-driven routines in combination with numerical calculations, a positive or negative confirmation can be made for each element present in the sample.

reverse Nier–Johnson double-focusing magnetic sector instrumentation The technology that all modern magnetic sector instrumentation is based on. Named after the scientists who developed it, Nier–Johnson geometry comes in two different designs, the "standard" and "reverse" Nier–Johnson geometry. Both these designs, which use the same basic principles, consist of two analyzers: a traditional electromagnet analyzer and an electrostatic analyzer (ESA). In the standard (sometimes called "forward") design, the ESA is positioned before the magnet, and in the reverse design, it is positioned after the magnet.

RF generator An alternative name for radio frequency generator.

right-angled ion lens design A recent development in ion-focusing optics, which utilizes a parabolic ion mirror to bend and refocus the ion beam at right angles to the ion source. The ion mirror incorporates a hollow structure that allows photons, neutrals, and solid particles to pass through it, while allowing ions to be deflected at right angles into the mass analyzer.

roughing pump Traditional mechanical roughing or oil-based pumps are used in ICP-MS to pump the interface region down to approximately 1–2 torr and also to back up the turbomolecular pump used in the ion optics region of the mass spectrometer.

ruby laser Ruby laser systems operate at 694 nm in the visible region of the electromagnetic spectrum.

S

S/B An abbreviation for **signal-to-background ratio**.

sample aerosol *Refer to* **aerosol**.

sample digestion The process of digesting a sample by traditional hot plate, fusion techniques, or microwave technology to get the matrix and analytes into solution.

sample dissolution The process of dissolving a sample by traditional hot plate, fusion techniques, or microwave technology to get the matrix and analytes into solution.

sample injector The central tube of the plasma torch that carries the sample aerosol mixed with the nebulizer gas. It can be a fixed part of the quartz torch or it can be separate (demountable) and be made from other materials, such as alumina, platinum, and sapphire, for the analysis of highly corrosive materials.

sample introduction system The part of the instrument that takes the liquid sample and puts it into the plasma torch as a fine-droplet aerosol. It comprises a nebulizer to generate the aerosol and a spray chamber to reject the larger droplets and allow only the smaller droplets into the plasma discharge.

sample preparation The entire process of preparing the sample for aspiration into the ICP mass spectrometer.

sample throughput The rate at which samples can be analyzed.

sample transport interferences A term used to describe a physical suppression of the analyte signal caused by matrix components in the sample. It is more exaggerated with samples having high levels of dissolved solids, because they are transported less efficiently through the sample introduction system than aqueous-type samples. *Also refer to* **physical interferences**.

sampler cone A part of the mass spectrometer interface region, where the ion beam from the plasma discharge first enters. The sampler cone, which is the first cone of the interface, is typically made of nickel or platinum and contains a small orifice of approximately 0.8–1.2 mm diameter, depending on the design. The sampler cone is much more pointed than the skimmer cone.

sampling accessories Customized sample introduction techniques optimized for a particular application problem or sample type. The most common types used today include the following: laser ablation/sampling (LA/S), flow injection analysis (FIA), electrothermal vaporization (ETV), desolvation systems, direct injection nebulizers (DIN), and chromatography separation techniques.

scan time The mass analyzer scan time is the time it takes to scan from one isotope to the next.

Scott spray chamber A sealed spray chamber with an inner tube inside a larger tube. The sample aerosol from the nebulizer is first directed into the inner tube. The aerosol then travels the length of the inner tube, where the larger droplets fall out by gravity into a drain tube and the smaller droplets return between the inner and outer tube, where they eventually exit into the sample injector of the plasma torch.

secondary discharge Another term used for the **pinch effect**.

secondary (side) reactions Reactions that occur in a collision/reaction cell that are not a part of the main interference reduction mechanism. If not anticipated and compensated for, secondary reactions can lead to erroneous results.

semiquant An abbreviated name used to describe **semiquantitative analysis**.

semiquantitative analysis A method for assessing the approximate concentration of up to 70 elements in an unknown sample. It is based on comparing the intensity of a small group of elements against known response tables stored in the instrument's calibration software. By correcting for common spectral interferences and applying heuristic, knowledge-driven routines in combination with numerical calculations, a positive or negative confirmation can be made for each element present in the sample.

settling time The time taken for the mass analyzer electronics to settle before a peak intensity measurement is taken for the operator-selected dwell time. The dwell time can usually be selected on an individual mass basis, but the settling time is normally fixed because it is a function of the mass analyzer and detector electronics.

shadow stop A grounded metal disk that stops particulate matter, neutral species, and photons from getting to the detector. It is considered a part of the ion optics and is sometimes called a photon stop.

side reactions Reactions that occur in a collision/reaction cell that are not a part of the main interference reduction mechanism. If not anticipated and compensated for, secondary reactions can lead to erroneous results.

signal-to-background ratio (S/B) The ratio of the signal intensity of an analyte to its background level at a particular mass. When considering the noise of the background signal (standard deviation of the signal), it is typically used as an assessment of the detection limit for that element. *Also refer to* **detection limit**, **background signal**, *and* **background noise**.

single-point peak hopping   A quantitation in which the quadrupole power supply is driven to a discrete position on the analyte mass (normally the maximum point), allowed to settle (settling time), and a measurement taken for a fixed amount of time (dwell time). The integration time for that peak is the dwell time multiplied by the number of scans (scan time). Multielement peak quantitation involves peak hopping to every mass in the multielement run. *Also refer to* **measurement duty cycle** *and* **peak measurement protocol**.

skimmer cone   A part of the mass spectrometer interface region where the ion beam from the plasma discharge first enters. The skimmer cone, which is the second cone of the interface, is typically made of nickel or platinum and contains a small orifice of approximately 0.5–0.8 mm diameter, depending on the design. The skimmer cone is much less pointed than the sampler cone.

solvent-based interferences   Spectral interferences derived from an elemental ion in the solvent (e.g., water, acid, etc.) combining with another ion from either the sample matrix or plasma gas (argon) to produce a polyatomic ion that interferes with the analyte mass.

space charge effect   A type of matrix-induced interference that produces a suppression of the analyte signal. This occurs mainly when low-mass analytes are being determined in the presence of larger concentrations of high-mass matrix components. It has the effect of defocusing the ion beam, and without compensation, the high-mass matrix element will dominate the ion beam, pushing the lighter elements out of the way, leading to low sensitivity and poor detection limits. The classical way to compensate for a space-charge matrix interference is to use an internal standard of similar mass to the analyte.

speciation analysis   In ICP-MS, it is the study and quantification of different species or forms of an element using a chromatographic separation device coupled to an ICP mass spectrometer. In this configuration, the instrument becomes a very sensitive detector for trace element speciation studies when coupled with high-performance liquid chromatography (HPLC), ion chromatography (IC), gas chromatography (GC), or capillary electrophoresis (CE). In these hybrid techniques, element species are separated on the basis of their chromatograph retention/mobility times and then eluted/passed into the ICP mass spectrometer for detection. The intensity of the eluted peaks is then displayed for each isotopic mass of interest in the time domain.

spectral interferences   A generic name given to interferences that produce a spectral overlap at or near the analyte mass of interest. In ICP-MS, there are two main types of spectral interference that have to be taken into account. Polyatomic spectral interferences (or molecular-based spectral interferences) are typically generated in the plasma by the combination of two or more atomic ions. They are caused by a variety of factors, but are usually associated with the argon plasma/nebulizer gas used, matrix components in the solvent/sample, other elements in the sample, or entrained oxygen/nitrogen from the surrounding air. The other type is an isobaric spectral interference, which is caused by different isotopes of other elements in the sample creating spectral interferences at the same mass as the analyte.

spray chamber The component of the sample introduction system that takes the aerosol generated by the nebulizer and rejects the larger droplets for the more desirable smaller droplets.

SRM An abbreviation for **standard reference materials**.

stability The ability of a measuring device to consistently replicate a measurement. In ICP-MS, it usually refers to the capability of the instrument to reproduce the signal intensity of the calibration standards over a fixed period of time without the use of internal standardization. Short-term stability is generally defined as the precision (as % RSD [relative standard deviation]) of 10 replicates of a single or multielement solution, whereas long-term stability is defined as the precision (as % RSD) of a fixed number of measurements over a 4–8 h time period of a single or multielement solution. However, stability in mass spectrometry can *also refer to* mass calibration stability, which is the ability of the mass spectrometer to repeatedly scan to the same mass position every time during a multielement analysis.

stability boundaries/regions The RF/DC boundaries of the Mathieu stability plot where an ion is stable as it passes through a quadrupole mass-filtering device. *Also refer to* **Mathieu stability plot**.

standard additions A method of calibration that provides an effective way to minimize sample-specific matrix effects by spiking samples with known concentrations of analytes. In standard addition calibration, the intensity of a blank solution is first measured. Next, the sample solution is "spiked" with known concentrations of each element to be determined. The instrument measures the response for the spiked samples and creates a calibration curve for each element for which a spike has been added. The calibration curve is a plot of the blank subtracted intensity of each spiked element against its concentration value. After creating the calibration curve, the unspiked sample solutions are then analyzed and compared to the calibration curve. Depending on the slope of the calibration curve and where it intercepts the X-axis, the instrument software determines the unspiked concentration of the analytes in the unknown samples.

standard reference materials (SRM) Well-established reference matrices that come with certified values and associated statistical data which have been analyzed by other complementary techniques. Their purpose is to check the validity of an analytical method, including sample preparation, instrument methodology, and calibration routines, to achieve sample results that are as accurate and precise as possible and can be defended under intense scrutiny.

standardization methods Refers to the different types of calibration routines available in ICP-MS, including quantitative analysis (external calibration and standard additions), semiquantitative analysis, isotope dilution, isotope ratio, and internal standardization methods.

T

thermoelectric cooling device Better known as a Peltier cooler, it generates a cold environment by creating a temperature gradient between two different materials. It uses electrical energy via a solid-state heat pump to transfer heat from a material on one side of the device to a different material on the other side, thus producing a temperature gradient across the device (similar to a household air conditioner).

time-of-flight mass spectrometry (TOFMS) A mass spectrometry technique based on the principle that the kinetic energy (KE) of an ion is directly proportional to its mass (m) and velocity (V), which can be represented by the equation $KE = \frac{1}{2}MV^2$. Therefore, if a population of ions with different masses is given the same KE by an accelerating voltage (U), the velocities of the ions will all be different, depending on their masses. This principle is then used to separate ions of different mass-to-charge (m/z) in the time (t) domain, over a fixed flight path distance (D), represented by the equation $m/z = 2Ut^2/D^2$. The simultaneous nature of sampling ions in TOF offers distinct advantages over traditional scanning (sequential) quadrupole technology for ICP-MS applications, where large amounts of data need to be captured in a short amount of time, such as the multielement analysis of transient peaks (laser ablation, flow injection, etc.).

time-of-flight (TOF) mass spectrometry (axial design) There are basically two different sampling approaches that are used in commercial TOF mass analyzers: the axial and orthogonal designs. In the axial design, the flight tube is in the same axis as the ion beam, whereas in the orthogonal design, the flight tube is positioned at right angles to the sampled ion beam. The axial approach applies an accelerating potential in the same axis as the incoming ion beam as it enters the extraction region. Because the ions are in the same plane as the detector, the beam has to be modulated using an electrode grid to repel the "gated" packet of ions into the flight tube. This kind of modulation generates an ion packet that is long and thin in cross section (in the horizontal plane), which is then resolved in the time domain according to the different ionic masses. *Also refer to* **time-of-flight mass spectrometry**.

time-of-flight (TOF) mass spectrometry (orthogonal design) There are basically two different sampling approaches that are used in commercial TOF mass analyzers, the axial and orthogonal designs. In the axial design, the flight tube is in the same axis as the ion beam, whereas in the orthogonal design, the flight tube is positioned at right angles to the sampled ion beam. With the orthogonal approach, an accelerating potential is applied at right angles to the continuous ion beam from the plasma source. The ion beam is then "chopped" by using a pulsed voltage supply coupled to the orthogonal accelerator to provide repetitive voltage "slices" at a frequency of a few kilohertz. The "sliced" packets of ions, which are typically tall and thin in cross section (in the vertical plane), are then allowed to "drift" into the flight tube, where the ions are temporally resolved according to their differing velocities. *Also refer to* **time-of-flight mass spectrometry**.

TOFMS An abbreviation for **time-of-flight mass spectrometry**.

torch design Refers to the different kinds of commercially available torch designs.

trace metal speciation studies *Refer to* **speciation analysis**.

transient signal (peak) A signal that lasts for a finite amount of time, compared to a continuous signal that lasts for as long as the sample is being aspirated. Transient peaks are typically generated by alternative sampling devices such as laser ablation, flow injection, or chromatographic separation systems where discrete amounts of sample are introduced into the ICP mass spectrometer.

turbomolecular pump (turbo pump) A type of vacuum pump used to maintain a high vacuum in the ion optics and mass analyzer regions of the ICP mass spectrometer. These pumps work on the principle that gas molecules can be given momentum in a desired direction by repeated collision with a moving solid surface. In a turbo pump, a rapidly spinning turbine rotor strikes gas (argon) molecules from the inlet of the pump towards the exhaust, creating and maintaining a vacuum. In the case of ICP-MS, two pumps are normally used, a large pump for the ion optic region, which creates a vacuum of approximately 10^{-3} torr, and another small pump for the mass analyzer region, which generates a vacuum of 10^{-6} torr. However, some designs use a twin-throated turbo pump, in which one powerful pump is used with two outlets, one for the ion optics and one for the mass analyzer region.

twin-throated turbomolecular pump In some designs of ICP mass spectrometer, a single twin-throated turbo pump is used instead of two separate pumps. In this design, one powerful pump is used with two outlets, one for the ion optics and one for the mass analyzer region.

U

ultrasonic nebulizer (USN) A type of desolvating nebulizer that generates an extremely fine-droplet aerosol for introduction into the ICP mass spectrometer. The principle of aerosol generation using this approach is based on a sample being pumped onto a quartz plate of a piezoelectric transducer. Electrical energy of 1–2 MHz is coupled to the transducer, which causes it to vibrate at high frequency. These vibrations disperse the sample into a fine-droplet aerosol, which is carried in a stream of argon. With a conventional ultrasonic nebulizer, the aerosol is passed through a heating tube and a cooling chamber, where most of the sample solvent is removed as a condensate before it enters the plasma. However, commercial ultrasonic nebulizers are also available with membrane desolvation systems.

USN An abbreviation for **ultrasonic nebulizer**.

UV laser A generic name given to a laser ablation system that works in the ultraviolet region of the electromagnetic spectrum. The three most common wavelengths used in commercial equipment are all UV lasers. They include the 266 nm (frequency-quadrupled) Nd:YAG laser, the 213 nm (frequency-quintupled) Nd:YAG laser, and the 193 nm ArF excimer laser system. *Also refer to* **excimer laser** *and* **Nd:YAG laser**.

V

vacuum chamber The region of the mass spectrometer that is under negative pressure created by a combination of roughing and turbomolecular pumps. As the ion beam moves from the plasma, which is at atmospheric pressure (760 torr), it enters the interface region between the sampler and skimmer cone (1–2 torr) before it is focused through the ion optic vacuum chamber region (10^{-3} torr) and eventually goes through the mass analyzer vacuum chamber (at 10^{-6} torr). *Also refer to* **turbomolecular pump**.

vacuum gauge Used to measure the pressure in the different vacuum chambers of the mass spectrometer.

vacuum pump A number of vacuum pumps are used to create the vacuum in an ICP mass spectrometer. Two roughing pumps are used, one for the interface region and another to back up the first turbomolecular pump of the ion optic region. Also, two turbomolecular pumps (or in some designs, one twin-throated pump) are used, one for the ion optics and another for the main mass analyzer region. *Also refer to* **roughing pump**, **turbomolecular pump**, *and* **twin-throated turbomolecular pump**.

visible laser A laser that operates in the visible region of the electromagnetic spectrum. An example is the ruby laser, which operates at 694 nm.

23 Useful Contact Information

The final chapter of the book is dedicated to providing you with useful contact information related to ICP mass spectrometry. It includes contact details for manufacturers of ICP-MS instrumentation, instrument consumables, sample introduction components, and alternative sources of sampling accessories, together with suppliers of laboratory chemicals, calibration standards, certified reference materials, high-purity gases, deionized water systems, and clean-room equipment. I have also included information about the major scientific conferences, professional societies, publishing houses, Internet discussion groups, and the most popular ICP-MS-related journals. It is sorted alphabetically by category. However, some vendors sell many different products, so they are listed under the category represented by their major product line.

1

Certified Reference Materials/Calibration Standards

National Research Council of Canada
1500 Montreal Road
Ottawa, Ontario, K1A 0R9, Canada
Phone: 800-668-1222
Fax: 613-952-8239
www.nrc.ca

NIST
100 Bureau Drive, Stop 200
Gaithersburg, MD 20899
Phone: 301-975-6776
Fax: 301-975-2183
www.nist.gov

VHG Labs
276 Abby Road
Manchester, NH 03103
Phone: 603-622-7660
Fax: 603-622-5180
www.vhglabs.com

High Purity Standards
P.O. Box 41727
Charleston, SC 29423
Phone: 843-767-7900
Fax: 843-767-7906
www.hps.net

Inorganic Ventures, Inc.
195 Lehigh Avenue, Suite 4
Lakewood, NJ 08701
Phone: 800-669-6799
Fax: 732-901-1903
www.ivstandards.com

SPEX Certiprep
203 Norcross Avenue
Metuchen, NJ 08840
Phone: 800-522-7739
Fax: 732-603-9647
www.spexcsp.com

2

Chemicals

Aldrich Chemicals
940 W. St. Paul Avenue
Milwaukee, WI 53233
Phone: 414-273-3850
Fax: 414-273-4979
www.sigma-aldrich.com

Eichrom Technologies, Inc.
8205 S. Cass Avenue
Darien, IL 60561
Phone: 800-422-6693
Fax: 630-963-1928
www.eichrom.com

Fisher Scientific, Inc.
2000 Park Lane
Pittsburgh, PA 15275
Phone: 412-490-8472
Fax: 412-809-1310
www.fishersci.com

J. T. Baker
222 Red School Lane
Phillipsburg, NJ 08865
Phone: 908-859-9315
Fax: 908-859-9385
www.jtbaker.com

3

Chromatographic Separation Equipment

Agilent Technologies
2850 Centerville Road
Wilmington, DE 19808
Phone: 302-633-8264
Fax: 302-633-8916
www.chem.agilent.com

PerkinElmer Life and Analytical Sciences
710 Bridgeport Avenue
Shelton, CT 06484
Phone: 800-762-4000
Fax: 203-944-4914
www.perkinelmer.com

Thermo Fisher Scientific
27 Forge Parkway
Franklin, MA 02038
Phone: 800-229-4087
Fax: 508-528-2127
www.thermoelemental.com

Varian, Inc.
2700 Mitchell Drive
Walnut Creek, CA 94598
Phone: 925-939-2400
Fax: 925-945-2102
www.varianinc.com

Dionex Corporation
500 Mercury Drive
P.O. Box 3603
Sunnyvale, CA 94088
Phone: 408-737-0700
Fax: 408-730-9403
www.dionex.com

Waters Corporation
34 Maple Street
Milford, MA 01757
Phone: 508-478-2000
Fax: 508-872-1990
www.waters.com

4

Clean Room Equipment

Cleanroom Consulting LLC
5396 Springview Drive
Fayetteville, NY 13066
Phone: 315-637-4030
Fax: 315-637-0928
www.cleanroomconsulting.com

Clestra Hauserman
259 Veterans Lane, Suite 201
Doylestown, PA 18901
Phone: 267 880 3700
Fax: 267 880 3705
www.clestra.com

Microzone Corp.
25F Northside Road, PO Box 11336
Ottawa, Ontario, K2H 7V1, Canada
Phone: 613-829-1433
Fax: 613-829-6331
www.microzone.com

5

Consumables (Detectors)

SGE, Inc.
2007 Kramer Lane
Austin, TX 78758
Phone: 800-945-6254
Fax: 512-836-9159
www.ecpsci.com

6

Consumables (Sample Introduction/Interface Components)

Burgener Research, Inc.
1680-2 Lakeshore Rd. W.
Mississauga, Ontario, L5J 1J5, Canada
Phone: 905-823-3535
Fax: 905-823-2717
www.burgenerresearch.com

CPI International
5580 Skylane Blvd.
Santa Rosa, CA 95403
Phone: 800-878-7654
Fax: 707-545-7901
www.cpiinternational.com

Elemental Scientific, Inc. (ESI)
2440 Cumming Street
Omaha, NE 68131
Phone: 402-991-7800
Fax: 402-997-7799
www.elementalscientific.com

Glass Expansion Pty.
4 Barlows Landing Road, Unit #2
Pocasset, MA 02559
Phone: 505-563-1800
Fax: 505-563-1802
www.geicp.com

Meinhard Glass Products
700 Corporate Circle, Suite A
Golden, CO 80401
Phone: 303-277-9776
Fax: 303-216-2649
www.meinhard.com

Precision Glassblowing
14775 E. Hindsdale Avenue
Engelwood, CO 80112
Phone: 303-693-7329
Fax: 303-699-6815
www.precisionglassblowing.com

SCP Science
21800 Clark Graham
Baie D'urfe, H9X 4B6, Canada
Phone: 800-361-6820
Fax: 514-457-4499
www.scpscience.com

Spectron, Inc.
2080 Sunset Drive
Ventura, CA 93001
Phone: 805-652-1992
Fax: 805-652-1994
www.spectronus.com

7

Expositions and Conferences

Eastern Analytical
P.O. Box 633
Montchanin, DE 19710
Phone: 610-485-4633
Fax: 610-485-9467
www.eas.org

FACSS
1201 Don Diego Avenue
Santa Fe, NM 87505
Phone: 505-820-1648
Fax: 505-989-1073
www.facss.org

Pittsburgh Conference
300 Penn Center Blvd., Suite 332
Pittsburgh, PA 15235
Phone: 412-825-3220
Fax: 412-825-3224
www.pittcon.org

Plasma Winter Conference
c/o Dr. Ramon Barnes
85 N. Whitney Street
Amherst, MA 01002-1869
Phone: 413-256-8942
Fax: 413-256-3746
www.uc.edu/plasmachem/taormina/
docments/2008_Winter_Conference_Information.pdf

8

Gases

Air Liquide
2700 Post Oak Blvd.
Houston, TX 77056
Phone: 800-248-1427
Fax: 281-474-8419
www.airliquide.com

Air Products and Chemicals, Inc.
7201 Hamilton Blvd.
Allentown, PA 18195
Phone: 800-654-4567
Fax: 800-880-5204
www.airproducts.com

Praxair Specialty Gases
7000 High Grove Blvd.
Burr Ridge, IL 60521
Phone: 877-772-9247
Fax: 630-320-4506
www.praxair.com/specialty gases

Scott Specialty Gases
6141 Easton Road, P.O. Box 310
Plumsteadville, PA 18949
Phone: 215-766-8861
Fax: 215-766-2476
www.scottgas.com

9

ICP-MS Instrumentation (Magnetic Sector Technology)

Thermo Finnigan
355 River Oaks Parkway
San Jose, CA 95134
Phone: 408-965-6000
Fax: 408-965-6010
www.thermofinnigan.com

Nu Instruments
Sales and Support
Newburyport, MA 01950
Phone: 978-465-2484
Fax: 978-465-2484
www.nu-ins.com

10

ICP-MS Instrumentation (Quadrupole Technology)

Agilent Technologies
2850 Centerville Road
Wilmington, DE 19808
Phone: 302-633-8264
Fax: 302-633-8916
www.chem.agilent.com

Thermo Fisher Scientific
27 Forge Parkway
Franklin, MA 02038
Phone: 800-229-4087
Fax: 508-528-2127
www.thermoelemental.com

PerkinElmer Life and Analytical Sciences
710 Bridgeport Avenue
Shelton, CT 06484
Phone: 800-762-4000
Fax: 203-944-4914
www.perkinelmer.com

Varian, Inc.
2700 Mitchell Drive
Walnut Creek, CA 94598
Phone: 925-939-2400
Fax: 925-945-2102
www.varianinc.com

11

ICP-MS Instrumentation (Time-of-Flight Technology)

GBC Scientific
3930 Ventura Drive, Suite 350
Arlington Heights, IL 60004
Phone: 800-445-1902
Fax: 847-506-1901
www.gbcsci.com

12

Internet Discussion Group

PLASMACHEM List Server
312 Heroy Geology Laboratory
University of Syracuse
Syracuse, NY 13244
Phone: 315-443-1261 (Michael Cheatham)
Fax: 315-443-3363
To subscribe:
Email: mmcheath@mailbox.syr.edu

European Virtual Institute for Speciation Analysis (EVISA)
A forum dedicated to trace element speciation analysis
http://www.speciation.net/

13

Journals/Magazines

American Laboratory
30 Control Drive
Shelton, CT 06484
Phone: 800-777-9009
Fax: 203-926-9310
www.iscpubs.com

Analytical Chemistry
1155 16th Street, NW
Washington, DC 20036
Phone: 202-872-4570
Fax: 202-872-4574
www.pubs.acs.org/ac

Applied Spectroscopy
201b Broadway Street
Frederick, MD 21701
Phone: 301-694-8122
Fax: 301-694-6860
www.s-a-s.org

JAAS (Royal Society of Chemistry)
Thomas Graham House
Science Park
Milton Rd.
Cambridge, CB4 4WF, England, UK
Phone: 44-1223-420066
Fax: 44-1223-420247
www.rsc.org

Spectroscopy Magazine (Editorial Office)
485 Route One South
Building F, First Floor
Iselin, NJ 08830
Phone: 732-596-0276
Fax: 732-225-0211
www.spectroscopyonline.com

14

Professional Societies/Services

American Chemical Society
1155 16th Street NW
Washington, DC 20036
Phone: 800-227-5558
Fax: 202-872-4615
www.pubs.acs.org

American Society for Mass Spectrometry
1201 Don Diego Avenue
Santa Fe, NM 87505
Phone: 505-989-4517
Fax: 505-989-1073
www.asms.org

American Society for Testing Materials
100 Barr Harbor Drive
West Conshohocken, PA 19428
Phone: 610-832-9605
Fax: 610-834-3642
www.astm.org

Chemical Abstract Services
2540 Olentangy Drive
Columbus, OH 43202
Phone: 800-753-4227
Fax: 614-447-3837
www.cas.org

Society for Applied Spectroscopy
201b Broadway Street
Frederick, MD 21701
Phone: 301-694-8122
Fax: 301-694-6860
www.s-a-s.org

Semiconductor Equipment & Materials International
3081 Zanker Road
San Jose, CA 95134
Phone: 408-943-6900
Fax: 408-428-9600
www.semi.org

15

Laser Ablation Equipment

SD Acquisition, Inc., dba CETAC Technologies
14306 Industrial Road
Omaha, NE 68144
Phone: 402-733-2829
Fax: 402-733-5292
www.cetac.com

New Wave Research
47613 Warm Springs Blvd.
Fremont, CA 94539
Phone: 510-249-1550
Fax: 510-249-1551
www.new-wave.com

16

Microwave Dissolution Equipment

CEM Corporation
3100 Smith Farm Road
Matthews, NC 62810
Phone: 800-726-3331
Fax: 704-821-5185
www.cem.com

Milestone, Inc.
160 B Shelton Road
Monroe, CT 06468
Phone: 203-261-6175
Fax: 203-261-6592
www.milestonesci.com

17

Publishers

CRC Press/Taylor and Francis
6000 Broken Sound Parkway NW,
  Suite 300
Boca Raton, FL 33487
Phone: 561-994-0555
Fax: 561-989-9732
www.crcpress.com

Elsevier Science Publishing
655 Avenue of the Americas
New York, NY 10010
Phone: 212-633-3756
Fax: 212-633-3112
www.elsevier.com

John Wiley and Sons
605 Third Avenue
New York, NY 10158
Phone: 212-850-6518
Fax: 212-850-6617
www.wiley.com

International Scientific Communications
30 Control Drive, P.O. Box 870
Shelton, CT 06484
Phone: 800-777-9009
Fax: 203-926-9310
www.iscpubs.com

18

Sample Introduction Delivery Systems
(USN, Cooled Spray Chambers, Desolvators, Autosamplers, Dilutors)

SD Acquisition, Inc., dba
  CETAC Technologies
14306 Industrial Road
Omaha, NE 68144
Phone: 402-733-2829
Fax: 402-733-5292
www.cetac.com

Elemental Scientific, Inc. (ESI)
2440 Cumming Street
Omaha, NE 68131
Phone: 402-991-7800
Fax: 402-997-7799
www.elementalscientific.com

19

Vacuum Pumps and Components

Leybold Vacuum (USA), Inc.
5700 Mellon Road
Export, PA 15632
Phone: 800-764-5369
Fax: 724-733-1217
www.leyboldvacuum.com

Varian Vacuum Technologies
121 Hartwell Avenue
Lexington, MA 02421
Phone: 781-861-7200
Fax: 781-860 5437
www.varianinc.com

Index

A

AA. *See* Atomic absorption
Abundance sensitivity specifications, 52–54
Acid digestion, 232, 234
Active film multipliers. *See* Channel electron
    multipliers
Addition calibration, 118
Aerosol generation, 13–15
Alkylated metals, 188–189
Alpha counting spectrometry, 296
Alternative sampling accessories, 36, 163–185,
    288
Alternative sampling devices. *See* Alternative
    sampling accessories
Analog counting, 97–100, 270f
Application segments. *See* Market segments
Applications, 203–236
    biomedical, 208–211
    electrothermal vaporization, 22, 174–178
    environmental, 204–208
    flow injection, 214–215 (*See also* Flow
        Injection)
    food, 234–236
    geochemical, 211–219
    laser ablation, 218–219 (*See also* Laser
        ablation)
    metallurgical, 229–231
    nuclear, 224–229
    organic samples, 231–234
    petrochemical, 231–234
    semiconductor, 219–224
    speciation, 187–201, 206–207
Argon gas flows, 15–16, 26–27f, 179–180f, 254
Argon gas purity, 147, 236
Argon-based interferences, 73
Ashing techniques, 140, 232
Atomic absorption (AA), 242f
Atomic emission (AE), 242f
Atomic mass, 8–9, 134
Atomic number, 8–9
Atomic structure, 9
Attenuation (of detector), 99–100, 270, 272
Autodilutors, 339
Autosamplers, 184–185, 339
Auxiliary gas flow, 24–25
Axial view, 243

B

Background equivalent concentration (BEC),
    263, 265, 277
Background noise, 95, 105–106, 109, 263–265
Background reduction, 221, 223, 265
Background signal, 82, 263, 265, 282
Bandpass filtering, 80–83
BEC. *See* Background equivalent concentration
Biomedical applications, 208–211
    analysis of body fluids, 209
    calibration routines, 210–211
    interferences, 210
    sample preparation, 209–210
    stability, 211
Biomolecules, 76, 187, 189t
By-product ions, 76, 78, 83–84, 222, 254,
    279–281

C

Calibration (standardization) routines, 115–123,
    142
Calibration curves. *See* Calibration
    (standardization) routines
Calibration standards suppliers, 333
Capacitive coupling, 32–34
Capillary (zone) electrophoresis, 187
CE. *See* Capillary (zone) electrophoresis
Cell (Collision/Reaction). *See* Collision and
    reaction cells
CEM. *See* Channel electron multipliers
Certified reference materials. *See* Standard
    reference materials
Channel electron multipliers, 93–94
Channeltron. *See* Channel electron multipliers
Chemical modifier, 176
Chemical reagents suppliers, 334
Chicane ion lens, 40
Chilled spray chambers. *See* Cooled spray
    chambers
Chromatographic separation devices, 334
Clean room equipment manufacturers, 334–335
Clean rooms, 145–146
Clinical applications. *See* Biomedical
    applications

Cold plasma technology. *See* Cool plasma
        technology
Cold Vapor Atomic Absorption (CVAA), 300
Collision cell technology, 85–89
Collision/reaction cell technology, 73–91, 209,
        279–281
Collision/reaction cells, 73–75, 131–132
Collision/reaction interface, 73, 75, 84–85, 90t,
        120, 197–199, 254, 280
Collisional damping, 301
Collisional focusing, 75
Collisional fragmentation, 75
Collisional mechanisms, 76–80
Commercial ion lens designs, 43–46
Comparing ICP-MS with other AS techniques,
        241–259
    dynamic range, 246, 249
    ease of use, 252
    precision, 249–250, 252
    running costs, 253–259
    sample throughput, 249–251
Concentric nebulizers, 16–18
Consumables suppliers, 335
Contact information, 333–339
Contamination sources, 137–149
Cool plasma technology, 73, 130–131, 276–279
Cooled spray chambers, 178–183, 339
Correction equations. *See* Mathematical
        correction equations
CRC. *See* Collision/reaction cell
CRI. *See* Collision/reaction interface
CRMs, Standard reference materials
Cross-flow nebulizers, 18
Curved fringe rods, 45
Cyclonic spray chamber, 21–22
Cylindrical quadrupoles, 47, 54
CZE. *See* Capillary (zone) electrophoresis

D

Data quality objectives, 107–113
DCPs. *See* Direct current plasmas
Dead time correction. *See* Detector dead time
Dead volume, 17, 195
Debye length, 35, 41
Deionized water quality, 125–126, 130–131,
        141–142
Desolvating microconcentric nebulizers, 303
Desolvating spray chambers, 339
Desolvation devices, 22, 178–183
Detection limits, 89–91, 263–264, 280–281
    comparison with AA and ICP-OES, 246,
        247–248t
    optimization, 105–113
    typical ICP-MS, 2f
Detector manufacturers, 335

Detectors, 4, 93–100, 159–160
    analog-counting, 97
    attenuation, 98–100
    cross calibration, 98
    dead time, 94
    digital-counting, 97
Digital counting, 97
DIHEN. *See* Direct injection high efficiency
        nebulizers
DIN. *See* Direct injection nebulizers
Direct current plasmas (DCPs), 23–24
Direct injection high efficiency nebulizers
        (DIHEN), 183–184
Direct injection nebulizers (DIN), 22, 183–184
Discrete dynode detectors (DDD), 4, 95–96
Double focusing magnetic sector instrument
        manufacturers, 336
Double pass spray chamber, 15f, 20
Double-focusing magnetic sector mass analyzers,
        57–63
Doubly-charged ions, 127
Drain system, 152
DRC. *See* Dynamic reaction cell technology
Droplet selection, 15–16
Duty cycle. *See* Measurement duty cycle
Dwell time. *See* Integration time
Dynamic range extension. *See* Extended dynamic
        range
Dynamic reaction cell (DRC) technology, 80
Dynamically-scanned ion lens, 44

E

EDR. *See* Extended dynamic range
Electron diffusion, 42f
Electron shells, 7
Electrons, 7
Electrostatic analyzer, 59–60
Electrothermal atomization, 1, 243, 250–259.
        See also Graphite furnace atomic
        absorption
Electrothermal vaporization, 22, 174–178
Elemental fractionation, 166
Endothermic reaction, 82f, 306
Environmental applications, 204–208
Environmental contamination, 145–146
Environmental Protection Agency ICP-MS
        Methodology, 204–208
EPA. *See* Environmental Protection Agency
ESA. *See* Electrostatic analyzer
ETA. *See* Electrothermal atomization
ETV. *See* Electrothermal vaporization
Evaluate ICP-MS (How to). *See* How to select an
        ICP-MS
Excimer laser, 165–166
Exothermic reaction, 306

Extended dynamic range, 96–100, 270
External standardization, 116–117
Extraction lens, 41, 44f, 45–46, 281–282

F

FAA. *See* Flame atomic absorption
Faraday collectors, 94–95
Faraday cup. *See* Faraday collectors
FIA. *See* Flow injection analysis
Flame atomic absorption (FAA), 1, 242, 249–259
Flight tube (TOF), 65
Flow Injection, 171–174, 214–215
Flow injection analysis (FIA), 19
Food applications, 234–236
Fringe rods, 45
Fusion mixtures, 140–141, 147, 212, 214, 268

G

Gamma counting spectrometry, 308
Gas chromatography, 187
Gas dynamics (of ion flow), 41
Gas flows (plasma), 175
Gases suppliers, 336
GC. *See* Gas chromatography
Geochemical applications, 211–219
GFAA. *See* Graphite furnace atomic absorption
Glassware (contamination in), 142–144
Graphite furnace atomic absorption (GFAA), 308
Grounded shadow stop, 44
Grounding (RF) mechanism, 4, 34

H

Half-life, 309
Heating zones. *See* Plasma heating zones
HEN. *See* High efficiency nebulizers
HEPA filters, 146
Hexapole, 74, 76, 80, 85, 279
HGAA. *See* Hydride generation atomic absorption
High efficiency nebulizers (HEN), 18–19. *See also* Microflow nebulizers
High performance liquid chromatography. *See* Liquid chromatography
High purity standards. *See* Calibration standards
High resolution (using a quadrupole), 51
High resolution mass analyzers, 132
High sensitivity interface, 46
Hollow ion mirror, 41f, 45
Hot plate digestion, 145
How to evaluate an ICP-MS system. *See* How to select an ICP-MS system

How to select an ICP-MS system, 261–293
    abundance sensitivity, 276
    accuracy, 271
    analytical performance, 262–289
    background levels, 265, 280
    BEC performance, 263, 265
    collision/reaction cell performance, 279–281
    cool plasma performance, 276–279
    detection capability, 263–267
    detectors, 272
    dynamic range, 272–274
    ease of use, 286–287
    evaluation objectives, 261–286
    extended dynamic range, 270
    financial considerations, 291–292
    installation requirements, 288–289
    interference reduction, 274–281
    isotope ratio precision, 269–271
    mass stability, 266–267, 279
    matrix-induced interferences, 281–283
    peak measurement protocol, 266
    precision, 267–269
    reliability, 289–290
    routine maintenance issues, 287–288
    sample throughput, 284–285
    sampling accessories, 288
    secondary discharge, 268, 277, 282–283
    service support, 290–291
    software, 286–287
    spectral interferences, 274–275
    speed of analysis, 277, 284–285
    stability, 267–269
    summary, 292–293
    technical support, 289
    transient capability, 285–286
    usability aspects, 286–293
HPLC. *See* Liquid chromatography
Hydride generation atomic absorption (HGAA), 310
Hyperbolic quadrupole rods, 47–48, 54

I

IC. *See* Ion chromatography
ICP. *See* Inductively-coupled plasma
ICP-OES. *See* Inductively-coupled plasma optical emission spectrometry
Inductively-coupled plasma, 23
Inductively-coupled plasma optical emission spectrometry, 1, 3, 243, 250–259
Infrared (IR) lasers, 165
Instrument (ICP-MS) manufacturers, 336–337
Instrument background levels, 146
Instrument comparisons. *See* How to select an ICP-MS system
Instrument detection limit, 263–264

Integration time, 107–113, 269
Interface region, 3–4, 31–37, 156–157
    cones, 31
    housing, 31, 156–157
    pressure, 41–42
Interference reduction, 274–281
    collision/reaction cell, 278–280
    collision/reaction interface, 84
    cool plasma technology, 277–278
    internal standardization, 281
    ion lens optimization, 281–282
    magnetic sector technology, 274–275
    mass analyzer resolution, 275–276
    mathematical equations, 128–130, 277
    RF coil grounding mechanisms, 277
Interferences, 251
    methods of compensation, 128–134
    overview, 125–135
Internal Standardization (IS), 123–124, 133–134
Ion
    chromatography, 187, 191–195
    energy spread, 33–35, 42, 69, 77–78
    flow, 41–42
    focusing guide, 45, 81
    focusing system, 39
    formation, 7–8
    kinetic energy, 34–35, 42, 65
    lens
        chamber, 41–42
        components, 41
        designs, 43–46
        voltages, 42–44, 158, 211, 226, 281
    mirrors, 45
    molecule chemistry, 80, 84
    optics, 39–41, 157–158, 268, 281–282
    packet, 66–67, 69–70
    repulsion, 43f
Ion focusing system, 39–46
Ionization source, 58, 156
Ionizing radiation counting techniques, 224
IR lasers. See Infrared lasers
Isobar
Isobaric interferences, 128–129t, 207, 210, 212,
        227, 262
Isotope dilution, 110, 120–122f
Isotope ratio calibration, 111–113, 123
Isotope ratio precision, 269–271
Isotopes, 8–11
Isotopic abundance. See Relative abundance of
        natural isotopes

K

KE. See Kinetic energy
KED. See Kinetic energy discrimination

Kinetic energy (KE), 65. See also Ion kinetic
        energy
Kinetic energy discrimination, in collision/
        reaction cells, 76–80, 83, 280–281

L

Laser ablation, 22, 164–171
Laser ablation applications, 218–219
Laser ablation manufacturers, 338
Laser absorption, 316
Laser sampling. See Laser ablation
Liquid chromatography, 18–19, 187, 190–191
Load coil, 33–34f
Low-temperature plasma. See Cool plasma
        technology

M

Magnetic sector mass analyzer technology,
        57–63, 336
    benefits, 63
    comparison with quadrupole technology,
        57, 60
    mass analyzers, 57–58
    precision, 63
    resolving power, 60–62
    transient peak capability, 63
Maintenance issues. See Routine maintenance
Manufacturers. See Instrument manufacturers
Market segments, 203–236. See also specific
        applications
Mass calibration, 103, 266–267
Mass discrimination, 123
Mass filtering discrimination, 80, 83. See also
        Collision/reaction cells
Mass filters, 47–48, 49f
Mass scanning, 102–105
Mass separation devices, 4, 47–48f
Mass spectrometers, 47, 57, 102, 287. See also
        specific technologies
Matching network (RF), 27–28
Mathematical correction equations, 128–130
Mathieu stability plot, 51f
Matrix interferences, 132–135, 281–283
Matrix separation, 73
Matrix suppression, 133–134, 163, 179, 210, 212,
        221, 223, 226, 281–282
Mattauch-Herzog magnetic sector design, 58
Measurement duty cycle, 69
Membrane desolvation, 179–183, 274
Metallurgical applications, 229–231
Microconcentric nebulizers. See Microflow
        nebulizers

Microflow nebulizers, 18. *See also* High efficiency nebulizers (HEN)
Microsampling, 19, 171–173, 214
Microwave digestion equipment manufacturers, 339
Microwave digestion/dissolution, 145, 210
Microwave induced plasma (MIP), 23–24
MIP. *See* Microwave induced plasma
Molecular spectral interferences. *See* Polyatomic interferences
Multichannel analyzers, 49, 103
Multichannel data acquisition, 44, 49, 50f, 103–106, 266–267
Multicomponent ion lens, 43–44f
Multipoles in collision/reaction cells, 45

N

Natural abundance, 11t
Natural isotopes. *See* Relative abundance of natural isotopes
Nd:YAG laser design, 165–168, 170, 218, 230
Nebulizer gas flow, 36, 127, 130, 147, 155, 179, 181, 197, 212, 221, 232–233, 264, 276, 282–283
Nebulizers, 16, 152–154
    Babington, 16
    concentric, 16–18
    cone spray, 16
    crossflow, 15f, 18
    high solids, 16
    microconcentric, 19
    microflow, 18–19, 181–183
    ultrasonic, 22, 178–181
Neodymium yttrium aluminum garnet lasers. *See* Nd-YAG lasers
Neutral species, 40
Neutron, 321
Nier-Johnson magnetic sector design, 58–59f
Nonionic species, 39–40
Nuclear applications, 224–229
    atom counting techniques, 224–225t
    characterization of nuclear waste, 227
    environmental monitoring, 227–228
    human health studies, 228–229
    production of nuclear materials, 226–227
    radiation counting techniques, 224–225

O

Octapole, 74, 80, 85, 89–90t, 279
Off-axis ion lens, 45
Organic samples, 36, 231–234
Oxide interferences, 127
Oxide ions, 321

Oxides, 36, 83, 127, 178–180, 218
Oxygen
    for analysis of organic solvents, 197
    in a collision/reaction cell, 198

P

Parabolic fields, 45
Particle counting techniques, 321
Peak hopping, 103–106, 111, 266–267
Peak integration. *See* Peak measurement protocol
Peak measurement protocol, 101–113, 199–200, 266
Peak quantitation, 106f, 266, 271. *See also* Quantitation methods
Petrochemical applications, 231–234
Photon stop, 40
Physical interferences, 322
Pinch effect. *See* Secondary discharge
Plasma, 23–29
    direct current, 23
    inductively coupled
        gas flow, 24
        heating zones, 28
        microwave induced, 23–24
        RF generators, 26–28
        source, 23–29
        torch, 3, 24–26, 155–156 (*See also* Radio frequency generators)
PlasmaChem internet discussion group, 289, 337
Polyatomic (spectral) interferences, 61t, 62, 125–126
Prescan (of detector), 97, 272–273, 286
Prescan methods, 272. *See also* Extended dynamic range
Proton transfer, 75
Publishers, 339. *See also* Supplier (vendor) contact information
Pulse counting, 270f

Q

Quadrupole instrument manufacturers, 337
Quadrupole mass analyzer technology, 47–54, 102–103, 111–113, 275–276
Quantitation methods, 115–124
Quantitative analysis, 103–106, 109–113, 115–118

R

Radio frequency (RF), 3
Radio frequency (RF) generators, 26–28

Radioactive isotopes, 314

Ramp scanning. *See* Multichannel data acquisition

Reaction cells, 80–83

Reaction interfaces. *See* Collision/reaction interface

Reactive gases, 76, 80–82, 86, 89, 91, 279–280

Reference materials. *See* Standard reference materials

Reference materials suppliers, 333

Relative abundance of natural isotopes, 2f, 8–10, 11t, 129t

Resolution, 50–52, 60–62, 69, 275–276

Response tables, 325

Reverse Nier-Johnson design, 59f

RF. *See* Radio frequency

RF generator. *See* Radio frequency generator

Right-angled ion lens design, 281–282, 326

Roughing pump, 158–159

Routine maintenance, 151–160
    detectors, 159–160
    filters, 159
    interface region, 156–157
    ion optics, 157–158
    mass analyzer, 160–161
    nebulizer, 152–154
    peristaltic pump, 152
    plasma torch, 155–156
    roughing pumps, 158–159
    sample introduction system, 152–156
    spray chamber, 154–155
    turbomolecular pumps, 160

S

S/B. *See* Signal-to-background ratio

Sample collection, 137–138

Sample delivery (autosamplers) manufacturers, 339

Sample digestion. *See* Sample dissolution

Sample dissolution, 139–141

Sample injectors, 14–15, 20, 24–29, 147, 155–156, 256, 268

Sample preparation contamination, 137–149

Sample preparation equipment, 142–145

Sample throughput optimization, 249–251, 284–285

Sample transport interferences, 132–133, 281

Sampler cone, 31, 33f

Sampling accessories. *See* Alternative sampling accessories

Sampling methods for ICP-MS, 163–185

Scan time (in integration time), 327

Scott spray chamber, 20

Secondary discharge, 4, 26, 33–36, 77, 155–156, 213, 268, 277, 282–283
    testing for one, 283

Secondary reactions in collision/reaction cells and interfaces, 75–76, 80, 82

Semiconductor applications, 219–224

Semiquant. *See* Semiquantitative analysis

Semiquantitative analysis, 118–120

Settling time (in integration time), 107–108, 269

Side reactions. *See* Secondary reactions

Signal-to-background ratio, 105–106f

Single ion lens, 46, 281

Single point, peak hopping. *See* Peak hopping

Skimmer cone, 16, 31, 33f, 40–42, 147, 157, 211, 256, 268, 283

Solvent-based interferences, 78, 178–180, 246

Space-charge effects, 39, 134–135

Speciation, 18

Speciation analysis, 187–201

Speciation applications, 206–207

Spectral interferences, 61–62, 73, 125–132, 274–275

Spray chambers, 20–22, 154–155, 178–179
    cooled, 22, 178, 197
    cyclonic, 21–22
    desolvating, 303
    double pass, 15f, 20
    impact bead, 310
    jacketed, 21
    Peltier-cooled, 178–179
    Scott-design, 20, 154

SRMs. *See* Standard reference materials

Stability regions of quadrupoles, 50–52

Standard additions, 117–118

Standard reference materials, 333

Standardization methods, 329. *See also* Calibration (standardization) routines

Standards. *See* Calibration standards

Supplier (vendor) contact information, 333–339
    autosamplers/auto dilutors, 339
    calibration standards, 333
    certified reference materials, 333
    chemicals and standards, 334
    chromatographic separation equipment, 334
    clean room equipment, 334–335
    consumables, 335
    detectors, 335
    expositions and conferences, 336
    gases, 336
    ICP-MS instrumentation, 336–337
    Internet discussion groups, 337
    journals/magazines, 337–338

laser ablation equipment, 338
microwave dissolution equipment, 339
professional societies, 338
publishers, 339
sample introduction/interface components,
335, 339
vacuum pumps and components, 339

T

Time-of-flight (TOF) mass analyzer technology,
65–72, 111, 337
Time-resolved peaks, 124, 198f
TOF. *See* Time-of-flight mass analyzer
technology
TOFMS. *See* Time-of-flight mass analyzer
technology
Torch design. *See* Plasma torch
Trace metal speciation studies. *See* Speciation
Training, 289
Transient signals, 71–72, 101, 113, 285–286
Twin-throated turbomolecular pumps

U

Ultrasonic nebulizers, 179–181, 339
USN. *See* Ultrasonic nebulizers
UV lasers, 165–167, 169

V

Vacuum chamber, 4
Vacuum gauge, 332
Vacuum pump, 45
Vacuum pump manufacturers, 339
Vendors. *See* Supplier (vendor) contact
information
Vessels and containers, 142–145
Volumetric ware, 144–145

W

Waste drain, 155
Water molecule interferences, 180
Water quality. *See* Deionized water quality